Transport of photoassimilates

Monographs and Surveys in the Biosciences

Plant Science

Main Editor
M. Wilkins, University of Glasgow

Submission of proposals for consideration
Suggestions for publication, in the form of outlines and representative
samples, are invited by the Editorial Board for assessment. Intending
authors should approach the main editor or another member of the
Editorial Board. Alternatively, outlines may be sent directly to the
publisher's offices. Refereeing is by members of the board and other
authorities in the topic concerned, throughout the world.

Longman Scientific & Technical
Longman House
Burnt Mill
Harlow, Essex, CM20 2JE, UK
Tel 0279 26721

Transport of photoassimilates

Editors

D. A. Baker
Department of Biological Sciences
Wye College (University of London)
Ashford, Kent, TN25 5AH
United Kingdom

and

J. A. Milburn
Department of Botany
University of New England
Armidale, N.S.W. 2351
Australia

Copublished in the United States with
John Wiley & Sons, Inc., New York

Longman Scientific & Technical
Longman Group UK Limited
Longman House, Burnt Mill, Harlow
Essex CM20 2JE, England
and Associated Companies throughout the world.

Copublished in the United States with
John Wiley & Sons Inc., 605 Third Avenue, New York, NY 10158

First published in 1989

British Library Cataloguing in Publication Data
Transport of photoassimilates. – (Monographs
    and surveys in the biosciences).
    1. Plants. Transport phenomena
    I. Baker, D. A. (Dennis Anthony), *1936-*
    II. Milburn, J. A. III. Series
    581.19′12

ISBN 0-582-46234-7

Library of Congress Cataloguing in Publication Data
Transport of photoassimilates.

    (Monographs and surveys in the biosciences)
    Includes bibliographies and indexes.
    1. Plant translocation. 2. Phloem. I. Baker, D. A.
II. Milburn, John A., 1936–   III. Series.
QK871.T725   1989   581.1′1   88–26863

ISBN 0-470-21386-8 (USA only)

Set in Lintron 202 10/12 pt Times
Printed and Bound in Great Britain
at the Bath Press, Avon

Contents

Preface

During the past ten years there has been a marked integration of approaches within the areas of study concerned with the translocation of assimilates in higher plants. Researchers from various areas – ranging from agronomists, horticulturalists, physiologists with interests in water relations, ion transport, and hormonal control, to anatomists, developmental morphologists, biophysicists and biochemists – have found common ground as the integrative role that phloem transport plays within the major processes of the plant has become increasingly apparent.

The plethora of investigations from such a wide range of disciplines has resulted in a greatly enhanced comprehension of the fundamental processes involved in the partitioning and compartmentation of photoassimilates between source and sink tissues within the plant. We are now able to formulate new concepts about these major processes, but because such concepts are still relatively inchoate, students and researchers in other fields are often unaware of the most recent developments. Some still regard phloem transport studies as a fringe area of research in which a few investigators hotly debate the pros and cons of the long-distance transport mechanisms, not knowing that this has ceased to be the case.

The aim of the present book is to present to advanced undergraduate and postgraduate students the major conceptual developments in the field of the transport of photoassimilates in plants. This objective is best achieved, we feel, by the multi-author approach and we have therefore invited acknowledged international experts to supply their specialist knowledge to provide the current synthesis. Topics included reflect the contemporary emphasis on regulatory processes within the plant. We have attempted the presentation of a balanced coverage although inevitably, in a subject area with many rapidly developing fields, some bias may be reflected in the relative emphasis and importance of specific topics.

Photoassimilate transport commences within the chloroplasts of the individual photosynthetic cell. The major movement of metabolites and the control mechanisms operating, which provide solutes for subsequent transport, are considered in Chapter 1. Related to these events is the intercellular movement of solutes between leaf cells, which is the subject of Chapter 2, where the contributions of symplasmic and apoplasmic transport to this movement of photoassimilates are evaluated. In Chapter 3 the pathway for the long-distance transport of the photoassimilates and other solutes is presented in a detailed description of the structure of the phloem. The composition of the solutes transported via the phloem pathway with reference to their destination and subsequent fate is discussed in Chapter 4, and illustrated by a number of case studies based on phloem sap composition. The mechanism whereby endogenous solutes are loaded into the phloem in the source tissues is considered in Chapter 5, where the possible control of this process is discussed and related to the potential utilization of this pathway for pesticide uptake. In Chapter 6 the events occurring in the sink tissues associated with the unloading of photoassimilates and the regulation of this process are appraised. The driving force for the long-distance transport of photoassimilates through the phloem is the topic of Chapter 7, where the various physiological studies which have resolved the pressure flow mechanism are critically evaluated. Finally, the overall relationships between source and sink organs and the subsequent regulation of compartmentation and partitioning within the plant are considered in Chapter 8.

We have thus covered in a relatively few chapters the major events associated with the translocation of solutes through the phloem. However, the subject is considerably larger than the coverage contained within this book and readers are encouraged to delve further into the extensive literature on the various topics which has been cited by the contributors.

We would like to thank our fellow scientists in the phloem transport field who have provided their detailed specialist knowledge and views in this book. Their presentation of the individual chapters has been of a very high standard and has made our task as editors a pleasant one. We are particularly indebted to Mrs Sue Briant for her patient and willing secretarial assistance throughout the production of the text.

Wye and Armidale, 1988   **D. A. B. and J. A. M.**

List of contributors

D. A. Baker,
Department of Biological Sciences,
Wye College, University of London,
Ashford,
Kent, TN25 5AH, England.

H.-D. Behnke,
Universität Heidelberg,
D-6900 Heidelberg,
Germany.

S. Delrot,
Charge de Recherches CNRS,
25, rue du Faubourg St Cyprien,
UA CNRS 575,
86000 Poiters, France.

W. Eschrich,
Forstbotanisches Institut,
Busgenweg 2,
D-3400 Göttingen,
Germany.

R. I. Grange,
Horticultural Research Institute,
Worthing Road,
Rustington,
Littlehampton, Sussex, BN16 3PU, England.

L. C. Ho,
Horticultural Research Institute,
Worthing Road,
Rustington,
Littlehampton, Sussex, BN16 3PU, England.

J. Kallarackal,
Botany Department,
University of New England,
Armidale,
NSW 2351, Australia.

R. C. Leegood,
Research Institute for Photosynthesis
University of Sheffield,
Sheffield, S10 2TN, England.

W. J. Lucas,
Department of Botany,
University of California,
Davis,
CA 95616, USA.

M. A. Madore,
Department of Botany and Plant Science,
Riverside,
CA 92521, USA.

J. A. Milburn,
Botany Department,
University of New England,
Armidale,
NSW 2351, Australia.

J. S. Pate,
Botany Department,
University of Western Australia,
Nedlands 6009, Western Australia.

A. F. Shaw,
Horticultural Research Station,
Worthing Road,
Rustington,
Littlehampton, Sussex, BN16 3PU, England.

M. N. Sivak,
Research Institute for Photosynthesis,
University of Sheffield,
Sheffield, S10 2TN, England.

List of contributors

D. A. Walker,
Research Institute for Photosynthesis,
University of Sheffield,
Sheffield, S10 2TN, England.

Abbreviations, symbols and units

ABA	abscisic acid
ADP	adenosine-5'-diphosphate
AFS	apparent free space
AMP	adenosine-5'-monophosphate
ATP	adenosine-5'-triphosphate
ATPase	adenosine triphosphatase
C_3	photosynthetic carbon reduction cycle
C_4	photosynthetic dicarboxylic acid pathway
CAM	Crassulacean acid metabolism
CCCP	carbonylcyanide m-chlorophenyl hydrazone
DAPI	4',6-diamidino-2-phenyl indole dihydrochloride
DCCD	N,N'-dicyclohexylcarbodiimide
DCMU	3'(3,4 dichlorophenyl)-1',1-dimethyl urea
DHAP	dihydroxyacetone phosphate
DNA	deoxyribonucleic acid
E4P	erythrose-4-phosphate
EDTA	ethyleneidiaminetetracetic acid
ER	endoplasmic reticulum
F6P	fructose-6-phosphate
FBP	fructose 1,6-bisphosphate
FC	fusicoccin
FCCP	carbonylcyanide p-trifluoromethoxyphenylhydrazone
FMN	flavin mononucleotide
$Fru2,6P_2$	fructose 2,6-bisphosphate
G6P	glucose-6-phosphate
GA	gibberellic acid
IAA	indolyl-3-acetic acid
LM	light microscope
MTs	microtubules
MW	molecular weight

NAD$^+$	nicotinamide adenine dinucleotide
NADH	reduced form of above
NADP$^+$	nicotinamide adenine dinucleotide phosphate
NADPH	reduced form of above
NEM	N-ethylmaleimide
OAA	oxaloacetate
PCMB	p-chloromercuribenzoic acid
PCMBS	p-chloromercuribenzene sulphonic acid
PCRC	photosynthetic carbon reduction cycle
PEP	phosphoenolpyruvate
PFK	phosphofructokinase
PGA	phosphoglyceric acid
PGRs	plant growth regulators
P$_i$	inorganic phosphate
PMF	proton motive force ($\triangle p$)
PP$_i$	inorganic pyrophosphate
P-protein	phloem protein
PSI	photosystem I
PSII	photosystem II
PVM	paraveinal mesophyll
RNA	ribonucleic acid
Ru5P	ribulose-5-phosphate
RUBISCO	ribulose bisphosphate/carboxylase-oxygenase
RuBP	ribulosebisphosphate
S7P	sedoheptulose-7-phosphate
SBP	sedoheptulose 1,7-bisphosphate
SEM	scanning electron microscope
SER	sieve element reticulum
SMT	specific mass transfer
SPS	sucrose phosphate synthase
TEM	transmission electron microscope
Xu5P	xylulose-5-phosphate

Symbols

Symbol	Description	Unit
A	area	m^2
a	activity	mol m^{-3} (mM)
C	capacity	F
c	concentration	mol m^{-3} (mM)
D	diffusion coefficient	m^2 s^{-1}
G	Gibbs' free energy	J
g	membrane conductance	S m^{-2}
I	current	A

J	generalized flow	
J_j	net flux	
J_s	solute flow	$mol\ m^{-2}\ s^{-1}$
J_v	volume flow	$m^3\ m^{-2}\ s^{-1}\ (m\ s^{-1})$
K_i	inhibitor constant	$mol\ m^{-3}$
K_m	Michaelis–Menten constant	$mol\ m^{-3}$
k	rate constant	
l	length	m
l_n	logarithm to base e	
log	logarithm to base 10	
n	number of moles	
P	permeability coefficient	ms^{-1}
P	turgor pressure	Pa
R	electrical resistance	ohm
T	temperature	$K\ or\ °C$
t	time	s
u	electrical mobility	$m^2\ V^{-1}\ s^{-1}$
V	volume	m^3
V_{max}	maximum reaction velocity	$mol\ s^{-1}$
v	velocity of reaction	$mol\ s^{-1}$
X	generalized force	
z	valency	
γ	activity coefficient	
ϵ	volumetric elastic modulus	Pa
μ	chemical potential	$J\ mol^{-1}$
μ	electrochemical potential	$J\ mol^{-1}$
π	osmotic pressure	Pa
ϕ	net flux	$mol\ m^{-2}\ s^{-1}$
ψ	water potential	Pa
ψ_π	osmotic potential	Pa
Ψ	electrical potential	V
[ ]	denotes concentration also used for isotopes	
$\triangle$	difference in	

Subscripts

c	cytoplasmic
i	inside
j	species
M	membrane
o	outside
t	tonoplast
v	vacuole

w   water
x   xylem

Chemical symbols (H, K, Cl) when used as subscripts specify the particular ion.

Constants

Constant	*Description*	
F	Faraday constant	96 490 coulombs mol^{-1}
R	gas constant	8.314 J mol^{-1} K^{-1}
RT/F		25.3 mV at 20°C

Units

The International System of Units (SI) is used wherever possible in this book. In some instances, where data are reproduced, other units are quoted with their SI equivalent values.

Prefixes

M	mega	10^6
k	kilo	10^3
c	centi	10^{-2}
m	milli	10^{-3}
μ	micro	10^{-6}
n	nano	10^{-9}
p	pico	10^{-12}

Unit	*Description*
A	ampere, electrical current
bar	unit of pressure, 10^5 Pa
°C	degree Celsius, temperature
Da	dalton
d	day
eq	equivalent mol × valency
F	farad, electrical capacitance
g	gram
h	hour
J	joule
K	kelvin (degrees)
l	litre
m	metre

min	minute, time
mol	mass equal to molecular weight in grams
osmol	sum of mole contribution to osmotic pressure
Pa	pascal, pressure
S	siemen, electrical conductance
s	second, time
V	volt, electrical potential difference
y	year
Ω	unit of resistance

1 Transport of photoassimilates within photosynthetic cells

M. N. Sivak, R. C. Leegood and D. A. Walker

1.1 Introduction

The major end-products of photosynthesis are starch and sucrose. Sucrose is exported to the rest of the plant whilst starch is accumulated in the leaf. How much mobile sugar will the leaf be able to supply? An understanding of how this supply is regulated requires an understanding of carbon metabolism in the photosynthetic cell and, in particular, of those processes that regulate the level of sucrose in the cytosol, both in the light and in the dark.

It is clear that the rate of photosynthesis does not depend on the amount of a single component, e.g. the activity of a particular enzyme. There is a wide range of possible regulatory factors, proven to exist *in vitro* but the importance of which *in vivo* has still to be determined. In particular, there is a multitude of factors affecting the activity of the enzymes involved, with pH, ions, coenzymes and metabolite effectors modulating the activity of every enzyme studied so far. Compartmentation is the other key factor. The role of metabolite transport in the cell, particularly between chloroplast and cytosol but also to and from mitochondria, vacuole and other organelles is now considered fundamental to the regulation of photosynthesis. In this chapter we will look at the factors now considered to be of major importance in the determination of the nature of the products of photosynthesis, with special emphasis on the characteristics and role of the movement of these photoassimilates in the photosynthetic cell. The intention is not to cover all the areas related to transport between the cell compartments but to deal only with those relevant to the movement of photoassimilates in the cell, and to refer the reader to the relevant literature when necessary.

1.2 The chloroplast as a transporting organelle

The chloroplast always seems to be faced with reconciling the irreconcilable. While evolving oxygen, it must simultaneously produce an intermediate more reducing than hydrogen. While reducing $NADP^+$ it can simultaneously generate ATP, an achievement which, when first reported, seemed almost as remarkable as making water flow uphill. No less striking in their own way are two seemingly conflicting roles in carbon metabolism. On the one hand, the chloroplast must operate its carbon cycle as an autocatalytic breeder reaction, while on the other, it must export elaborated carbon and chemical energy to its cellular environment. In order to export, it must produce more than it uses, however it can only do this by returning newly synthesized intermediates to the cycle; but in order to satisfy the needs of the cell, it must release newly made products to the cytoplasm. Clearly, these processes could not be efficiently accomplished unless it were possible to strike a delicate balance between recycling, export and internal storage.

1.2.1 The development of the experimental study of metabolite translocation in chloroplasts

The classic equation for photosynthesis is often written

$$6CO_2 + 6H_2O \rightarrow C_6H_{12}O_6 + 6O_2 \tag{1.1}$$

although it is conceded that the empirical $C_6H_{12}O_6$ might be something other than free glucose. Originally starch seemed the most likely end-product, and certainly the historic observations of Sachs (1862, 1887), Pfeffer and Godlewski established an intimate relationship between photosynthesis and the process and location of starch accumulation in green leaves (see Rabinowitch, 1945). Even in Sachs' day it was known that some plants seem incapable of accumulating starch within their leaves in any circumstances. Moreover, an almost insoluble polysaccharide could quite clearly not be moved about the plant unchanged. As sucrose emerged as the major compound to be transported within the plant (e.g. Thomas *et al.*, 1973), it became increasingly accepted as the real end-product of photosynthesis (Rabinowitch, 1956), whereas starch was relegated to the role of a temporary storage compound. Even when it became evident that sugar phosphates played a central role in photosynthetic carbon metabolism (Benson *et al.*, 1950, 1952; Benson & Calvin, 1950; Bassham & Calvin, 1957), there was no suggestion that these compounds should be regarded as end-products rather than intermediates. The percentage of radioactive carbon in sucrose extrapolated to zero at zero time, and thereafter it increased in a progressive manner in precisely the way which might have been predicted for an end-product awaiting shipment to other parts of the plant. The first real hint that sucrose was not necessarily synthesized within the chloroplast prior to export came from Heber & Willenbrink (1964). Using freeze-drying and non-aqueous fractionation, these workers produced good evidence that a large part of sucrose synthesis occurred not inside the chloroplast but within the cytoplasm.

Almost all workers in the field were also slow to realize the significance of the use of sugars in the isolation of chloroplasts capable of relatively fast rates of photosynthesis. Sucrose had been used as an osmoticum by Hill, as it had been by Engelmann *et al.* before him (see Hill, 1965), but the correlation between chloroplast intactness and the ability to assimilate carbon, while recognized by Whatley *et al.* (1956), was not related to envelope integrity until the mid-1960s (Leech, 1964; James & Leech, 1964; Walker, 1965a). In retrospect it is clear from this work (see e.g. Heber, 1974; Walker, 1974) that if sucrose is an effective osmoticum for isolated chloroplasts it cannot readily penetrate the chloroplast envelope. This has also been established independently by Heldt and Rapley (1970a) and Heldt and Sauer (1971), whose results showed that it is the inner of the two envelopes which is impermeable.

By 1965 then, there were already good indications that neither sucrose nor reducing power in the form of NADP and ATP could move freely across both chloroplast envelopes. Conversely there was at least presumptive evidence that the chloroplast might be more readily permeated by some of the intermediates of the Benson–Calvin cycle. Since this time the picture which had emerged is that the principal imports are CO_2 and orthophosphate and that the principal export is triose phosphate (Heber, 1974; Walker, 1974). This view is based largely on evidence obtained by the sorts of methods described below. Similar methods are employed in the study of other organelles such as mitochondria (for a review see Douce, 1985).

1.2.2 Methods

1.2.2.1 Distribution *in vivo*

Radioactive carbon has been used extensively in the study of metabolic pathways. The use of $^{14}CO_2$ (supplied to an illuminated leaf) to determine the partition of photosynthetic product between chloroplast and cytosol, however, has to take into account two main difficulties. Firstly, a transport metabolite must ultimately be used in further metabolism and, since processes such as the RPP pathway and glycolysis share common intermediates, it is clearly difficult to decide whether compound X has been transported from chloroplast to cytoplasm as such, or whether it is converted into Y for transport back to X in the first stage of subsequent metabolism. Secondly, chloroplasts and other subcellular organelles are fragile and breakage during fractionation would make results very difficult to interpret.

1.2.2.1.1 Non-aqueous techniques Non-aqueous techniques rest on the concept that if a leaf is quickly frozen, the distribution of compounds within its tissues at that moment will remain unchanged throughout subsequent non-aqueous fractionation. This approach has been used in enzyme studies and it is here that its credibility is most suspect because of the very real probability that, for example, denatured cytoplasmic protein can be precipitated on to chloroplasts and then separated with this fraction. Nevertheless, when applied with the scrupulous attention to detail which has been exercised in some laboratories, it is a useful technique, especially in relation to the distribution of metabolites.

1.2.2.1.2 Aqueous separation To avoid redistribution, aqueous separation often depends on rapidity of separation and on the assumption, as above, that fractionation can be achieved without cross-contamination.

Before the development of procedures better suited to the maintenance of envelope integrity, chloroplasts were found to be particularly leaky and it was concluded that major damage must occur during isolation in aqueous media. The separation of really active chloroplasts, however, indicates that massive damage can be avoided. Even so there is no real doubt that chloroplasts isolated in aqueous media may be contaminated by enzymes from other organelles. Similarly, any smaller molecules or ions which are free to cross the envelope are likely to do so to some extent, however rapidly separation is achieved. However, careful washing may remove contaminating enzymes and there seems every prospect that the development of increasingly sophisticated types of centrifugation (Robinson *et al.*, 1987) could lead to considerable advances in this field.

1.2.2.2 Distribution *in vitro*

An alternative to the fractionation of whole tissues is to isolate functional organelles, to see what they produce *in vitro* and what they export to the surrounding medium. Direct analysis is used to determine the ways in which various compounds partition themselves between isolated chloroplasts (and chloroplast compartments) and the suspending medium under different conditions. The isolation of functional chloroplasts is not in itself an easy task and nearly 30 years elapsed before CO_2-dependent oxygen evolution by isolated chloroplasts was demonstrated at rates comparable to the release of O_2 observed in the presence of artificial hydrogen acceptors, by Robert Hill in the 1930s.

1.2.2.2.1 Centrifugal filtration This procedure, first devised by Werkheiser & Bartley (1957), and used extensively with mitochondria (see e.g. Klingenberg & Pfaff, 1967), has been applied to chloroplasts with great success by Heldt and his colleagues (see e.g. Heldt *et al.*, 1972). It

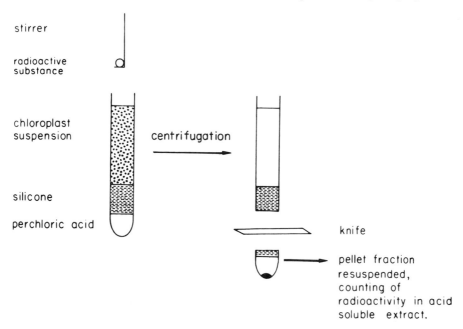

stirrer

radioactive substance

chloroplast suspension   centrifugation

silicone

perchloric acid   knife

pellet fraction resuspended, counting of radioactivity in acid soluble extract.

Fig. 1.1 The centrifugal filtration procedure for the measurement of metabolite transport (after Heldt, 1976).

usually involves loading chloroplasts with a labelled metabolite and following the redistribution of label between the plastid and a bathing solution after centrifugal acceleration through a filtering layer of silicone oil into a denaturing agent such as perchloric acid (Fig. 1.1). Its disadvantage, if it has any, is that it has not yet proved possible to apply the method to chloroplasts which are actively engaged in photosynthesis under relatively normal conditions. In an interesting improvement of this technique, uptake is *initiated* by centrifugation of the chloroplasts through a Percoll layer containing the labelled metabolite in silicone oil microfuge tubes (Howitz & McCarty, 1985a).

The availability of isolated protoplasts allowed the application of this approach to the rapid separation of chloroplasts from other cellular fractions and has greatly extended the usefulness of *in vitro* studies. Protoplasts, after illumination or other pre-treatments, are ruptured by centrifugation through a net. The cytoplasm is held above a silicone oil layer, whilst the chloroplasts are filtered through it into the denaturating solution (Fig. 1.2). This technique allows rapid examination of the partitioning of metabolites between these two fractions during photosynthesis or respiration.

1.2.2.2.2 Chromatographic analysis Mixtures containing intact isolated chloroplasts are analysed during and after photosynthesis under a variety

5

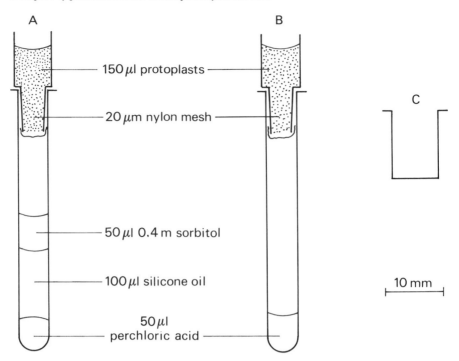

Fig. 1.2 A technique for the rapid separation of cytoplasm and chloroplasts from isolated protoplasts. The latter are suspended in the top container where they may be illuminated or otherwise pre-treated. At zero time they are ruptured by centrifugation through a 20 μm net. The extra-chloroplastic components are held above the silicone oil, the chloroplasts filtered through it into a denaturating solution.

of conditions. This is a natural extension of the work on *Chlorella* on which the Benson–Calvin cycle is based (Benson & Calvin, 1950; Bassham & Calvin, 1957). Its principal disadvantage is that it is extremely laborious, which inevitably restricts the number of observations on which conclusions are based. The advent of high performance liquid chromatography (HPLC) and the extensive use of spectrometric assays, however, has improved this aspect; but, like other procedures involving isolated chloroplasts, it cannot readily distinguish between events which take place inside the chloroplast and those which may occur in the surrounding medium.

1.2.2.2.3 Identification of specific translocators Once a specific inhibitor of transport of a given metabolite is found (this inhibition must be prevented by the presence of the natural substrates) differential labelling of the inhibitor allows the identification of the polypeptide(s) involved in transport. This method has been used in the identification of the phosphate translocator (see Sect. 1.2.3.1) and the glycollate transporter of the chloroplast (see Sect. 1.2.5.1).

1.2.2.2.4 Equilibrium-density gradient (isopycnic) centrifugation This method is useful for separating organelles and studying their enzyme composition. Peroxisomes were found to contain some key enzymes of the glycollate pathway, including glycollate oxidase, catalase and glyoxylate reductase, i.e. the enzymes needed to convert glycollate to glycine. The mitochondria contain the enzymes which decarboxylate glycine to serine + CO_2. This aspect will be discussed in detail in Sect. 1.3.2.

1.2.2.3 Indirect methods

These are often based on attempts to influence the course of photosynthesis in intact isolated chloroplasts, protoplasts or leaves by the addition of exogenous agents. If, for example, an added metabolite can produce a rapid and profound effect, the simplest explanation is that it does so following penetration. The advantage of this approach is that many experiments can be carried out in a short time, so that a large body of information can be elicited, and that the experiments relate to chloroplasts engaged in active photosynthesis. The disadvantages derive from the fact that interpretation is largely presumptive. Clearly it does not necessarily follow that an additive must penetrate the chloroplast envelope in order to produce an effect, or that if it does penetrate, it does so unchanged.

1.2.2.3.1 Catalysis by intact and ruptured chloroplasts Techniques developed with the aim of separating intact chloroplasts from leaf tissues (Walker, 1971) yield preparations containing, on average, some 70–80% of Class A chloroplasts (Hall, 1972). These may be readily stripped of their envelopes by brief osmotic shock, and reaction mixtures can be prepared which differ only in the degree of envelope integrity displayed by the chloroplasts within them. Envelope rupture exposes cryptic catalytic sites to reagents which do not cross the envelope, and this has been used for some years as a criterion of permeability (see e.g. Walker, 1965b; Heber *et al.*, 1967a; Mathieu, 1967). Osmotic shock abolishes the ability of intact chloroplasts to assimilate carbon at rapid rates (see e.g. Walker, 1965b), principally because of dilution of stromal coenzymes and cofactors. Conversely it accelerates the Hill reaction with non-penetrating oxidants such as ferricyanide (Heber & Santarius, 1970; Cockburn *et al.*, 1967b) and dark CO_2-fixation with ribulose biphosphate or ribose-5-phosphate + ATP as substrates.

1.2.2.3.2 Osmotic volume changes Intact chloroplasts shrink in hypertonic media and swell in hypotonic media in which the osmoticum is a non-penetrating compound such as sucrose (Nobel & Wang, 1970; Heldt & Sauer 1971; Wang & Nobel, 1971). Thus sorbitol causes a fast and irreversible increase in light-scattering, as measured by apparent absorbance at 535 nm, indicating chloroplast shrinkage brought about by exosmosis of water. However, a rapid exchange process would not necessarily bring about an osmotic response.

1.2.2.4 Other aspects of work with functional chloroplasts

The question of rate and chloroplast intactness is not unimportant in this context. If an isolated chloroplast is capable of rapid electron transport and photophosphorylation but has lost the ability to assimilate CO_2 when illuminated in a suitable reaction mixture, then it has obviously been exposed to damage or irreversible inhibition during isolation. Because there is often a clear correlation between envelope integrity and function, results obtained with relatively inactive chloroplasts must be viewed with more suspicion than those obtained with chloroplasts capable of achieving the rates of photosynthesis of the order displayed by the parent tissue.

1.2.2.5 Use of mutants

Mutants (generated by azide treatment) have been useful in the study of several metabolic routes, particularly photorespiration. Many mutants of *Arabidopsis thaliana* and barley have been characterized. Experiments with a photorespiratory mutant of *Arabidopsis thaliana* showed that dicarboxylate transport, which catalyses the exchange of 2-oxoglutarate and glutamate, is an essential step in the photorespiratory nitrogen metabolism (see below). One of the air-sensitive lines (unable to grow in air but which grow under high CO_2 concentrations which suppress photorespiration) from barley has a lesion which blocks the uptake of 2-oxoglutarate (Wallsgrove *et al.*, 1986).

1.2.3 Specific transport

Chloroplasts are enclosed by two membranes. The outer membrane is freely permeable to small molecules (up to about 10 000 Da), due to the presence of a *porin*, and the inner membrane is the osmotic barrier and the site where specific transport occurs. The specificity of envelope permeability is strikingly highlighted by P_i and PP_i, the former amongst the most rapidly translocated molecules, the latter amongst those to which the envelope is relatively impermeable.

Carrier-mediated anion transport can be classified as: (a) electroneutral, involving exchange of one anion with another of equal charge; (b) electroneutral proton compensated, where the different charge is compensated by co-transport of a proton; and (c) electrogenic, involving an exchange between anions of different charges (which requires energy as membrane potential or proton electrochemical gradient).

1.2.3.1 The transport of orthophosphate and the phosphate translocator

In whole plants and isolated chloroplasts, photosynthesis does not reach its full rate immediately upon illumination (unlike photosynthetic electron tranport), but only after a lag or induction period which is believed to represent the time taken for Benson–Calvin cycle intermediates to build up to the steady state concentration dictated by the prevailing light intensity (Rabinowitch, 1956; Walker, 1973, 1975). This lag can be shortened by certain intermediates (see e.g. Walker *et al.*, 1967) and shortening of induction has been taken as an indication of penetration (Walker & Crofts, 1970; Gibbs, 1971; Heber, 1974; Walker, 1974). Similarly the initial lag can be lengthened almost indefinitely by exogenous P_i, and reversal of P_i inhibition of metabolites has therefore also been interpreted as evidence of entry (Cockburn *et al.*, 1967a,b; Cockburn *et al.*, 1968; Walker, 1969). This sort of work led to the proposal of the existence (Walker & Crofts, 1970) and identification, of the P_i translocator (see e.g. Heldt & Rapley, 1970b).

Isolated chloroplasts are essentially P_i consuming organelles, and if illuminated in P_i-free media will cease to photosynthesize as the endogenous P_i is consumed (Cockburn *et al.*, 1967b). Under the conditions usually employed, the release of P_i associated with starch synthesis is insufficient, even in the short term, to offset the P_i incorporated into sugar phosphates and lost to the medium in this form. *In vivo*, P_i which is lost to the cytoplasm in this way is then released in processes such as sucrose synthesis and is constantly recycled. *In vitro*, a constant supply of P_i may be maintained by the inclusion of PP_i in the reaction mixture (Jensen & Bassham, 1966; Walker, 1971). Chloroplasts contain an active pyrophosphatase and, in the presence of exogenous Mg^{2+}, there is normally enough of this enzyme released from damaged chloroplasts to supply P_i at a noninhibitory concentration (Lilley *et al.*, 1973; Schwenn *et al.*, 1973).

If photosynthesis by isolated chloroplasts has been allowed to cease for lack of P_i, it can then be restarted without appreciable delay by the addition of P_i (Cockburn *et al.*, 1967c). In the short term there is then a rough stoichiometry of three molecules of O_2 evolved or three molecules of CO_2 for each molecule of P_i added. This is consistent with observations (Baldry *et al.*, 1966) which show that ^{32}P is initially incorporated into triose phosphate according to the overall equation:

$$3CO_2 + P_i + 2H_2O \rightarrow 1 \text{ triose phosphate} + 3O_2 \qquad (1.2)$$

As the P_i concentration in the external medium is increased, an optimum is reached at about $10^{-4}M$ and thereafter increasing concentrations at first extend the induction period and then depress the maximal rate until photosynthesis is ultimately almost entirely suppressed within the first hour of illumination (Cockburn *et al.*, 1967a,b). The precise nature of the response to P_i depends on the activity of the individual chloroplast preparation and the pre-treatment of the parent tissue. Thus very active chloroplasts from pre-illuminated leaves are difficult to inhibit, whereas at the other extreme, poor chloroplasts from dark-stored leaves are easy to inhibit. Inhibition is reversed, often completely, by the addition of certain cycle intermediates (Cockburn *et al.*, 1968). The response to PGA is very

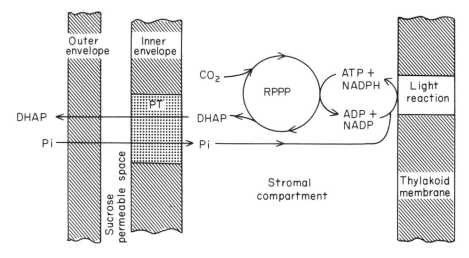

Fig. 1.3 The P_i translocator. This is located in the inner envelope and permits a strict stoichiometric exchange of external P_i with internal DHAP. Other compounds such as PGA will also exchange with PGA or DHAP. Movements may occur in either direction according to the conditions.

rapid and triose phosphates also produce a very fast response. Ribose-5-phosphate and fructose-1,6-bisphosphate produce reversal after a lag. Ribulose-1,5-bisphosphate is ineffective. P_i inhibition is seen as an exaggeration of the normal process in which newly synthesized triose phosphate is exchanged for external P_i. If high external P_i encourages export, low internal P_i should aid retention of product within the chloroplast. Orthophosphate may also be sequestered within the parent tissue by feeding mannose, with consequent increase in starch synthesis. When mannose is fed to leaves, it does not penetrate the chloroplast envelope but it is rapidly converted to mannose phosphate in the cytosol. Particularly in some species (e.g. spinach beet) which are unable to further metabolize mannose phosphate, mannose feeding leads to a sequestration of cytosolic phosphate. As the concentration of P_i in the cytosol falls, exchange of P_i and triose phosphate across the chloroplast envelope diminishes. This results in decreased photosynthesis but increased starch synthesis (see Sect. 1.8).

The phosphate triose phosphate phosphoglycerate translocator of the chloroplast is located in the inner envelope (Fig. 1.3) and catalyses the strict counter-exchange (for one molecule transported inwards, one is transported outwards; Fig. 1.4) of P_i, triose phosphate (DHAP and glyceraldehyde phosphate) and 3-PGA, providing the cytosol with triose phosphate. It also catalyses the export of DHAP in exchange for 3-PGA, providing the cytosol with extra ATP and NADH (when conversion is via NAD-glyceraldehyde dehydrogenase and phosphoglycerate kinase) or

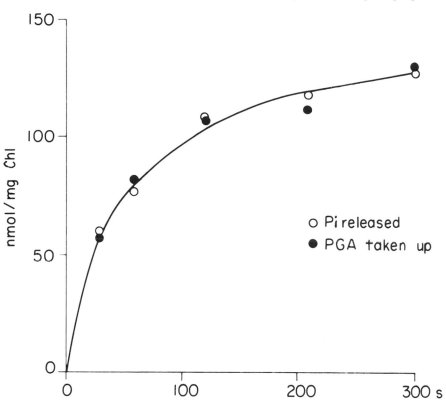

Fig. 1.4 Stoichiometry of P_i release and PGA uptake via the P_i translocator (after Heldt, 1976).

NADPH (if catalysed by NADP glyceraldehyde phosphate dehydrogenase), required for biosynthetic processes.

The existence of the phosphate translocator was predicted from indirect methods (Sect. 1.2.2.3) and demonstrated when silicone-layer filtering was used to investigate the permeability properties of isolated intact chloroplasts (Fig. 1.1 and see Sect. 1.2.2.2.1.). The phosphate translocator protein is the largest fraction of a chloroplast membrane preparation, about 15%, and its activity is very high (5000 min^{-1}). The phosphate translocator of C_3 plants binds either P_i or a P_i molecule attached to the end of a 3-carbon chain (2-PGA and PEP are poorly transported) and studies of pH dependence indicated that it accepts double negative charged anions. For this reason, the H^+ gradient across the chloroplast membrane has major effects on the transport by the P_i translocator of the three main metabolites. Inhibitor experiments indicated that a lysine group and an arginine group are involved in substrate binding and that a very reactive sulfhydryl group is essential (although it is not located in the binding site). Labelling experiments (with pyridoxal 5′-phosphate, an inhibitor of the phosphate translocator) implicate a polypeptide of MW 29 000 present in the membrane fraction in the phosphate translocator activity. The protein, MW 61 000, would be a dimer, coded by the nuclear genome and, after translation, imported into the chloroplast envelope membrane where it is converted to its mature conformation (Flügge & Heldt, 1984).

Although, as mentioned above, the maximum capacity of the translocator is very high, it has been proposed that, because of its competition between substrates, it may limit the overall rate of photosynthetic carbon assimilation in some circumstances (Lilley *et al.*, 1977; Portis 1983). As noted above, it seems that the translocator transports only the doubly ionized species, and the pK_a's of the three metabolites transported differ considerably. The actual ratio inside the stroma depends on stromal metabolism and the cytosolic concentrations but also on the cytosolic and stromal pH. It seems that pH differences may have an effect not only on the concentration of the transported species but also on the maximal velocity and affinity of the translocator.

1.2.4 The transport of cycle intermediates

At the time when rapid rates of CO_2-dependent oxygen evolution by isolated chloroplasts were first demonstrated (Walker & Hill, 1967), it was also shown that PGA would act as a Hill oxidant and subsequently (Cockburn *et al.*, 1967b) that this ability was lost following osmotic shock, implying that PGA must cross the envelope at least twice as fast as the rate of the observed O_2 evolution (i.e. at about 80 μmol mg^{-1} chlorophyll h^{-1}). This was in accord with a previous report of PGA reduction by intact chloroplasts (Urbach *et al.*, 1965).

After a relatively brief lag, all of the pentose monophosphates of the Benson–Calvin cycle will reverse P_i inhibition (Cockburn *et al.*, 1967a; Cockburn *et al.*, 1968; Walker, 1969), and R5P has also been shown to shorten induction (Bucke *et al.*, 1966; Baldry *et al.*, 1966; Walker *et al.*, 1967; Walker & Crofts, 1970; Gibbs, 1971; Walker, 1973) and to reverse inhibition by iodoacetate (Schachter *et al.*, 1968). Chromatographic analysis also indicates release of pentose monophosphates to the medium (Bassham & Jensen, 1967; Bassham *et al.*, 1968).

Fructose-6-phosphate (F6P) and glucose-6-phosphate (G6P) appear fairly quickly in the cytoplasm of leaves illuminated in the presence of $^{14}CO_2$ (Heber & Willenbrink, 1964; Heber *et al.*, 1967b), indicating export from the chloroplast, but (as with FBP) there is ambiguity because of the possibility of external synthesis from exported triose phosphate. Chromatography of isolated chloroplasts (Bassham & Jensen, 1967; Bassham *et al.*, 1968) indicates a low permeability to F6P and G6P. F6P has been reported to bring about a moderately fast reversal of P_i inhibition under conditions in which G6P gave no detectable response (Cockburn *et al.*,

1968). Subsequently, smaller (sometimes negligible and even inhibitory) responses have also been observed (Walker, unpublished) and the possibility that F6P actually moves, as such, at significant rates seems increasingly doubtful. There seems no real likelihood that F6P could be converted externally to FBP and triose phosphate but, in the presence of appreciable transketolase from ruptured chloroplasts, it might donate 2-carbon units to exported triose phosphate and enter as erythrose-4-phosphate and ribose-5-phosphate.

Sedoheptulose-7-phosphate(S7P) does not appear rapidly in the cytoplasm (Heber & Willenbrink, 1964; Heber *et al.*, 1967b) or in the medium (Bassham & Jensen, 1967; Bassham *et al.*, 1968), suggesting that it is slow to leave the chloroplast. Similarly, 6-phosphogluconate (which is known both to stimulate and to inhibit RuBP carboxylase, according to concentration) (Tabita & McFadden, 1972; Chu & Bassham, 1972, 1973, 1974; Buchanan & Schürmann, 1973) apparently fails to cross the envelope because it docs not inhibit CO_2 fixation by intact chloroplasts, nor does it reverse P_i inhibition or interfere with the subsequent reversal of P_i inhibition by triose phosphate (Cockburn *et al.*, 1968), or facilitate the release of PGA, etc. (Heldt & Rapley, 1970b).

There is general agreement that RuBP docs not move through the envelope. It does not appear in appreciable quantities in the cytoplasm (Heber & Willenbrink, 1964; Heber *et al.*, 1967b) or in the medium (Bassham & Jensen, 1967; Bassham *et al.*, 1968). It neither shortens the lag (Walker, 1969) nor reverses orthophosphate inhibition (Cockburn *et al.*, 1968) and only slightly affects inhibition by arsenate, etc. (Schachter *et al.*, 1968).

Both SBP and FBP are formed in relatively large quantities in illuminated chloroplasts *in situ* but SBP does not pass readily into the cytoplasm (Heber & Willenbrink, 1964; Heber *et al.*, 1967b) nor from isolated chloroplasts into the medium (Bassham *et al.*, 1968). Apparent movement of FBP is considerable (Heber, 1967; Heber *et al.*, 1967a) but is questioned in view of the possibility of external condensation of freely permeable triose phosphates (Bassham & Jensen, 1967; Bassham *et al*, 1968; Cockburn *et al.*, 1968). FBP shortens induction (Baldry *et al.*, 1966), but again external lysis followed by endodiffusion of the products cannot be discounted and it may be supposed that the envelope is as impermeable to FBP as it is to RuBP and SBP. Direct measurements (Heldt & Rapley, 1970b) support this view.

1.2.4.1 The transport of free sugars

There is clear evidence that sucrose does not cross the inner envelope of the chloroplast. Intact chloroplasts have also been successfully isolated in solutions in which a variety of sugars and sugar alcohols have been used to maintain the osmotic pressure, and it may be concluded, therefore, that the inner envelope is also largely impermeable to glucose, fructose, sorbitol and mannitol. This, however, is not the case for several pentoses and hexoses, such as D-xylose, D-ribose, D-glucose, D-mannose and D-fructose, which are transported across the envelope. The explanation for this apparent contradiction is that, at least for the glucose translocator (with a K_m of 20 mol m^{-3} and which also appears to transport maltose; Herold *et al.*, 1981), high concentrations of the substrate inhibit its transport. The glucose and maltose transporter would participate in the transport of these sugars produced by breakdown of stromal starch (Herold *et al.*, 1981). Thus although the glucose transporter has a lower activity than the phosphate translocator, it is active enough to play a substantial role in the export of hexoses formed during the degradation of chloroplast starch at night (Stitt *et al*, 1987).

1.2.5 The transport of carboxylic acids

1.2.5.1 Glycollate and glycerate

Photorespiration involves the oxidation of glycollate, but glycollate oxidase is located in peroxisomes whereas glycollate itself derives from the Benson–Calvin cycle. The photorespiratory metabolites glycollate and D-glycerate cross the chloroplast envelope on the same carrier. Studies with uncouplers indicate that glycerate uptake is driven by a proton gradient

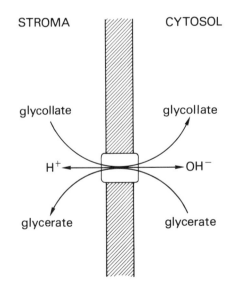

Fig. 1.5 Exchange of glycerate for glycollate at the chloroplast envelope. Note that during photorespiration the ratio of chloroplast glycollate export to glycerate import is 2:1. Therefore half the glycollate efflux would occur by glycollate/H$^+$ symport or glycollate/OH$^-$ antiport (see Howitz & McCarty, 1986).

established across the chloroplast envelope in the light (Robinson, 1984). The overall stoichiometry of this transport, during photorespiration, has been proposed to be two glycollates plus one proton leaving the chloroplast and one D-glycerate entering it (Fig. 1.5). The substrates of the glycollate transporter appear to be limited to 2- or 3-carbon 2-hydroxy-monocarboxylates.

1.2.5.2 Transport of dicarboxylates

The dicarboxylates, malate, succinate, oxaloacetate, 2-oxoglutarate, aspartate and glutamate are transported through the chloroplast envelope in a carrier-mediated mode (Lehner & Heldt, 1978; Proudlove & Thurman, 1981; Day & Hatch 1981b; Proudlove *et al.*, 1984). The dicarboxylate translocator transports compounds such as malate, succinate and fumarate, which undergo counter-exchange, and the K_m for uptake of a particular dicarboxylic acid is similar to the K_i for the inhibition of uptake by other dicarboxylates; for example, the K_i (0.3 mol m^{-3}) for oxaloacetate inhibition of malate transport is comparable to the K_m (0.4 mol m^{-3}) for malate uptake. However, such a system would be unsuitable for catalysing oxaloacetate uptake in a system where oxaloacetate concentrations are several orders of magnitude less than malate concentrations. Hatch *et al.* (1984) have provided evidence for a very active oxaloacetate carrier (K_m (OAA) 45 mmol m^{-3} in maize, 9 mmol m^{-3} in spinach) which is little affected by malate (K_i (malate) 7.5 mol m^{-3} in maize, 14 mol m^{-3} in spinach chloroplasts). Whether this transporter exists in C$_4$ plants such as *Amaranthus edulis*, which accumulate large quantitites of oxaloacetate (Leegood & von Caemmerer, 1988) is not known.

Chloroplasts photoreduce oxaloacetate *via* NADP-malate dehydrogenase located in the stroma. In C$_3$ plants this process may be part of a mechanism for transferring reducing equivalents across the chloroplast envelope. Somerville and Ogren (1983) and Somerville and Somerville (1985) isolated a mutant of *Arabidopsis thaliana* deficient in the chloroplast dicarboxylate translocator and have suggested that the fact that it grows normally under non-photorespiratory conditions relegates a role for the dicarboxylate transporter in shuttling reducing equivalents to one of minor physiological importance. However, in NADP-malic enzyme-type C$_4$ plants such as maize, the reduction of oxalocetate is a critical step in the pathway of CO$_2$ assimilation.

A detailed study of dicarboxylate transport by Woo *et al.* (1987) has provided evidence that two translocators operate in conjunction (Fig. 1.6). The 2-oxoglutarate translocator is apparently specific for the transport of only 2-oxoglutarate, malate, succinate, fumarate and glutamate. In contrast, the dicarboxylate translocator apparently transports all these dicarboxylates (including 2-oxoglutarate) and aspartate. Note that a crucial role is played by malate, so that the system operates in a cascade-like

15

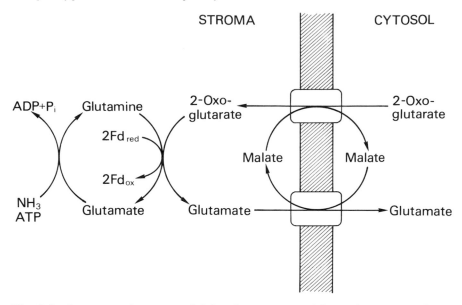

Fig. 1.6 A two-translocator model for the transport of 2-oxoglutarate on the 2-oxoglutarate translocator and of glutamate on the dicarboxylate translocator during ammonia assimilation in the light. Malate is the counter-ion for both the import of 2-oxoglutarate and the export of glutamate (redrawn from Woo *et al.*, 1987).

manner, and that by this means rapid transport of 2-oxoglutarate and glutamate can be maintained for photorespiratory NH_3 assimilation and recycling in C_3 leaves *in vivo*. This two-translocator model helps to resolve a number of discrepancies which have been reported in studies of dicarboxylate transport into chloroplasts, including: (a) the overlapping specificity of dicarboxylate transport (Lehner & Heldt, 1978); (b) the presence of a distinct site for 2-oxoglutarate transport (Proudlove *et al.*, 1984; Woo *et al.*, 1984); (c) the effect of malate, succinate, fumarate and glutamate in stimulating NH_3 + 2-oxoglutarate-dependent O_2 evolution (Dry & Wiskich, 1983; Woo, 1983; Woo & Osmond, 1982) and its decline after an initial rapid phase (Woo & Osmond, 1982); and (d) malate stimulation of NH_3 + 2-oxoglutarate-dependent O_2 evolution in intact chloroplasts but not of glutamate + 2-oxoglutarate-dependent O_2 evolution in a reconstituted chloroplast system (Anderson & Walker, 1984).

Studies with intact and lysed peroxisomes show that the peroxisomal membrane does not impose a barrier to the entry of glycollate, serine or O_2, but that it can limit the entry of NADH (to which the membrane appears impermeable) and the entry of glyoxylate, hydroxypyruvate, alanine and glutamate (Liang & Huang, 1983; Schmitt & Edwards, 1982), although nothing is known of the nature of any carriers that may be involved.

Mitochondria from leaves of C_3 plants convert glycine to serine at high rates (Journet *et al.*, 1981). Glycine oxidation by isolated mitochondria is accompanied by an efflux of serine, but to date there is no evidence for a specific glycine-serine antiport or a common carrier. Glycine transport into the mitochondrion may have both carrier-mediated and diffusional components (see Hanson, 1985).

1.2.5.3 Amino acids

Conversion of exogenous aspartate and 2-oxoglutarate to glutamate and oxaloacetate by intact chloroplasts is accelerated from *c.* 10 to 40 μmol mg^{-1} chlorophyll h^{-1} following osmotic shock, indicating that the intact envelope limits transport of the slowest of these reactants to the lower rate (Heber *et al.*, 1967a).

Nobel *et al.* (Nobel & Wang, 1970; Wang & Nobel, 1971; Nobel & Cheung, 1972) proposed two carriers on the basis of light-dependent shrinkage in chloroplasts caused by osmotic extraction of water (cf. Packer *et al.*, 1965; Packer & Crofts, 1967). It was proposed (Nobel & Cheung, 1972) that one carrier transports glycine, L-alanine, L-leucine, L-isoleucine and L-valine, and the other L-serine, L-threonine and L-methionine, but Gimmler *et al.* (1974) found no evidence for the existence of specific carriers which might bring about the rapid uptake of neutral amino acids. The slow diffusion which they observed, however, was more than sufficient to permit the 0.1 μmol mg^{-1} chlorophyll h^{-1} uptake demanded by protein turnover. Since the rate of neutral amino acid uptake was linear in respect to concentration, it was concluded that they permeate the envelope by simple diffusion.

1.2.5.4 Adenylates

There is considerable variation in the capacities of chloroplasts to transport adenylates. Transport of ATP and ADP across the chloroplast envelope is by counter-exchange, is very slow, and is highly specific for external ATP. It would not contribute much to the export of ATP to the cytosol but could act mainly to import ATP into the stroma, providing the chloroplast with ATP (produced by glycolysis or respiration) during the night. In spinach chloroplasts direct adenine nucleotide transport proceeds by counter-exchange at low maximum velocities (less than 5 μmol mg^{-1} chlorophyll h^{-1} at 20°C; Heldt, 1969). In chloroplasts of young peas and in wheat, transport is much more rapid, while mesophyll chloroplasts of *Digitaria sanguinalis*, an NADP^{+}-malic enzyme-type C$_4$ plant, have a high capacity for ATP transport (greater than 40 μmol mg^{-1} chlorophyll h^{-1}) (Huber & Edwards 1976). PP$_i$ is also transported on the pea chloroplast adenine nucleotide transporter (Robinson & Wiskich, 1976, 1977; Stankovic & Walker, 1977) and an exchange between ATP and PEP was recently reported for mesophyll chloroplasts of both maize and pea (Woldegiorgis *et al.*, 1983).

17

1.2.6 The PGA/DHAP shuttle

If the permeability characteristics of the chloroplast envelope preclude rapid export of reducing power in the shape of ATP and NADPH, can this be achieved indirectly? More or less simultaneously and independently several authors (see e.g. Heber, 1970; Walker & Crofts, 1970) postulated that existence of 'shuttles', which would serve this purpose. The first evidence in support of these exercises in paper biochemistry came from Stocking & Larson (1969), who observed external reduction of NAD by chloroplasts in the presence of PGA, PGA kinase and triose phosphate dehydrogenase.

The export of DHAP by the phosphate translocator provides fixed carbon for the synthesis of sucrose in the cytosol (by exchanging DHAP for P_i), but also (by exchanging DHAP with 3-PGA) a shuttle results by which the chloroplast provides the cytosol with ATP and NADPH. The DHAP is converted into PGA in the cytosol either via NAD-glyceraldehyde dehydrogenase and phosphoglycerate kinase (generating NADH and ATP) or via the non-phosphorylating and non-reversible NADP glyceraldehyde phosphate dehydrogenase (generating NADPH).

1.3 The fate of triose phosphate

1.3.1 Regeneration of RuBP

The entire Benson–Calvin cycle can be divided into three phases. The initial carboxylation is followed by reduction to triose phosphate, and in the third phase of these C_3 molecules are rearranged to regenerate $3(C_5)$ molecules of CO_2-acceptor (Fig. 1.7). It should be noted that the cycle consumes 9 molecules at ATP and 6 molecules of NADPH in the formation of one triose phosphate product (which can also feed back into the cycle to promote autocatalytic acceleration). In total, 5 molecules of H_2O are consumed in the cycle proper and 3 are released in the generation of assimilatory power. If the triose phosphate product were hydrolysed to give free triose in a reaction consuming 1 molecule of H_2O, there would be no net P_i consumption and the entire sequence would simplify to the classic overall equation for photosynthesis.

1.3.2 Photorespiration and the regeneration of RuBP

At its most simple, photorespiration is a process involving light-stimulated CO_2 release and O_2 uptake (see e.g. Goldsworthy, 1970; Jackson & Volk, 1970; Gibbs, 1971; Zelitch, 1971, 1973). It is a characteristic of C_3 plants which have CO_2 compensation points of about 50 ppm or greater. Factors

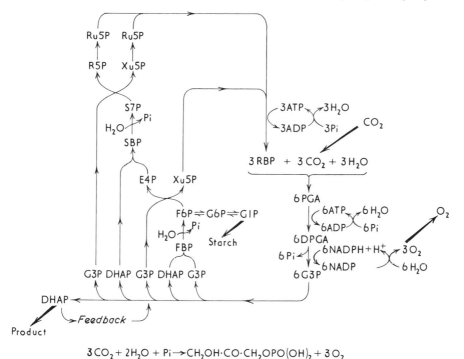

$$3\,CO_2 + 2H_2O + Pi \rightarrow CH_2OH\cdot CO\cdot CH_2OPO(OH)_2 + 3\,O_2$$

Fig. 1.7 (after Edwards & Walker, 1983) The reactions that lead to the regeneration of RuBP. On the right 3 molecules of RuBP combine with 3 molecules of CO_2 and 3 molecules of water to give 6 molecules of PGA. These are phosphorylated at the expense of ATP and the resulting DPGA is reduced by NADPH to G3P. The major part of this is converted to its isomer DHAP. Aldol condensation of these 2 triose phosphates gives a molecule of FEB which undergoes hydrolysis to F6P. This hexose phosphate is also the precursor of G6P and G1P which, after further transformation, give rise to starch. The F6P also enters the first transketolase reaction donating a 2-carbon unit to G3P to form Xu5P and E4P. The process of condensation, phosphorylation and 2-carbon transfer is repeated yielding SBP, S7P and 2 more molecules of pentose phosphate respectively. All 3 molecules of pentose monophosphate are finally converted to Ru5P, which is phosphorylated to RuBP.

which favour photorespiration (e.g. high light intensity, high temperature and low CO_2) also favour glycollate production. Photorespiration refers to the fact that C_3 plants evolve CO_2 when illuminated in CO_2-free air and this is one of the many methods used to measure this process. Photorespiration is apparently absent in C_4 plants (see Sect. 1.6). Glycollate is considered the primary substrate of this 'light-respiration'. RuBP carboxylase-oxygenase is a branch point between photorespiratory and photosynthetic metabolism. Oxygen, reacting with RuBP, leads to glycollate synthesis and photorespiration, whilst CO_2, reacting with RuBP, leads to photosynthesis.

INTRACELLULAR MOVEMENT OF PHOTORESPIRATORY METABOLITES

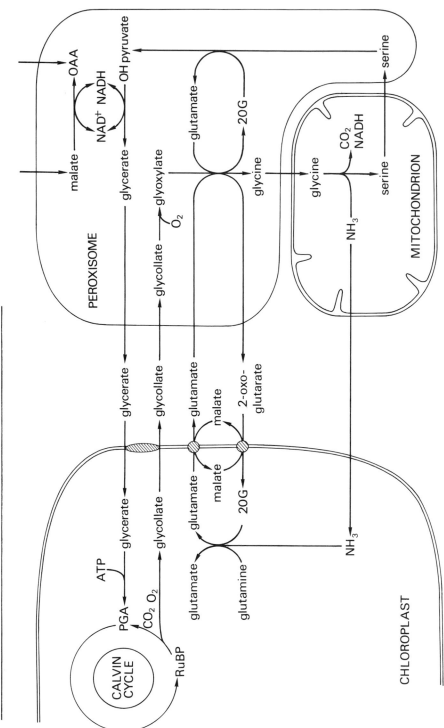

The synthesis of glycollate and its subsequent metabolism depends on partial reactions of the pathway in chloroplasts, peroxisomes and mitochondria and therefore on substantial metabolite transport between them (Fig. 1.8). It has been proposed that part of this transport constitutes a malate-oxaloacetate shuttle, by which redox equivalents are transferred from the mitochondial matrix to the peroxisomes (see Sect. 1.2.5.2).

1.3.3 Starch metabolism

The evidence that starch is synthesized within the chloroplast is, of course, visual and unequivocal (Sachs, 1862, 1887; Rabinowitch, 1945). Stromal starch is believed to be formed primarily from triose phosphate released from the RPP pathway. If triose phosphate is retained within the chloroplast the initial reactions are the same as those involved in sucrose synthesis in the cytoplasm, i.e. a proportion of the triose phosphate undergoes aldol condensation to FBP, which is then hydrolysed to F6P. This is then converted to its isomer by hexose phosphate isomerase. In the reaction catalysed by phosphoglucomutase, G6P is converted to G1P. At equilibrium, a mixture of the above enzymes would yield hexose phosphates in the proportions of approximately F6P (10) → G6P (50) → G1P (1). Although these reactions are considered freely reversible (but see Dietz, 1985) the overall equilibrium may still be important in determining the distribution of carbon between starch and pentose monophosphate in the illuminated chloroplast. Both F6P and triose phosphate are substrates for the first transketolase reaction and this will, in turn, influence the amount of triose phosphate entering the second aldolase condensation and the second transketolase reaction. An active sink for G1P would therefore tend to deflect carbon towards starch (the sink could be provided by ADP-glucose pyrophosphorylase, which catalyses the reaction: glucose-1-P + ATP → ADP-glucose + PP_i, when low external P_i decreases triose phosphate and PGA export).

A stromal endoamylase is believed to initiate starch degradation in the chloroplast. With the help of R-enzymes and D-enzymes this produces substrates which readily undergo phosphorolysis. G1P is then converted to triose phosphate:

$$G1P \rightarrow G6P \rightarrow F6P \rightarrow FBP \rightarrow G3P + DHAP$$

This triose phosphate can be exported from the chloroplast.

Fig. 1.8 The photorespiratory pathway, showing the major shuttles which may operate between the chloroplasts, peroxisomes and the mitochondria. Note that the diagram is non-stoichiometric, but since the pathway involves the conversion of 2 molecules of glycine to 1 of serine, a double transamination is shown for the conversion of glyoxylate to glycine. For the purpose of illustration, re-assimilation of ammonia is assumed to be entirely chloroplastic. Re-reduction to malate of oxaloacetate (or aspartate) exported from the peroxisomes may occur in either the mitochondria or the chloroplasts.

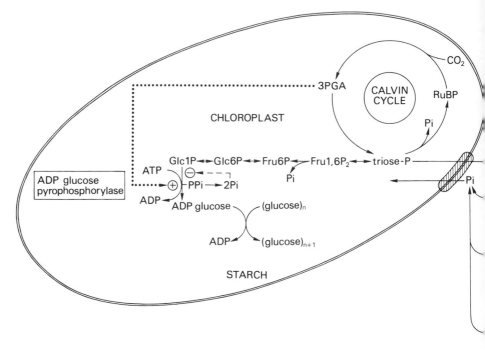

Fig. 1.9 The fine control of sucrose and starch synthesis and the role of metabolite modulation, including that by fructose 2,6-bisphosphate (after Heldt & Stitt, 1986).

1.3.4 Synthesis of sucrose; the site of sucrose synthesis

Experimental evidence indicates that (in both photosynthetic and non-photosynthetic cells) sucrose is synthesized via sucrose phosphate synthetase. Sucrose phosphate synthetase catalyses the reaction: UDP-glucose + fructose-6-phosphate $\rightarrow$ sucrose-6-phosphate + UDP, and the sucrose phosphate formed is hydrolysed by a specific phosphatase to give free sucrose. The formation of UDP-glucose is analogous to the formation of ADP-glucose in starch synthesis (Fig. 1.9).

1.3.5 Metabolism of sucrosyl-fructans

Accumulation of fructans is not limited to reserve tissues (e.g. Jerusalem artichoke) but can also be found in leaves of temperate grasses. It has been proposed that these play an important role in the ability of temperate species to grow in northern latitudes, by providing storage of fixed carbon in a readily accessible state under conditions of low temperatures and short daylength (Pollock *et al.*, 1980). Two enzymes are, together, capable of

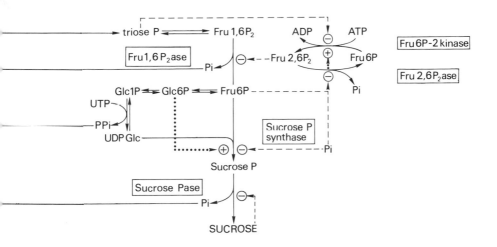

synthesizing fructan. The first, sucrose-sucrose 1-fructosyl transferase (SST) catalyses the irreversible synthesis of the trisaccharide 1^F-fructosyl sucrose (also known as isokestose or 1-kestose) from two molecules of sucrose:

$$G\text{-}1,2\text{-}F + G\text{-}1,2\text{-}F \rightarrow G\text{-}1,2\text{-}F\text{-}1,2\text{-}F + glucose$$

Synthesis of polymers is catalysed by the action of β-(2,1)-fructan 1-fructosyl transferase (FFT) in this reaction: $G\text{-}F\text{-}(F)_n + G\text{-}F\text{-}(F)_m \rightarrow G\text{-}F\text{-}(F)_{n-1} + G\text{-}F\text{-}(F)_{m+1}$

This enzyme mediates the transfer of terminal β-1-2-linked fructosyl residues between two oligosaccharide molecules, with the size and concentration of the various sugars determining which is the donor and which is the acceptor. There is strong evidence that all the major events of fructan metabolism occur in the vacuole. Fructans are accumulated in the vacuole (Wagner *et al.*, 1983) and it has been suggested (Edelman & Jefford, 1968) that FFT acts vectorially across the tonoplast, transferring fructosyl residues from trisaccharides in the cytoplasm to acceptors of higher degree of polymerization in the vacuole. Hydrolysis of fructan is presumed to occur in the vacuole, through the action of enzymes, fructan hydrolases, catalysing the stepwise hydrolysis of the terminal fructoside linkage:

$$\text{G-F-(F)}_n + H_2O \rightarrow \text{G-F-(F)}_{n-1} + F$$

By a succession of these reactions, fructan is converted irreversibly to a mixture of fructose and sucrose, with hydrolysis presumed to occur in the vacuole and the products passing into the cytosol for subsequent metabolism.

1.4 Control of carbohydrate metabolism

During the day, rates of starch and sucrose synthesis and the rate of photosynthetic carbon assimilation have to be co-ordinated. There is a clear need to determine how much assimilated carbon can be diverted to sucrose and starch synthesis without unduly decreasing the amount that returns to the Benson–Calvin cycle. Conversely, when sucrose accumulates in the cytosol because the rate of export diminishes (and/or photosynthesis increases) starch begins to accumulate inside the chloroplast. During the night, the sucrose accumulated in the vacuole during the day and the starch accumulated in the chloroplast are remobilized to be used to support the metabolism of the leaf itself or to be exported as sucrose (Stitt *et al.*, 1987). In this way, photosynthates are available 'around the clock'. The importance of these remobilization mechanisms is highlighted when they are disturbed. For example, mutants of the crucifer *Arabidopsis thaliana* which are unable to synthesize starch but can still synthesize sucrose will grow at the same rate as the wild type under continuous light, but the growth rate will be drastically diminished if grown in a day-night regime (Caspar *et al.*, 1985).

1.4.1 The role of compartmentation and transport

The chloroplast uses light to transform CO_2, P_i and water to triose phosphate. As noted above, the major metabolites transported through the phosphate translocator of the chloroplast envelope are P_i, PGA and DHAP. Besides the effects that these metabolites might have on sucrose and starch metabolism through their effects on the enzymes involved, levels of these metabolites could be a primary means of co-ordinating chloroplast and cytosol.

Plastids play an essential role in the biosynthetic activity of photosynthetic (as well as non-photosynthetic) cells. For example, fatty acid biosynthesis occurs exclusively in plastids, as does the biosynthesis of the majority of amino acids. Fructan synthesis occurs in the vacuole. It seems that most of a plastid's enzymes are encoded in the nuclear genome, translated in the cytosol and transported into the plastid. When similar reactions occur in the cytosol and plastids, these are catalysed by isoenzymes, with different properties and regulatory characteristics and coded for by different genes. Concentrations of the key metabolites of photosynthesis are central in the co-ordination of cell metabolism. Transport between compartments is as important as the activities of the key enzymes in the determination of these levels. The main limitation in assessing accurately the role of transport in the control of cell metabolism is the lack of reproducible techniques for measuring metabolite concentrations in the different compartments, and detailed information on the kinetic characteristics is only available for a few of the translocators.

1.4.2 *The fine control of sucrose and starch synthesis: the role of fructose 2,6-bisphosphate*

The role of metabolite effectors in the regulation of the key enzymatic reactions in photosynthesis has been the object of great attention during the last few years. Fructose 2,6-bisphosphate (and the enzymes that, in turn, regulate its level) plays a major role in the fine regulation of starch and sucrose synthesis, as it does in the regulation of glycolysis and gluconeogenesis in animals. This effector (it is a signal metabolite rather than an intermediate in the pathways concerned) was first discovered in liver but it is also present in leaves and other plant tissues at micromolar concentrations. It modifies the activity of cytosolic fructose-1,6-bisphosphatase (FBPase) and the PP_i-F6P phosphofructophosphotransferase (PFPase). It is synthesized and degraded by specific enzymes, the fructose-6-phosphate, 2-kinase (F6P,2-kinase) and the $Fru2,6P_2$ phosphatase, respectively. The activity of these enzymes is, in turn, affected by the concentration of other metabolites such as DHAP, PGA, P_i and F6P (Stitt, 1987; Stitt *et al.*, 1987; Fig. 1.9).

According to present understanding, $Fru2,6P_2$ is involved in the fine regulation of sucrose synthesis by 'sensing' the availability of substrates (e.g. DHAP) and adjusting sucrose synthesis accordingly. For example, high DHAP increases the amount of substrate available to the FBPase, because it is converted, in reactions which are near to equilibrium, into $Fru1,6P_2$. But increased DHAP would have a second effect on the enzyme by decreasing the concentration of $Fru2,6P_2$ (because it inhibits the activity of F6P,2-kinase, the enzyme which synthesizes $Fru2,6P_2$). This fall in $[Fru2,6P_2]$ would, in turn, release the inhibition of FBPase, which has a critical role in sucrose synthesis, because it catalyses the first irreversible reaction during the conversion of triose phosphate to sucrose in the cytosol. $Fru2,6P_2$ will thus be involved in the fine regulation of the partitioning between sucrose and starch.

25

1.5 Transport of photoassimilates into the vacuole

Giaquinta (1978) postulated that sink leaves store part of the imported sucrose in the vacuoles of the mesophyll cells. Although in the leaf most of the products of photosynthesis are usually translocated through the phloem to other parts of the plant, part of them is stored in the mesophyll cells, where they will be consumed during the subsequent dark period. Storage can be through accumulation of starch in the chloroplast stroma or accumulation of starch and malate in the extrachloroplast compartment. Two distinct sucrose pools exist in the palisade parenchyma cells of *Vicia faba* leaves, the cytosolic and the vacuolar (Fisher & Outlaw, 1979). During the day, a 40-fold increase in vacuolar sucrose in spinach leaves has been measured (Gerhardt & Heldt, 1984). The relatives sizes of these stores vary with species and environmental conditions. In spinach leaves, 280, 100 and 25 μatom C mg^{-1} chlorophyll of starch, sucrose and malate have been measured after the end of a 9 h illumination period, and all of these were almost depleted at the end of the following dark period (Gerhardt & Heldt, 1984).

Understanding of the intracellular transport of sucrose requires the study of the vacuolar component in the traffic of the key metabolites. This has long been a subject for speculation but the fragility of the vacuole has precluded, until relatively recently, experimental studies of the kind performed on isolated chloroplasts.

Most of the studies on vacuolar transport have been performed on protoplasts, which cannot export assimilates (Kaiser *et al.*, 1979; Giersch *et al.*, 1980). After 10 min $^{14}CO_2$ fixation about 90% of the total soluble label is outside the chloroplasts (Kaiser *et al.*, 1979; Giersch *et al.*, 1980), but if the newly synthesized sucrose were retained by the cytosol it would decrease osmotic potential and result in swelling of the cytosol, in metabolite dilution and in drastic changes in cytosol metabolism. The alternative was that in photosynthesizing protoplasts, sucrose and other soluble photosynthetic products were transported from the cytosol into the vacuole compartment. Kinetic work involving rapid isolation of vacuoles from photosynthesizing barley protoplasts showed that transport of photosynthetic products from the cytosol, across the tonoplast and into the vacuoles is a rapid process which may approach the rate of synthesis (Kaiser *et al.*, 1982). Sucrose, malate, citrate, glutamate, glutamine and alanine are among the metabolites transported. Further improvement in techniques allowed the isolation of intact vacuoles from barley protoplasts and the study of sucrose transport across the tonoplast (Kaiser & Heber, 1984). Sucrose transport is carrier-mediated as shown by substrate saturation, the effect of metabolic inhibitors and competitive substrates; the apparent K_m for sucrose was 21 mol m^{-3}. Transport was not energy-dependent and efflux experiments with pre-loaded vacuoles indicated that it is mediated by the same carrier which catalyses uptake (Kaiser & Heber, 1984).

Conversely, transport of malate is energy-dependent (Martinoia *et al.*, 1985) with uptake being markedly enhanced in the presence of MgATP (V_{max} increased from 2.6 to 8 μmol mg^{-1} chlorophyll h^{-1}; the K_m remained constant at 2.5 mol m^{-3}). This energy-dependent transport could occur against transmembrane concentration gradients and is probably coupled to the action of a largely specific H$^+$-translocating ATPase, different to those present in the mitochondrial and plasma membrane ATPases (for a review see Lin & Wagner, 1985).

1.5.1 *Centrifugal filtration*

Methods for isolation of vacuoles from protoplasts are essentially similar to those used with chloroplasts (see Sect. 1.2.2). Protoplasts are disrupted by squeezing them through the needle (5 cm × 0.1 mm) of a 250 μl syringe. The plasmalemma of about 40% of the protoplasts is ruptured, but most of the large central vacuoles remain intact and can then be purified by flotation using a preformed silicone oil/sorbitol or silicone oil/Percoll gradient (Kaiser *et al.*, 1982).

1.5.2 *Non-aqueous procedures*

Non-aqueous procedures can also been used (as with chloroplasts), and it has been demonstrated that both sucrose and malate are mainly located in the vacuole, although the modes of transport (inferred from the relative concentrations present in the different cellular compartments) seem to be different (Gerhardt & Heldt, 1984).

1.5.3 *Nuclear magnetic resonance*

Nuclear magnetic resonance spectroscopy (NMR) can be applied to higher plant tissues to follow long-term intracellular changes, to follow the uptake of small molecules and spectroscopic probes into the cytoplasmic and vacuolar compartments of the cell. NMR studies on plant systems are, so far, limited to ^{1}H, ^{13}C, ^{14}N and ^{31}P. Phosphorus-31 NMR, for example, can provide information on the relative concentration of phosphorus-containing molecules in the different cellular compartments. Contribution of the cytoplasm and vacuole can be distinguished because of the pH-dependence of the ^{31}P chemical shifts (the nuclear magnetic resonance of a nucleus is modified by its chemical environment). This technique has been used in the study of the mechanism of mobilization of stored carbo-hydrate in cells of sycamore (see e.g. Journet *et al.*, 1986) and of ortho-phosphate (Foyer *et al.*, 1981; Kime *et al.*, 1982; Rebeille *et al.*, 1983). So far, the requirements of the technique, which is relatively insensitive, limit

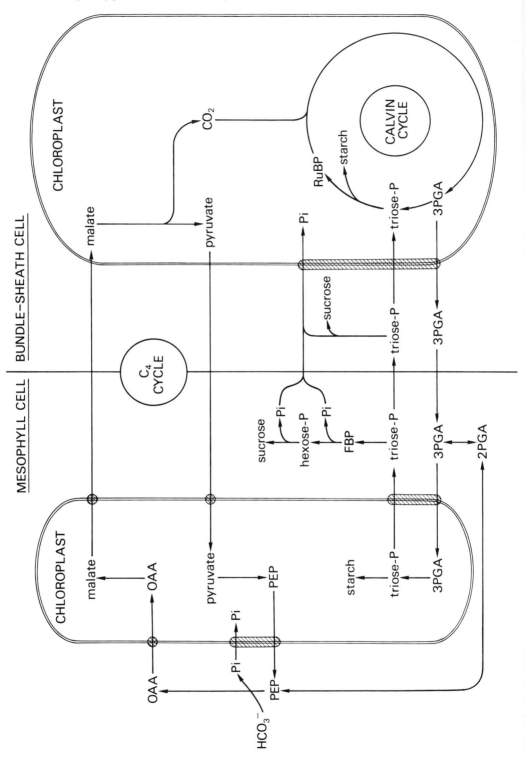

its usefulness to relatively long-term changes (hours or days rather than minutes) and a few plant materials, but recent progress suggests that in the future it might be possible to apply it to the more rapid changes which occur during illumination (for a review see Foyer & Spencer, 1987).

1.6 Transport of photoassimilates in C_4 photosynthesis

The C_4 cycle can be viewed simply as an ATP-dependent CO_2 pump which delivers CO_2 from the mesophyll cells to the bundle-sheath cells and thereby suppresses photorespiration (Hatch & Osmond, 1976; Fig. 1.10). However, the development of the C_4 syndrome has resulted in considerable modifications of inter- and intra-cellular transport processes. Perhaps the most striking development with regard to the formation of assimilates is that sucrose and starch formation are not only compartmented within cells, but in C_4 plants may be largely compartmented between mesophyll and bundle-sheath cells. This has been achieved together with a profound alteration of Calvin cycle function, in that 3PGA reduction is shared between the bundle-sheath and mesophyll chloroplasts in all the C_4 subtypes. Moreover, since C_4 plants are polyphyletic in origin, several different metabolic and structural answers have arisen in response to the same problem of how to concentrate CO_2. C_4 plants have three distinct mechanisms based on decarboxylation by $NADP^+$-malic enzyme, by NAD^+-malic enzyme, or by PEP carboxykinase in the bundle-sheath (Hatch & Osmond, 1976).

1.6.1 The phosphate translocator in C_4 plants

Several features distinguish chloroplastic transport of phosphorylated intermediates in C_4 plants from that in C_3 plants. During photosynthesis in C_4 plants, the mesophyll chloroplasts *import* 3PGA and *export* triose-P, and the bundle-sheath chloroplasts *export* 3PGA and *import* triose-P which has been reduced by the mesophyll chloroplasts. Thus most of the metabolite fluxes are opposed to those which occur in chloroplasts of a C_3 plant. This results from the fact that up to 50% of the 3PGA generated by the Calvin

Fig. 1.10 Inter- and intra-cellular transport processes during photosynthesis in an $NADP^+$-malic enzyme-type C_4 plant such as maize. Hatched areas indicate identified carrier-mediated transporters of metabolites across the chloroplast envelope. While starch and sucrose synthesis are shown to occur in both compartments, it should be noted that sucrose synthesis predominates in the mesophyll, while starch synthesis predominates in the bundle-sheath cells (see text). Malate, pyruvate, 3PGA and triose-P all move between the cytosols of the two cell-types along diffusion-driven concentration gradients. Phosphate presumably moves in a similar fashion.

cycle in the bundle-sheath is reduced by the mesophyll chloroplasts. The main function of this 3PGA export (which occurs in all C_4 decarboxylation types) may be to reduce O_2 evolution in the bundle-sheath, and so further the creation of a favourable CO_2/O_2 ratio which suppresses photorespiration. The triose-P which is formed in the mesophyll then has two possible fates. The first is conversion to sucrose (or starch) in the mesophyll cells (see below), while the second fate is to return to the bundle-sheath to regenerate RuBP. In species such as maize, where 50% of the 3PGA generated in the bundle-sheath is reduced in the mesophyll, then a minimum of two-thirds of the triose-P must be returned to the bundle-sheath to maintain pools of Calvin cycle intermediates.

In addition to the exchange of 3PGA, triose-P and P_i, the mesophyll chloroplasts also catalyse the export of PEP in exchange for P_i. PEP is formed in the chloroplast by the action of pyruvate P_i dikinase and is then utilized by PEP carboxylase in the carboxylation step of the C_4 cycle. Although the phosphate translocator of C_4 mesophyll chloroplasts has not been studied to the extent that it has in C_3 chloroplasts (and has not yet been isolated), there is evidence which indicates that exchange of PEP/P_i and 3PGA/triose-P occurs on a single common translocator in the chloroplast envelope (Day & Hatch, 1981a; Hallberg & Larsson, 1983a; Rumpho & Edwards, 1984, 1985, Huber & Edwards, 1977a). In contrast, the phosphate translocator in spinach chloroplasts appears both to recognize, and transport, PEP poorly (Fliege *et al.*, 1978). Studies on metabolite transport in C_4 mesophyll chloroplasts show that 3PGA uptake is competitively inhibited by PEP and P_i (Day & Hatch, 1981a,b; Rumpho & Edwards, 1984). More direct evidence has been produced by the use of DIDS (4-4'-diisothiocyano-2,2'-disulphonic acid stilbene) which, at micromolar levels, irreversibly inhibits 3PGA-dependent O_2 evolution in C_4 mesophyll chloroplasts. This inhibition can be prevented by pre-incubation with PEP, 3PGA or P_i. DIDS also inhibits 3PGA-dependent O_2 evolution in barley chloroplasts, which can be prevented by pre-incubation with PEP (Rumpho & Edwards, 1985). Rumpho *et al.* (1987) have also shown that the translocator in C_3 mesophyll chloroplasts can only recognize certain organic phosphates with phosphate in the C-3 position, while the translocator in C_4 mesophyll chloroplasts can recognize certain organic phosphates with phosphate in the C-2 or C-3 position. Thus, in addition to PEP, phosphoglycollate and 2PGA are effective inhibitors of 3PGA-dependent O_2 evolution in C_4 mesophyll chloroplasts but have no influence on 3PGA-dependent O_2 evolution in barley chloroplasts. The practical consequences, in terms of metabolic regulation, of having intermediates of the C_3 and C_4 pathways carried on a common translocator cannot easily be predicted, especially since we are ignorant of the subcellular compartmentation of these metabolites under varying environmental conditions.

Virtually nothing is known of the properties of the bundle-sheath chloroplast phosphate translocator, but bundle-sheath strands and chloroplasts from all the C_4 sub-groups metabolize Calvin cycle substrates in ways which suggest that the phosphate translocator may be modified (Boag & Jenkins, 1985). Bundle-sheath chloroplasts are also unusual in catalysing 3PGA export. In C_3 plants, 3PGA is not exported by chloroplasts either *in vitro* or *in vivo* to any great extent because 3PGA is transported by the translocator as the $3PGA^{2-}$ ion, whereas $3PGA^{3-}$ is the form which predominates at the pH which obtains in the illuminated stroma (Flügge & Heldt, 1984). Whether this means that the stromal pH is lower in bundle-sheath chloroplasts, or whether $3PGA^{3-}$ is indeed exported, or whether it simply reflects build-up of 3PGA within the chloroplast to such high levels that transport out of the chloroplast is inevitable, is not known.

1.6.2 Synthesis of starch and sucrose in C_4 plants

Although in a species such as maize the synthesis of sucrose appears to occur largely in the mesophyll cells, while the synthesis of starch occurs largely in the bundle-sheath cells, it is clear that there is a good deal of flexibility both within maize and between C_4 plants in general. For example, while the mesophyll tissue of maize grown under normal conditions contains no detectable starch, growth of plants in continuous light induces starch formation in the mesophyll (Downton & Hawker, 1973). On the other hand *Digitaria* spp. (which, like maize, are also $NADP^+$-malic enzyme-type C_4 plants) synthesize both sucrose and starch in the mesophyll compartment (Mbaku *et al.*, 1978; Hallberg & Larsson, 1983b).

A number of pieces of evidence support the contention that sucrose synthesis is confined to the mesophyll cells in leaves of at least some C_4 plants. Bucke & Oliver (1975) and Furbank *et al.* (1985) found that the majority of the sucrose-phosphate synthesis is located in the mesophyll of maize, *Pennisetum purpureum* and *Muhlenbergia montana*. Other studies have shown the cytosolic fructose bisphosphatase to be largely confined to the mesophyll (Furbank *et al.*, 1985) as well as F6P, 2-kinase, fructose-2,6-bisphosphatase (Soll *et al.*, 1983) and fructose-2,6-bisphosphate itself (Stitt & Heldt, 1985). However, Ohsugi & Huber (1987) have shown that sucrose-phosphate synthetase activity is present in both mesophyll and bundle-sheath cells in all C_4 sub-types, including maize. In addition, the response of the enzyme to light was different in the two compartments, with the bundle-sheath enzyme requiring higher irradiance for activation. Ohsugi and Huber (1987) suggest that sucrose-phosphate synthetase may function in both mesophyll cells and bundle-sheath cells for sucrose synthesis in the light, particularly at high light intensity, while in the dark the major function of bundle-sheath cell sucrose-phosphate synthetase may be in sucrose formation following starch degradation, a function which has been largely overlooked. Perhaps the safest conclusion to make at present is that starch and sucrose synthesis may predominate in one or other compartment in maize, but that this compartmentation may readily be overridden when environmental conditions (e.g. high light intensity) require it.

1.6.3 *The regulation of sucrose synthesis in the mesophyll cells*

When triose-P is exported from the chloroplast in the cell of a C_3 plant and is converted to sucrose, the P_i release re-enters the chloroplast in exchange for more triose-P. This provides a direct link between the provision of triose-P and its utilization in sucrose synthesis. Fine control of the enzymes of sucrose synthesis by triose-P, P_i, Fru2,6P_2, etc., then allows a sensitive adjustment of the rate of sucrose synthesis to the availability of fixed carbon (Stitt *et al.*, 1983). As we have seen, the situation in maize is rather more complicated. The triose-P which is available for sucrose synthesis in the mesophyll contains P_i which was incorporated into 3PGA in the bundle-sheath. Hence P_i released in sucrose synthesis in the mesophyll must be recycled back to the bundle-sheath, but this P_i will nevertheless play a part in regulating the transport of 3PGA, triose-P and PEP across the mesophyll chloroplast envelope.

The movement of metabolites such as 3PGA, triose-P and C_4 acids between the mesophyll and bundle-sheath cells is accomplished by diffusion which is dependent upon the generation of relatively large intercellular concentration gradients. In the case of triose-P this results in concentrations of this compound of the order of 10–15 mol m^{-3} in the mesophyll cells and 1 mol m^{-3} in the bundle-sheath cells (Leegood, 1985; Stitt & Heldt, 1985). Most of this triose-P must return to the bundle-sheath and not more than one-third can be siphoned off for sucrose synthesis, despite the high concentrations of triose-P present. Consequently the regulation of the cytosolic fructose-1,6-bisphosphatase from maize has been modified in showing a much higher K_m for FBP (3 mmol m^{-3} in spinach, 20 mmol m^{-3} in maize; and in the presence of Fru2,6P_2: 20 mmol m^{-3} in spinach, 3–5 mol m^{-3} in maize; Stitt & Heldt, 1985). In addition, although the F6P,2-kinase from maize is inhibited by triose-P and 3PGA, as in spinach, higher concentrations of triose-P are needed to achieve inhibition (Soll *et al.*, 1983). The enzyme is also inhibited by PEP and OAA in maize (Soll *et al*, 1983), which may facilitate co-ordination between the C_3 and C_4 cycles (i.e. Fru2,6P_2 will build up and inhibit sucrose synthesis when both C_3 and C_4 metabolites are low (see Leegood & von Caemmerer, 1988)).

1.6.4 *The regulation of starch synthesis in C_4 plants*

A number of studies have shown that the activities of starch synthase, of the branching enzyme and of ADP-glucose pyrophosphorylase are higher in the bundle-sheath than in the mesophyll. On the other hand, the enzymes of starch degradation, starch phosphorylase and amylase are more evenly distributed and slightly higher in the mesophyll (Huber *et al.*, 1969; Downton & Hawker, 1973; Echeverria & Boyer, 1986; Spilatro & Preiss, 1987). ADP-glucose pyrophosphorylase from spinach is activated by 3PGA and inhibited by P$_i$, with ratios of 3PGA/P$_i$ for half maximal activation typically being less than 1.5 (Ghosh & Preiss, 1966). ADP-glucose pyrophosphorylase from maize leaves requires very much higher ratios of 3PGA to P$_i$ for half-maximal activation, with the enzyme from the mesophyll cells requiring a higher ratio (9–16) than the enzyme from the bundle-sheath (7–10) (Spilatro & Preiss, 1987). Measurements show that 3PGA may be as high as 15 to 16 mol m^{-3} in the bundle-sheath and 5 to 7 mol m^{-3} in the mesophyll (Leegood, 1985; Stitt & Heldt, 1985), due to the requirement for metabolite gradients during photosynthesis. Thus 3PGA/P$_i$ ratios in the mesophyll are likely to be considerably lower than in the bundle-sheath. This factor, and the relatively low activities of enzymes of starch synthesis in the mesophyll, would appear to be one factor which limits synthesis of starch in the mesophyll relative to the bundle-sheath, but is a relationship which could readily be modified with fluctuations in physiological and developmental conditions and could therefore account for variations in the capacity of the mesophyll to make starch.

1.6.5 *Movement of other carbon compounds during photosynthesis in C$_4$ plants*

The transport of intermediates of the C$_4$ cycle has a direct bearing upon the intracellular transport of Calvin cycle intermediates, because although they are not, in a strict sense, assimilates, they may nevertheless be interconverted with pools of compounds such as 3PGA, via 2PGA and PEP, in the reactions catalysed by phosphoglycerate mutase and enolase (Furbank & Leegood, 1984).

Although transport processes across the envelope of mesophyll chloroplasts are now adequately characterized, virtually nothing is known of transport into the bundle-sheath chloroplasts, largely because these cannot easily be isolated in appreciable quantities from any C$_4$ plant, particularly maize. However, earlier work in which aspartate was shown to stimulate malate decarboxylation in bundle-sheath strands (Chapman & Hatch, 1981) has been shown by Boag and Jenkins (1985) to occur at the level of the isolated chloroplast. They suggest that the bundle-sheath chloroplasts of maize may possess a carrier specific for malate uptake which depends upon the presence of aspartate for maximum activity.

The transport of dicarboxylates into mesophyll chloroplasts of C$_4$ plants has already been mentioned in Sect. 1.2.5.2, but transport of another C$_4$

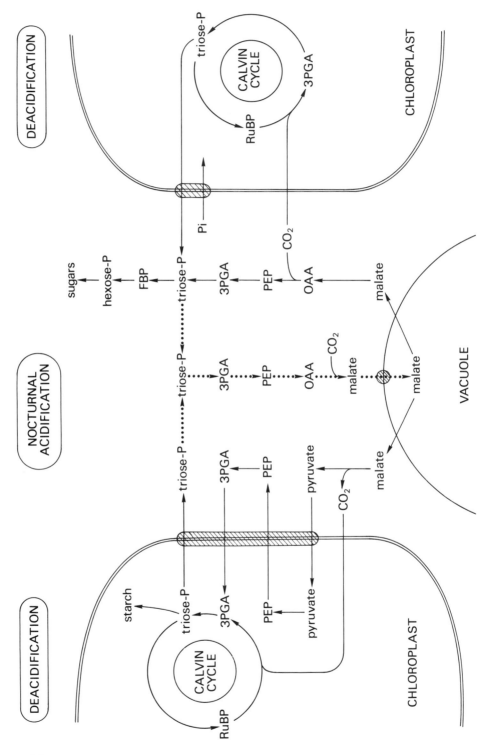

pathway intermediate, pyruvate, appears to occur on a separate carrier in mesophyll chloroplasts of both C_3 and C_4 chloroplasts. Huber and Edwards (1977b) were the first to demonstrate carrier-mediated uptake of pyruvate into chloroplasts of the $NADP^+$-malic enzyme species *Digitaria sanguinalis*. They studied transport in the dark and found that internal concentration of pyruvate did not rise above that in the medium. More recently, Flügge *et al.*, (1985) demonstrated light-dependent pyruvate uptake in maize and Ohnishi & Kanai (1987) have shown the same in mesophyll chloroplasts of *Panicum miliaceum*. In both cases pyruvate is concentrated within the stroma. In the bundle-sheath, by contrast, a much slower carrier-mediated uptake of pyruvate had very similar properties to the carrier in wheat (Ohnishi & Kanai, 1987) and pea (Proudlove & Thurman, 1981) chloroplasts and was not light-stimulated.

1.7 Movement of metabolites in CAM photosynthesis

Metabolism in CAM plants involves the transfer of large amounts of carbon between two storage pools, malic acid and storage carbohydrates (Fig. 1.11). While malate is invariably stored in the vacuole, the carbohydrate store may be either chloroplastic (e.g. starch-storers such as *Bryophyllum tubiflorum, Kalanchoe diagremontiana*) or extrachloroplastic (as in *Ananas comosus* (pineapple) and *Aloe arborescens*). This transfer of carbon involves glycolytic carbohydrate breakdown in the dark and gluconeogenic carbohydrate synthesis during the light. In starch-formers, large amounts of carbon must therefore enter the chloroplast during deacidification.

Fahrendorf *et al.* (1987) have proposed, as a working hypothesis, that in CAM plants which are primarily starch-formers: (a) malic enzyme is the principal decarboxylase, while pyruvate P_i dikinase is present in amounts sufficient to convert the pyruvate formed back to PEP; and (b) the capacity for conversion of FBP to F6P in the cytosol is low and levels of $Fru2,6P_2$ are low during deacidification. It is suggested that this apparent paradox may be due to a very low affinity of the FBPase for FBP, as occurs in the

Fig. 1.11 Possible patterns of carbon flow and metabolite transport in CAM plants. Nocturnal acidification, following breakdown of starch or sugars to form triose-P, is shown by the dotted line and involves active uptake of malate into the vacuole. Possible pathways of diurnal deacidification are shown in a starch-former on the left and in a soluble sugar-former on the right. Carbon traffic across the envelope would involve the phosphate translocator, with additional uptake of P_i in exchange for PEP as well as uptake of P_i in exchange for triose-P formed during starch breakdown (unless hexoses or hexose-P are exported across the chloroplast envelope) and in exchange for triose-P formed during CO_2 fixation. It is not known whether pyruvate uptake by CAM chloroplasts is carrier-mediated.

CAM but not in the C_3 form of *Mesembryanthemum crystallinum* (Keiller *et al.*, 1987) nor in C_4 plants such as maize (Stitt & Heldt, 1985). On the other hand, in CAM plants which store soluble, extrachloroplastic sugars: (a) PEP carboxykinase is the principal decarboxylase (and pyruvate P_i dikinase is virtually absent); and (b) the cells possess high activities of PP_i-dependent phosphofructokinase (PFP) and large amounts of Fru2,6P_2. PFP is then proposed to operate in a gluconeogenic direction.

Since *Bryophyllum* and *Mesembryanthemum* generate PEP inside the chloroplasts during deacidification, and since the enzymes which convert PEP to PGA appear to be exclusively cytoplasmic (Spalding *et al.*, 1979; Winter *et al.*, 1982), it is likely that, in a manner analogous to the exchange which occurs in C_4 plants, PEP is exported in exchange for cytoplasmic 3PGA (or possibly triose-P). Chloroplasts from CAM-*Mesembryanthemum* possess a P_i translocator which, in common with that described in C_4 plants, can exchange PEP, 3PGA, triose-P and P_i (Holtum *et al.*, 1987). There is an additional possibility that hexose-P could be manufactured in the cytoplasm and taken up by the chloroplasts, since chloroplasts of *Sedum* show a large stimulation of CO_2 fixation by F6P (Piazza & Gibbs, 1982). Chloroplasts from starch-forming CAM plants such as *Sedum* spp. are also remarkable in showing a very high P_i optimum for CO_2 fixation (i.e. it is difficult to enforce export of fixed carbon from the chloroplasts (Spalding & Edwards, 1980; Nishida & Sanada, 1977). Whether this results from altered affinities of the P_i translocator for its substrates, as compared with chloroplasts from a C_3 plant, or whether it results, for example, from a chloroplastic FBPase with a very high affinity for FBP, or from some other mechanism, is not known.

1.8 Measurement of photosynthesis *in vivo* in the study of metabolite flow between compartments

Measurement of different aspects of photosynthesis *in vivo* (CO_2 and O_2 exchange, light-scattering, chlorophyll fluorescence and its quenching components) led to the characterization of a broad range of symptoms caused by experimental limitation of P_i supply to the chloroplast (Sivak & Walker, 1987). These findings led to re-evaluation of data reported in the literature and symptoms of inadequate P_i supply were also recognized in various circumstances of physiological relevance. Several phenomena which can be observed *in vivo* and sometimes *in vitro* are strongly modified by P_i availability, e.g. oscillatory behaviour, the slow increase in photosynthesis displayed by many leaves after re-illumination, the 'CO_2 gulp' and associated 'O_2 burst', chloroplast energization and the response to decreased O_2 concentration.

1.8.1 Limitation of photosynthesis in vivo by P_i supply

The P_i requirement by the chloroplast depends on the supply of the other substrates, light and CO_2, and on the rates of the partial processes of photosynthesis in which these substrates are utilized. For example, the relative requirement for orthophosphate diminishes with photoinhibition (Walker & Osmond, 1986). At low temperatures, the $[P_i]$ optimum for isolated spinach chloroplasts increases (Leegood & Walker, 1983), implying that an unchanged P_i supply to the chloroplast which was adequate at 20°C would be inadequate at 10°C. It is self-evident that the relative importance of the P_i-recycling processes will vary with the plant material and environmental conditions. Decreased temperature has similar effects to sequestration of cytosolic P_i (and opposite to those of P_i feeding and increased temperature) in terms of oscillatory behaviour and the effect of oxygen (Sivak, 1987). Isolated chloroplasts require higher $[Pi]$ for maximum photosynthesis at low temperatures, and the lower thresholds in $[CO_2]$ needed to initiate oscillatory behaviour at low temperatures suggest that this is also true for chloroplasts *in situ*. Alternatively (or possibly, as well as) the overall rate of P_i recycling (by starch and sucrose synthesis, and movement from the vacuole) is slower at these temperatures.

The plant cell can adjust its P_i supply up to a degree by increasing the rates of one or more of the processes which recycle P_i (Walker & Sivak, 1985; Sivak & Walker, 1986). Sometimes, however, P_i recycling is insufficient and symptoms resembling those shown under experimental sequestration of P_i can be observed (Walker & Sivak, 1985). However, if the leaf can adjust the rate of the processes that recycle P_i to the new demands (e.g. increasing sucrose or starch synthesis following an increase in $[CO_2]$), the symptoms are only transitory (Sivak & Walker, 1986; Sivak, 1987).

Characterization of the syndrome led to re-evaluation of data reported in the literature, and the references below are just examples. One or more of the symptoms of inadequate P_i supply have been described in leaves of plants grown under limited P_i in nutrient solutions (Brooks, 1986; Prinsley & Leegood, 1986; Rao *et al.*, 1986), or grown in warm conditions and tranferred to relatively low temperatures (Cornic & Louason, 1980; Leegood & Furbank, 1986; Sivak & Walker, 1986; Labate & Leegood, 1988), in leaves photosynthesizing in saturating light and $[CO_2]$ (Walker & Sivak, 1985), in plants infected by pathogens (Walters & Ayres, 1983), or photosynthesizing in optimum conditions after a long period of illumination (Azcon-Bieto, 1983; Sharkey *et al.*, 1986), and in plants grown under high $[CO_2]$ (see Kramer, 1981).

There are, therefore, many circumstances in which leaves (even from plants grown in complete nutrient supply) may show symptoms of transient or permanent inadequate P_i supply leading to misinterpretations (e.g. 'O_2 insensitivity' has been explained as a change in the carboxylase/oxygenase ratio of RuBP carboxylase, a ratio believed to be extremely constant for a given species). If unusual responses of the kind described above are observed, inadequate P_i supply should be suspected and, if orthophosphate feeding via the petiole or through the edges of leaf pieces eliminates the syndrome restoring the 'normal' pattern (Walker & Sivak, 1985; Leegood & Furbank, 1986, ·Labate & Leegood, 1988), the diagnosis will then be confirmed. In addition, limitation of photosynthesis by inadequate P_i supply could be expected to cause secondary effects. For example, the observed decrease in the activation status of RuBP carboxylase-oxygenase in plants transferred to relatively low temperatures (Schnyder *et al.*, 1984) or grown in low P_i (Brooks, 1986) could be interpreted in this way. Long-term effects may also include photoinhibition (Azcon-Bieto, 1983) and accelerated senescence (Sivak, unpublished). Inadequate P_i supply during long periods could well result in slower plant growth than the rate predicted from 'optimum' environmental conditions, including high CO_2 (for a review see Kramer, 1981).

Whatever the primary events, experimental evidence has accumulated which indicates that RuBP regeneration cannot be separated from P_i recycling or supply. This has implications for basic research which seeks to increase photosynthesis and plant growth by improving RuBP carboxylase and decreasing photorespiration. Decreased photorespiration should decrease the light and $[CO_2]$ thresholds at which P_i becomes limiting, firstly because less P_i will be made available through photorespiration, and secondly because increased carboxylation will make higher demands on the supply of P_i, itself a substrate of photosynthesis. Symptoms of inadequate P_i may be displayed by leaves photosynthesizing under conditions which impose higher P_i demands than those the plant was likely to experience under the conditions in which it was grown, e.g. higher light intensities and CO_2 concentrations. Characterization of the syndrome, however, now makes it easier to recognize inadequate P_i supply in other circumstances which are more physiologically relevant. Thus, if limited P_i supply could be recognized as limiting photosynthesis and growth during the growth of cereals at relatively low temperatures or in plants growing in enriched CO_2, improved understanding of photosynthetic regulation could be used for the genetic or chemical improvement of the crop.

References

Anderson, J. W. & Walker, D. A. (1984). Ammonia assimilation and oxygen evolution by a reconstituted chloroplast system in the presence of 2-oxo-glutarate and glutamate. *Planta* **159**, 247–53.

Azcon-Bieto, J. (1983). Inhibition of photosynthesis by carbohydrates in wheat leaves. *Plant Physiology* **73**, 681–6.

Baldry, C. W., Walker, D. A. & Bucke, C. (1966). Calvin-cycle intermediates in relation to induction phenomena in photosynthetic carbon dioxide fixation by isolated chloroplasts. *Biochemical Journal* **101**, 641–6.

Bassham, J. A. & Calvin, M. (1957). *The Path of Carbon in Photosynthesis*. Englewood Cliffs, N.J., Prentice-Hall Inc.

Bassham, J. A. & Jensen, R. G. (1967). Photosynthesis of carbon compounds. In San Pietro, A., Greer, F. A. & Army, T. J. (eds.) *Harvesting The Sun*, pp. 79–110. New York, Academic Press.

Bassham, J. A., Kirk, M. & Jensen, R. G. (1968). Photosynthesis by isolated chloroplasts. I. Diffusion of labelled photosynthetic intermediates between isolated chloroplasts and suspending medium. *Biochimica et Biophysica Acta* **153**, 211–18.

Benson, A. A., Bassham, J. A., Calvin, M., Goodale, T. C., Haas, V. A. & Stepka, W. (1950). The path of carbon in photosynthesis. V. Paper chromatography and radioautography of the products. *Journal of American Chemical Society* **72**, 1710–18.

Benson, A. A., Bassham, J. A., Calvin, M., Hall, A. G., Hirsch, H. E., Kawaguchi, S., Lynch, V. & Tolbert, N. E. (1952). The path of carbon in photosynthesis. XV. Ribulose and sedoheptulose. *Journal of Biological Chemistry* **196**, 703–16.

Benson, A. A. & Calvin, M. (1950). Carbon dioxide fixation by green plants. *Annual Review of Plant Physiology* **1**, 25–42.

Bergmeyer, H. U. (1974). *Methods of Enzymic Analysis* Vol. 2, Ed. 3. Weinheim, Verlag Chemie.

Beutling, H. O. (1978). Enzymatische Bestimmung von Starke in Lebensmitteln mit Hilfe der Hexokinase-Methode. In *Starch/Starke*, pp. 309–12, Boehringer Publication 30/9.

Boag, S. & Jenkins, C. L. D. (1985). CO_2 Assimilation and malate decarboxylation by isolated bundle sheath chloroplasts from *Zea mays*. *Plant Physiology* **79**, 165–70.

Brooks, A. (1986). Effect of phosphorus nutrition on ribulose-1,5-biphosphate carboxylase activation, photosynthetic quantum yield and amounts of some Calvin-cycle metabolites in spinach leaves. *Australian Journal of Plant Physiology* **13**, 221–37.

Buchanan, B. B. & Schürmann, P. (1973). Regulation of ribulose-1,5-diphosphate carboxylase in the photosynthetic assimilation of carbon dioxide. *Journal of Biological Chemistry* **248**, 4956–64.

Bucke, C. & Oliver, I. R. (1975). Location of enzymes metabolising sucrose and starch in the grasses *Pennisetum purpureum* and *Muhlenbergia montana*. *Planta* **122**, 45–52.

Bucke, C., Walker, D. A. & Baldry, C. W. (1966). Some effects of sugars and sugar phosphates on carbon dioxide fixations by isolated chloroplasts. *Biochemical Journal* **101**, 636–41.

Caspar, T., Somerville, C. & Huber, S. (1985). Alterations in growth, photosynthesis and respiration in a starchless mutant of *Arabidopsis thaliana* (L.) deficient in chloroplast phosphoglucomutase activity. *Plant Physiology* **79**, 11–17.

Chapman, K. S. R. & Hatch, M. D. (1981). Aspartate decarboxylation in bundle sheath cells of *Zea mays* and its possible contribution to C_4 photosynthesis. *Australian Journal of Plant Physiology* **8**, 237–48.

Chu, D. K. & Bassham, J. A. (1972). Inhibition of ribulose-1,5-diphosphate carboxylase by 6-phospho-gluconate. *Plant Physiology* **50**, 224–7.

Chu, D. K. & Bassham, J. A. (1973). Activation and inhibition of ribulose-1,5-diphosphate carboxylase by 6-phosphogluconate. *Plant Physiology* **52**, 373–9.

Cockburn, W., Baldry, C. W. & Walker, D. A. (1967a). Photosynthetic induction phenomena in spinach chloroplasts in relation to the nature of the isolating medium. *Biochimica et Biophysica Acta* **143**, 603–13.

Cockburn, W., Baldry, C. W. & Walker, D. A. (1967b). Some effects of inorganic phosphate on O_2 evolution by isolated chloroplasts. *Biochimica et Biophysica Acta* **143**, 614–24.

Cockburn, W., Baldry, C. W. & Walker, D. A. (1967c). Oxygen evolution by isolated chloroplasts with carbon dioxide as the hydrogen acceptor. A requirement for orthophosphate or pyrophosphate. *Biochimica et Biophysica Acta* **131**, 594–6.

Cockburn, W., Walker, D. A. & Baldry, C. W. (1968). Photosynthesis by isolated chloroplasts. Reversal of orthophosphate inhibition by Calvin cycle intermediates. *Biochemical Journal* **197**, 89–95.

Cornic, G. & Louason, G. (1980). The effects of O_2 on net photosynthesis at low temperature. *Plant, Cell and Environment* **3**, 149–57.

Day, D. A. & Hatch, M. D. (1981a). Dicarboxylate transport in maize mesophyll chloroplasts. *Archives of Biochemistry and Biophysics* **211**, 738–42.

Day, D. A. & Hatch, M. D. (1981b). Transport of 3-phosphoglyceric acid, phosphoenolpyruvate, and inorganic phosphate in maize mesophyll chloroplasts, and the effect of 3-phosphoglyceric acid on malate and phosphoenolpyruvate production. *Archives of Biochemistry and Biophysics* **211**, 743–9.

Dietz, K.-J. (1985). A possible rate-limiting function of chloroplast hexosemonophosphate isomerase in starch synthesis of leaves. *Biochimica et Biophysica Acta* **839**, 240–8.

Douce, R. (1985). *Mitochondria in higher plants. Structure, Function and Biogenesis.* pp. 1–327. Orlando, Florida, Academic Press.

Downton, W. J. S. & Hawker, J. S. (1973). Enzymes of starch and sucrose metabolism in *Zea mays* leaves. *Phytochemistry* **12**, 1551–6.

Dry, I. B. & Wiskich, J. T. (1983). Characterization of dicarboxylate stimulation of ammonia, glutamine and 2-oxoglutarate-dependent O_2 evolution in isolated pea chloroplasts. *Plant Physiology* **72**, 291–6.

Echeverria, E. & Boyer, C. D. (1986). Localization of starch biosynthetic and degradative enzymes in maize leaves. *American Journal of Botany* **73**, 167–71.

Edelman, J. & Jefford, T. J. (1968). The mechanism of fructosan metabolism in higher plants as exemplified in *Helianthus tuberosus*. *New Phytologist* **67**, 517–31.

Fahrendorf, T., Holtum, J. A. M., Muckerjee, U. & Latzko, E. (1987). Fructose 2,6-biphosphate, carbohydrate partitioning, and Crassulacean acid metabolism. *Plant Physiology* **84**, 182–7.

Fisher, D. B. & Outlaw, H. W. (1979). Sucrose compartmentation in the palisade parenchyma of *Vicia faba*. *Plant Physiology* **64**, 481–3.

Fliege, R., Flügge, U., Werdan, K. & Heldt, H. W. (1978). Specific transport of inorganic phosphate, 3-phosphoglycerate and triose-phosphate across the inner membrane of the envelope in spinach chloroplasts. *Biochimica et Biophysica Acta* **502**, 232–47.

Flügge, U. I. & Heldt, H. W. (1984). The phosphate-triose phosphate-phosphoglycerate translocator of the chloroplast. *Trends in Biochemical Science* **9**, 530–3.

Flügge, U. I., Stitt, M. & Heldt, H. W. (1985). Light-driven uptake of pyruvate into mesophyll chloroplasts from maize. *FEBS Letters* **183**, 335–9 number 2.

Foyer, C. H. & Spencer, C. M. (1987). NMR as a probe of biochemical and physiological processes in plants, *Annual Report Research Institute for Photosynthesis, University of Sheffield*, pp. 16–18.

Foyer, C., Walker, D. A., Spencer, C. & Mann, B. (1981). Observations on the phosphate status and intracellular pH of intact cells, protoplasts and chloroplasts from photosynthetic tissue using phosphorus-31 NMR. *Biochemical Journal* **202**, 429–34.

Furbank, R. T. & Leegood, R. C. (1984). Carbon metabolism and gas-exchange in leaves of *Zea mays* L. Interaction between the C_3 and C_4 pathways during photosynthetic induction. *Planta* **162**, 457–62.

Furbank, R. T., Stitt, M. & Foyer, C. H. (1985). Intercellular compartmentation of sucrose synthesis in leaves of *Zea mays*. *Planta* **164**, 172–8.

Gerhardt, R. & Heldt, H. W. (1984). Measurement of subcellular metabolite levels in leaves by fractionation of freeze stopped material in nonaqueous medium. *Plant Physiology* **75**, 542–7.

Ghosh, H. P. & Preiss, J. (1966). Adenosine diphosphate glucose pyrophosphorylase. A regulatory enzyme in the biosynthesis of starch in spinach leaves chloroplasts. *Journal of Biological Chemistry* **241**, 4491–504.

Giaquinta, R. (1978). Source and sink leaf metabolism in regulation of phloem translocation. Carbon partitioning and enzymology. *Plant Physiology* **61**, 380–5.

Gibbs, M. (1971) (ed.) *Structure and Function of Chloroplasts*, pp. 169–214. Berlin, Heidelberg, New York, Springer-Verlag.

Giersch, C., Heber, U., Kaiser, G., Walker, D. A. & Robinson, S. P. (1980). Intracellular metabolite gradients and flow of carbon during photosynthesis of leaf protoplasts. *Archives of Biochemistry and Biophysiology* **205**, 246–59.

Gimmler, H., Schafter, G., Kraminer, H. & Huber, U. (1974). Amino acid permeability of the chloroplast envelope as measured by light scattering, volumetry and amino acid uptake. *Planta* **120**, 47–61.

Goldsworthy, A. (1970). Photorespiration. *Botanical Review* **36**, 321–40.

Hall, D. O. (1972). Nomenclature for isolated chloroplasts. *Nature New Biology* **235**, 125–6.

Hallberg, M. & Larsson, C. (1983a). Highly purified intact chloroplasts from mesophyll protoplasts of the C_4 plant *Digitaria sanguinalis*. Inhibition of phosphoglycerate reduction by orthophosphate and by phosphoenolpyruvate. *Physiologia Plantarum* **57**, 330–8.

Hallberg, M. & Larsson, C. (1983b). Metabolism of labelled 3-phosphoglycerate with mesophyll protoplasts and purified mesophyll chloroplasts from the C_4 plant *Digitaria sanguinalis*. In Sybesma, C. (ed.) *Proceedings of the VIth International Congress on Photosynthesis* Vol. III. pp. 417–20. The Hague, M. Nijhoff/W. Junk.

Hanson, J. B. (1985). Membrane transport systems of plant mitochondria. In Douce, R. & Day, D. A. (eds.). *Higher Plant Cell Respiration. Encyclopedia of Plant Physiology 18*, pp. 248–80. Berlin, Springer-Verlag.

Hatch, M. D., Droscher, L., Flügge, U. I. & Heldt, H. W. (1984). A specific translocator for oxaloacetate transport in chloroplasts. *FEBS Letters* **178**, 15–19.

Hatch, M. D. & Osmond, C. B. (1976). Compartmentation and transport in C_4 photosynthesis. In Stocking, C. R. & Heber, U. (eds.). *Transport in Plants III. Encyclopedia of Plant Physiology New Series*, pp. 144–84. New York, Springer-Verlag.

Heber, U. (1970). Flow of metabolites and compartmentation phenomena in chloroplasts. In Mothes, K., Muller, E., Nelles, A. & Nedumann, D. (eds.). *Transport and Distribution of Matter in Cells of Higher Plants*, pp. 152–84. Berlin, Akademie Verlag.

Heber, U. (1974). Metabolite exchange between chloroplasts and cytoplasm. *Annual Review of Plant Physiology* **25**, 393–421.

Heber, U., Hallier, U. W. & Hudson, M. A. (1967a). Lokalisation von enzymen des reduktiven und oxydativen pentosephosphat-zyklus in den chloroplasten und permeabilitat der chloroplastenmembran gegenüber metaboliten. *Zeitschrift Naturforschung* **22b**, 1200–15.

Heber, U. & Santarius, K. A. (1970). Direct and indirect transport of ATP and ADP across the chloroplast envelope. *Zeitschrift Naturforschung* **25b**, 718–78.

Heber, U., Santarius, K. A., Hudson, M. A. & Hallier, U. W. (1967b). Intracellularer transport von zwischenprodukten der photosynthese in photosynthese-gleichgewicht und im dunkel-licht-weschel. *Zeitschrift Naturforschung* **22b**, 1189–99.

Heber, U. & Willenbrink, J. (1964). Sites of synthesis and transport of photosynthetic products within the leaf cell. *Biochimica et Biophysica Acta* **82**, 313–24.

Heldt, H. W. (1969). Adenine nucleotide translocation in spinach chloroplasts. *FEBS Letters* **5**, 11–14.

Heldt, H. W. (1976). Metabolite transport in intact spinach chloroplasts. In Barber, J. (ed.) *Topics in Photosynthesis Vol. 1. The Intact Chloroplast*, pp. 215–34. Amsterdam, Elsevier.

Heldt, H. W. & Rapley, L. (1970a). Unspecific permeation and specific uptake. of substances in spinach chloroplasts. *FEBS Letters* **7**, 139–42.

Heldt, H. W. & Rapley, L. (1970b). Specific transport of inorganic phosphate, 3-phosphoglycerate and dihydroxyacetonephosphate, and of dicarboxylates across the inner membrane of spinach chloroplasts. *FEBS Letters* **10**, 143–8.

Heldt, H. W. & Sauer, F. (1971). The inner membrane of the chloroplast envelope as the site of specific metabolite transport. *Biochimica et Biophysica Acta* **234**, 83–91.

Heldt, H. W., Sauer, F. & Rapley, L. (1972). Differentiation in the permeability properties of the two membranes of the chloroplast envelope. In Forti, G., Avron, M. & Melandria, A. (eds.). *Progress in Photosynthesis. Proceedings of the 2nd International Congress of Photosynthesis Research*, pp. 13345–55. The Hague, W. Junk.

Heldt, H. W. & Stitt, M. (1986). The regulation of sucrose synthesis in leaves. In J. Biggins (ed.). *Progress in Photosynthesis Research*. pp. 675–90.

Herold, A., Leegood, R. C., McNeil, P. H. & Robinson, S. P. (1981). Accumulation of maltose during photosynthesis in protoplasts isolated from spinach leaves treated with mannose. *Plant Physiology* **67**, 85–8.

Hill, R. (1965). The biochemists' green mansions: The photosynthetic electron-transport chain in plants. *Essays in Biochemistry* **1**, 121–51.

Holtum, J. A. M., Fahrendorf, T., Neuhaus, H. E., Mukherjee, M. & Latzko, E. (1987). Carbohydrate storage strategy, F-2,6-P_2 and CAM. In Biggins, J. *Proceedings of the VIIth International Congress on Photosynthesis*. Vol. III. pp. 735–8. The Hague, M. Nijhoff/W. Junk.

Howitz, K. T. & McCarty, R. E. (1982). pH dependence and kinetics of glycolate uptake by intact pea chloroplasts. *Plant Physiology* **70**, 949–52.

Howitz, K. T. & McCarty, R. E. (1983). Evidence for a glycolate transporter in the envelope of pea chloroplasts. *FEBS Letters* **154**, 339–42.

Howitz, K. T. & McCarty, R. E. (1985a). Kinetic characteristics of the chloroplast envelope glycolate transporter. *Biochemistry* **24**, 2645–52.

Howitz, K. T. & McCarty, R. E. (1985b). Substrate specificity of the pea chloroplast glycolate transporter. *Biochemistry* **24**, 3645–50.

Howitz, K. T. & McCarty, R. E. (1986). D-glycerate transport by the pea chloroplast glycolate carrier. Studies on $(1-^{14}C)$D-glycerate uptake and D-glycerate dependent O_2 evolution. *Plant Physiology* **80**, 390–5.

Huber, S. C. & Edwards, G. E. (1976). A high activity ATP translocator in mesophyll chloroplasts of *Digitaria sanguinalis*, a plant having the C_4 acid pathway of photosynthesis. *Biochemica et Biophysica Acta* **440**, 675–87.

Huber, S. C. & Edwards, G. E. (1977a). Transport in C_4 mesophyll chloroplasts. Evidence for an exchange of inorganic phosphate and phosphoenolpyruvate. *Biochimica et Biophysica Acta* **462**, 603–12.

Huber, S. C. & Edwards, G. E. (1977b). Transport in C_4 mesophyll chloroplasts. Characterization of the pyruvate carrier. *Biochimica et Biophysica Acta* **462**, 583–602.

Huber, W., DeFekete, A. & Ziegler, H. (1969). Enzymes of starch metabolism in chloroplasts of the bundle-sheath and palisade cells of *Zea mays*. *Planta* **87**, 360–4.

Jackson, W. A. & Volk, R. J. (1970). Photorespiration. *Annual Review of Plant Physiology* **21**, 385–432.

James, W. O. & Leech, R. M. (1964). The cytochromes of isolated chloroplasts. *Proceedings of the Royal Society, London B* **160**, 13–24.

Jensen, R. G. & Bassham, J. A. (1966). Photosynthesis by isolated chloroplasts. *Proceedings of the National Academy of Sciences, USA* **56**, 1095–101.

Journet, E.-P., Bligny, R. & Douce, R. (1986). Biochemical changes during sucrose deprivation in higher plant cells. *Journal of Biological Chemistry* **261**, 3193–9.

Journet, E.-P., Neuberger, M. & Douce, M. (1981). Role of glutamate-oxaloacctatc transaminase and malate dehydrogenase in the NAD^+ for glycine oxidation by spinach leaf mitochondria. *Plant Physiology* **67**, 467–9.

Kaiser, G. & Heber, U. (1984). Sucrose transport into vacuoles isolated from barley mesophyll protoplasts. *Planta* **161**, 562–8.

Kaiser, G., Martinoia, E. & Wiemken, A. (1982). Rapid appearance of photosynthetic products in the vacuoles isolated from barley mesophyll protoplasts by a new fast method. *Zeitschrift für Pflanzenphysiologie* **107**, 103–13.

Kaiser, W. M., Paul, J. S. & Bassham, J. A. (1979). Release of photosynthates from mesophyll cells *in vivo* and *in vitro*. *Zeitschrift für Pflanzenphysiologie* **94**, 377–85.

Keiller, D. R., Paul, M. J. & Cockburn, W. (1987) Regulation of reserve carbohydrate metabolism in *Mesembryanthemum crystallinum* exhibiting C_3 and CAM photosynthesis. *New Phytologist* **107**, 1–13.

Kime, M. J., Ratcliffe, R. G. & Loughman, B. C. (1982). The application of ^{31}P nuclear magnetic resonance to higher plant tissue. II. Detection of intracellular changes. *Journal of Experimental Botany* **33**, 670–81.

Klingenberg, M. & Pfaff, E. (1967). Means of terminating reactions. *Methods in Enzymology* **10**, 680–4.

Kramer, P. J. (1981). Carbon dioxide concentration, photosynthesis, and dry matter production. *BioScience* **21**, 29–33.

Labate, C. A. & Leegood, R. C. (1988). Limitation of photosynthesis by changes in temperature. Factors affecting the response or carbon dioxide assimilation to temperature in barley leaves. *Planta* **173**, 519–27.

Leech, R. M. (1964). The isolation of structurally intact chloroplasts. *Biochimica et Biophysica Acta* **79**, 637–9.

Leegood, R. C. (1985). The inter-cellular compartmentation of metabolites in leaves of *Zea mays*. *Planta* **164**, 163–71.

Leegood, R. C. & Furbank, R. T. (1986). Simulation of photosynthesis by 2% O_2 at low temperatures is restored by phosphate. *Planta* **168**, 84–93.

Leegood, R. C. & von Caemmerer, S. (1988). The relationship between contents of photosynthetic metabolites and the rate of photosynthetic carbon assimilation in leaves of *Amaranthus edulis*. *Planta* **174**, 253–62.

Leegood, R. C. & Walker, D. A. (1983). The role of transmembrane solute flux in regulation of CO_2 fixation in chloroplasts. *Biochemical Society Transactions* **11**, 74–6.

Lehner, K. & Heldt, H. W. (1978). Dicarboxylate transport across the inner membrane of the chloroplast envelope. *Biochimica et Biophysica Acta* **501**, 531–44.

Liang, Z. & Huang, A. H. C. (1983). Metabolism of glycollate and glyoxylate in intact spinach leaf peroxisomes. *Plant Physiology* **73**, 147–52.

Lilley, R.McC., Chon, C. J., Mosbach, A. & Heldt, H. W. (1977). The distribution of metabolites between spinach chloroplasts and medium during photosynthesis *in vitro*. *Biochimica et Biophysica Acta* **460**, 259–72.

Lilley, R.McC., Schwenn, J. D. & Walker, D. A. (1973). Inorganic pyrophosphatase and photosynthesis by isolated chloroplasts. II. The controlling influence of orthophosphate. *Biochimica et Biophysica Acta* **325**, 596–604.

Lin, W. & Wagner, G. J. (1985). Isolation, properties and functions of tonoplast ATPase from higher plants. In Marin, B.P. (ed.). *Biochemistry and Function of Vacuolar Adenosine-triphosphatase in Fungi and Plants*, pp. 67–76. Berlin, Heidelberg, New York, Tokyo, Springer-Verlag.

Martinoia, E., Flügge, U. I., Kaiser, G., Heber, U. & Heldt, H. W. (1985). Energy-dependent uptake of malate into vacuoles isolated from barley mesophyll protoplasts. *Biochimica et Biophysica Acta* **806**, 311–19.

Mathieu, Y. (1967). Sur l'isolement, en milieu aqueux, de chloroplastes 'intacts' a partir de feuilles de plantules d'Orge. *Photosynthetica* **1**(1–2), 57–63.

Mbaku, S. B., Fritz, G. J. & Bowes, G. (1978). Photosynthetic and carbohydrate metabolism in isolated leaf cells of *Digitaria pentzii*. *Plant Physiology* **62**, 510–15.

Nishida, K. & Sanada, Y. (1977). Carbon dioxide fixation in chloroplasts isolated from CAM plants. *Plant and Cell Physiology* (special issue) **3**, 341–6.

Nobel, P. S. & Cheung, Y. S. (1972). Two amino acid carriers in pea chloroplasts. *Nature* **237**, 207–8.

Nobel, P. S. & Wang, C. T. (1970). Amino acid permeability of pea chloroplasts as measured by osmotically determined reflection coefficients. *Biochimica et Biophysica Acta* **211**, 79–87.

Ohnishi, J. & Kanai, R. (1987). Pyruvate uptake by mesophyll and bundle sheath chloroplasts of a C_4 plant, *Panicum miliaceum* L. *Plant Cell Physiology* **28**, 1–10.

Ohsugi, R. & Huber, S. C. (1987). Light modulation and localization of sucrose phosphate synthase activity between mesophyll cells and bundle sheath cells in C_4 species. *Plant Physiology* **84**, 1096–101.

Packer, L. & Crofts, A. R. (1967). The energised movement of ions and water by chloroplasts. *Current Topics in Bioenergetics* **2**, 23–64.

Packer, L., Siegenthaler, P. A. & Nobel, P. S. (1965). Light induced volume changes in spinach chloroplasts. *Journal of Cell Biology* **26**, 593–9.

Piazza, G. J. & Gibbs, M. (1982). Photosynthetic CO_2 assimilation by *Sedum* chloroplasts. In Ting, I. P. & Gibbs, M. (eds.). *Crassulacean Acid Metabolism*, pp. 128–152. Rockville, Maryland, American Society of Plant Physiology.

Pollock, C. J., Riley, G. J. P., Stoddart, J. L. & Thomas, H. (1980). The biochemical basis of plant response to temperature limitations. *Annual Report of Welsh Plant Breeding Station for 1979*, pp 227–46.

Portis, A. R. (1983). Analysis of the role of the phosphate translocator and external metabolites in steady-state chloroplast photosynthesis. *Plant Physiology* **71**, 936–43.

Prinsley, R. T. & Leegood, R. C. (1986). Factors affecting photosynthetic induction in leaves. *Biochimica et Biophysica Acta* **849**, 244–53.

Proudlove, M. O. & Thurman, D. A. (1981). The uptake of 2-oxoglutarate and pyruvate by isolated pea chloroplasts. *New Phytologist* **88**, 255–64.

Proudlove, M. O., Thurman, D. A. & Salisburg, J. (1984). Kinetic studies on the transport of 2-oxoglutarate and L-malate into isolated pea chloroplasts. *New Phytologist* **96**, 1–8.

Rabinowitch, E. I. (1945). *Photosynthesis and Related Processes*, Vol. 1. New York, Wiley Interscience.

Rabinowitch, E. I. (1956). *Photosynthesis and Related Processes*, Vol. II, Part 2. New York, Wiley Interscience.

Rao, I. M., Abadia, J. & Terry, N. (1986). Leaf phosphate status and photosynthesis *in vivo*: changes in light scattering and chlorophyll fluorescence during photosynthetic induction in sugar beet leaves. *Plant Science* **44**, 133–7.

Rebeille, F., Bligny, R., Marin, J.-B. & Douce, R. (1983). Relationship between the cytoplasm and the vacuole phosphate pool in *Acer pseudoplatanus* cells. *Archives of Biochemistry and Biophysics* **225**, 143–8.

Robinson, S. P. (1982). Transport of glycerate across the envelope membrane of isolated spinach chloroplasts. *Plant Physiology* **70**, 1032–8.

Robinson, S. P. (1984). Lack of ATP requirement for light stimulation of glycerate transport into intact isolated chloroplasts. *Plant Physiology* **75**, 425–30.

Robinson, S. P., Cerovic, Z. & Walker, D. A. (1987). Isolation of intact chloroplasts: General principles and criteria of integrity. *Methods in Enzymology*, Vol. 148 (in press).

Robinson, S. P. & Wiskich, J. T. (1976). Stimulation of carbon dioxide fixation in isolated pea chloroplasts by catalytic amounts of adenine nucleotides. *Plant Physiology* **58**, 156–62.

Robinson, S. P. & Wiskich, J. T. (1977). Pyrophosphate inhibition of carbon dioxide function in isolated pea chloroplasts by uptake in exchange for endogenous adenine nucleotides. *Plant Physiology* **59**, 422–7.

Rumpho, M. E. & Edwards, G. E. (1984). Inhibition of 3-phosphoglycerate-dependent O_2 evolution by phosphoenolpyruvate in C_4 mesophyll chloroplasts of *Digitaria sanguinalis* L. Scop. *Plant Physiology* **76**, 711–18.

Rumpho, M. E. & Edwards, G. E. (1985). Characterization of 4,4'-diisothiocyano-2,2'-disulfonic acid stilbene inhibition of 3-phosphoglycerate-dependent O_2 evolution in isolated chloroplasts. Evidence for a common binding site on the C_4 phosphate translocator for 3-phosphoglycerate, phosphoenolpyruvate and inorganic phosphate. *Plant Physiology* **78**, 537–44.

Rumpho, M. E., Wessinger, M. E. & Edwards, G. E. (1987). Influence of organic phosphates on 3-phosphoglycerate dependent O_2 evolution in C_3 and C_4 mesophyll chloroplasts. *Plant Cell Physiology* **28**, 805–13.

Sachs, J. (1862). Über den einfluss des lichtes auf die bildung des amylums in den chlorophyllkornern. *Botanisches Zeitschrift* **20**, 365–73.

Sachs, J. (1887). *Lectures on the Physiology of Plants*. Translated by Ward, H. M., pp. 1–309. Oxford, Clarendon Press.

Schachter, B., Eley, J. H. Jr. & Gibbs, M. (1968). Effect of sugar phosphates and photosynthesis inhibitors on CO_2 fixation and O_2 evolution by chloroplasts. *Plant Physiology* **43**, 8–30.

Schmitt, M. & Edwards, G. E. (1982). Isolation and purification of intact peroxisomes from green leaf tissues. *Plant Physiology* **70**, 1213–17.

Schnyder , H., Machler, F. & Nosberger, J. (1984). Influence of temperature and O_2 concentration on photosynthesis and light-activation of ribulose bisphosphate carboxylase in intact leaves of white clover (*Trifolium repens* L.). *Journal of Experimental Botany* **35**, 147–56.

Schwenn, J. D., Lilley, R.McC. & Walker, D. A. (1973). Inorganic pyrophosphatase and photosynthesis by isolated chloroplasts. I. Characterisation of chloroplast pyrophosphatase and its relation to the response to exogenous pyrophosphate. *Biochimica et Biophysica Acta* **325**, 586–95.

Sharkey, T. D., Stitt, M. N., Heineke, D., Gerhardt, R., Raschke, K. & Heldt, H. (1986). Limitation of photosynthesis by carbon metabolism. II. O_2 insensitive CO_2 uptake results from limitation of triose phosphate utilization. *Plant Physiology* **81**, 1123–9.

Sivak, M. N. (1987). The effect of oxygen on photosynthetic carbon assimilation and hypothesis concerning the mechanisms involved. *Photobiochemistry and Photobiophysics*, Supplement, 141–56.

Sivak, M. N. & Walker, D. A. (1986). Photosynthesis *in vivo* can be limited by phosphate supply *New Phytologist* **102**, 499–512.

Sivak, M. N. & Walker, D. A. (1987). Oscillations and other symptoms of limitation of *in vivo* photosynthesis by inadequate phosphate supply. *Plant Physiology and Biochemistry* **25**, 635–48.

Soll, J., Worzer, C. & Buchanan, B. B. (1983). Fructose-2,6-biphosphate and C_4 plants. In Sybesma, C. (ed.). *Advances in Photosynthesis Research* 3, pp. 485–8.

Somerville, S. C. & Ogren, W. L. (1983). An *Arabidopsis thaliana* mutant defective in chloroplast dicarboxylate transport. *Proceedings of National Academy of Sciences, USA* **80**, 1290–4.

Somerville, S. C. & Somerville, C. R. (1985). A mutant *Arabidopsis* deficient in chloroplast dicarboxylate transport is missing an envelope protein. *Plant Science Letters* **37**, 317–30.

Spalding, M. H. & Edwards, G. E. (1980). Photosynthesis in isolated chloroplasts of the Crassulacean acid metabolism plant *Sedum prealtum. Plant Physiology* **65**, 1044–8.

Spalding, M. H., Schmidtt, M. R., Ku, M. S. B. & Edwards, G. E. (1979). Intercellular localization of some key enzymes of Crassulacean acid metabolism in *Sedum prealtum. Plant Physiology* **63**, 738–43.

Spilatro, S. R. & Preiss, J. (1987). Regulation of starch synthesis in the bundlesheath and mesophyll of *Zea mays* L. Intercellular compartmentation of enzymes of starch metabolism and the properties of the ADPglucose pyrophosphorylases. *Plant Physiology* **83**, 621–7.

Stankovic, Z. S. & Walker, D. A. (1977). Photosynthesis by isolated pea chloroplasts. Some effects of adenylates and inorganic pyrophosphate. *Plant Physiology* **59**, 428–32.

Stitt, M. N. (1987). Fructose, 2.6-bisphosphate and plant carbohydrate metabolism. *Plant Physiology* **84**, 201–4.

Stitt, M. N. & Heldt, H. W. (1985). Control of photosynthetic sucrose synthesis by fructose-2,6 bisphosphate. *Planta* **164**, 179–88.

Stitt, M. N., Huber, S. C. & Kerr, P. (1987). Regulation of photosynthetic sucrose synthesis. In Hatch, M. D. & Boardman, N. K. (eds.). *The Biochemistry of Plants*, Vol. 13, pp. 327–409. New York, Academic Press (in press).

Stitt, M. N., Wirtz, W. & Heldt, H. W. (1983). Regulation of sucrose synthesis by cytoplasmic fructose bisphosphatase and sucrose phosphate synthase during photosynthesis in varying light and CO_2. *Plant Physiology* **72**, 767–74.

Stocking, C. R. & Larson, S. (1969). A chloroplast cytoplasmic shuttle and the reduction of extraplastid NAD. *Biochemical and Biophysical Research Communications* **37**, 278–82.

Tabita, F. R. & McFadden, B. A. (1972). Regulation of ribulose-1,5-diphosphate carboxylase by 6-phospho-D-gluconate. *Biochemical and Biophysical Research Communations* **48**, 1153–9.

Thomas, M., Ranson, S. L. & Richardson, J. A. (1973). *Plant Physiology*. London, Longman.

Urbach, W., Hudson, M. A., Ullrich, W., Santarius, K. A. & Heber, U. (1965). Verteilung und wanderung von phosphglycerat zwischen den chloroplasten und dem zytoplasma während der photosynthese. *Zeitschrift Naturforschung* **20b**, 890–8.

Wagner, W., Keller, F. & Wiemken, A. (1983). Fructan metabolism in cereals: induction in leaves and compartmentation in protoplasts and vacuoles. *Zeitschrift fur Pflanzenphysiologie* **112**, 359–72.

Walker, D. A. (1965a). Photosynthetic activity of isolated pea chloroplasts. In Goodwin, T. W. (ed.). *Biochemistry of the Chloroplast Vol.II*, pp. 53–9, Proceedings of NATO Advanced Study Institute, Aberystwyth. New York, Academic Press.

Walker, D. A. (1965b). Correlation between photosynthetic activity and membrane integrity in isolated pea chloroplasts. *Plant Physiology* **40**, 1157–61.

Walker, D. A. (1969). Permeability of the chloroplast envelope. In Metzner, H. (ed.). *Progress in Photosynthesis Vol. 1*, pp. 250–7, Proceedings of 1st International Congress for Photosynthetic Research, Freudenstadt 1968. Union Biological Science, Tübingen.

Walker, D. A. (1971). Chloroplasts (and grana) – Aqueous (including high carbon fixation ability). In San Pietro, A. (ed.). *Methods in Enzymology Vol. XXIII*, pp. 211–20. London, New York, Academic Press.

Walker, D. A. (1973). Photosynthetic induction phenomena and the light activation of ribulose diphosphate carboxylase. *New Phytologist* **72**, 209–35.

Walker, D. A. (1974). Chloroplast and cell. Concerning the movement of certain key metabolites etc. across the chloroplast envelope. In Northcote, D. H. (ed.). *Med. Tech. Public. International Review of Science Biochemistry Series I*, Vol XI, pp. 1–49. London, Butterworths.

Walker, D. A. (1976). Photosynthetic induction and its relation to transport phenomena in chloroplasts. In Barber, J. (ed.). *The Intact Chloroplast: Topics in Photosynthesis, Vol. 1*. Amsterdam, Elsevier.

Walker, D. A., Cockburn, W. & Baldry, C W. (1967). Photosynthetic oxygen evolution by isolated chloroplasts in the presence of carbon cycle intermediates. *Nature* **216**, 597–9.

Walker, D. A. & Crofts, A. R. (1970). Photosynthesis. *Annual Review of Biochemistry* **39**, 389–428.

Walker, D. A. & Hill, R. (1967). The relation of oxygen evolution to carbon assimilation with isolated chloroplasts. *Biochimica et Biophysica Acta* **131**, 330–8.

Walker, D. A. & Osmond, C. B. (1986). Measurement of photosynthesis *in vivo* using a leaf disc electrode: correlations between light dependence of steady-state photosynthetic O_2 evolution and chlorophyll *a* fluorescence transients. *Proceedings of The Royal Society, London* B **227**, 267–80.

Walker, D. A. & Sivak, M. N. (1985). Can phosphate limit photosynthesis *in vivo*? *Physiologie Végétale* **23**, 829–41.

Wallsgrove, R. M., Kendall, A. C., Hall, N. P., Turner, J. C. & Lea, P. J. (1986). Carbon and nitrogen metabolism in barley (*Hordeum vulgare* L.) mutants with impaired chloroplast dicarboxylate transport. *Planta* **168**, 324–9.

Walters, D. R. & Ayres, P. G. (1983). Changes in nitrogen utilization and enzyme activities associated with CO_2 exchanges in healthy leaves of powdery mildew-infected barley. *Physiological Plant Pathology* **23**, 447–59.

Wang, C-T, & Nobel, P. S. (1971). Permeability of pea chloroplasts to alcohols and aldoses as measured by reflection coefficients. *Biochimica et Biophysica Acta* **241**, 200–12.

Werdan, K. & Heldt, H. W. (1972). The phosphate translocator of spinach chloroplasts. In Forti, G., Avron, M. & Melandri, A. (eds.). *Progress in Photosynthesis Research*. Proceedings of 2nd International Conference of Photosynthetic Research, pp. 1337–44. The Hague, W. Junk.

Werkheiser, W. C. & Bartley, W. (1957). The study of steady-state concentrations of internal solutes of mitochondria by rapid centrifugal transfer to a fixation medium. *Biochemical Journal* **66**, 79–91.

Whatley, F. R. Allen, M. B., Rosenberg, L. L., Capindale, J. B. & Arnon, D. I. (1956). Photosynthesis by isolated chloroplast. V. Phosphorylation and carbon dioxide fixation by broken chloroplasts. *Biochimica et Biophysica Acta* **20**, 462–8.

Winter, K., Foster, J. G., Edwards, G. E. & Holtum, J. A. M. (1982). Intracellular localization of enzymes of carbon metabolism in *Mesembryanthemum crystallinum* exhibiting C_3 photosynthetic characteristics or performing crassulacean acid metabolism. *Plant Physiology* **69**, 300–7.

Woldegiorgis, G., Voss, S., Shrago, E., Werner-Washburne, M. & Keegstra, K. (1983). An adenine nucleotide-phosphoenolpyruvate counter-transport system in C_3 and C_4 plant chloroplasts. *Biochemical and Biophysical Research Communications* **116**, 945–51.

Woo, K. C. (1983). Effect of inhibitors on ammonia, 2-oxoglutarate- and oxaloacetate-dependent O_2 evolution in illuminated chloroplasts. *Plant Physiology* **71**, 112–17.

Woo, K. C., Flügge, I. U. & Heldt, H. W. (1984). Regulation of 2-oxoglutarate and dicarboxylate transport in spinach chloroplasts by ammonia in the light. In Sybesma, C. (ed.). *Advances in Photosynthesis Research Vol. III*, pp. 685–8. The Hague, M. Nijhoff/W. Junk.

Woo, K. C., Flügge, I. U. & Heldt, H. W. (1987). A two-translocator model for the transport of 2-oxoglutarate and glutamate in chloroplasts during ammonia assimilation in the light. *Plant Physiology* **84**, 624–32.

Woo, K. C. & Osmond, C. B. (1982). Stimulation of ammonia and 2-oxoglutarate-dependent O_2 evolution in isolated chloroplasts by dicarboxylates and the role of the chloroplast in photorespiratory nitrogen recycling. *Plant Physiology* **69**, 591–6.

Zelitch, I. (1971). *Photosynthesis, Photorespiration and Plant Productivity*. New York, Academic Press.

Zelitch, I. (1973). The biochemistry of photorespiration. *Current Advances in Plant Science* **6**, 44–54.

2 Transport of photoassimilates between leaf cells

M. A. Madore and W. J. Lucas

2.1 Introduction

The assignment of photosynthetic and absorptive capacities to the different above- and below-ground plant parts necessitates long-distance transport systems by which photoassimilated or absorbed solutes can be relayed, exchanged and recirculated between photosynthetic and non-photosynthetic tissues for use in various metabolic reactions. In leaf tissues, the study of photoassimilate export and its relationship to the physiology and biochemistry of photosynthesis has been the particular focus of many recent scientific investigations. Details of compartmentation within the photosynthetic cells and the accumulation of solutes within phloem tissues are dealt with elsewhere in this volume (Chs. 1 and 5) and will not be discussed at length in this chapter. Instead, we will confine our attention to the interface between these two systems, focusing on the short-distance relay of photoassimilates *between* the leaf mesophyll and the phloem.

We should perhaps emphasize at this point that, by imposing this relatively narrow focus, we do not intend to imply that the processes of photosynthesis, cellular compartmentation, short-distance transport or phloem loading can, or should, be regarded as mutually exclusive metabolic events. Indeed, such a view would defeat the purpose of this volume. However, in the years since this subject was last reviewed (Gunning, 1976), many exciting advances have been made in the field of cell-to-cell solute movement in plant tissues. A chapter devoted exclusively to the transport of photoassimilates between leaf cells is, therefore, both warranted and timely.

2.2 Symplasmic transport

As early as 1914, Haberlandt drew attention to the apparently deliberate arrangement of leaf mesophyll cells as chains of cells emanating from the vasculature. He was quick to view this arrangement as one that would facilitate 'expeditious translocation' of photoassimilates from the sites of synthesis in the chlorophyllous tissues to the vascular phloem tissues. Later, Münch (1930) hypothesized that plasmodesmata, which interconnect virtually all leaf cells, could allow for continuity of the cytosolic compartment and thus provide a potential pathway for solute movement between these cells. However, many decades later Gunning, in his 1976 review of this subject, reported that the evidence for a role for plasmodesmata in the short-distance transport of assimilates between the mesophyll and the phloem was still 'both meagre and circumstantial'. Nonetheless, the prevailing view to date is that solute movement, at least between mesophyll cells, does occur via a symplasmic route through plasmodesmata (Giaquinta, 1983).

Over the past five years, new evidence for the functionality of the symplasmic pathway has been accumulating for tissues and organs of many plant species. This new information on intercellular communication in plant tissues has been unquestionably a direct result of the many advances made during the past decade in techniques for study of intercellular communication in animal tissues. Techniques for micro-injection of non-toxic, membrane-impermeable dyes (e.g. Lucifer Yellow CH), which have been used extensively for fluorescent tracing of neurological interconnections (Stewart, 1981) and gap junctions (Loewenstein & Rose, 1978), have found, with only minor refinements, direct applicability to a large number of plant tissues. In addition, the ease of synthesis of fluorescent peptide probes of known specific molecular weight and radius, developed to probe the size exclusion limits of gap junctions (Simpson *et al.*, 1977; Flagg-Newton *et al.*, 1979), has also been exploited by plant scientists to establish the extent of symplasmic permeability in plant tissues. Perhaps of more importance, though, is the effect of the rapidly growing understanding of the molecular basis for control of solute flow through gap junctions. Major advances in this area are providing us with more ideas about possible mechanisms for similar control of solute flow through plasmodesmata. (In view of this, a brief digression into recent advances in the molecular biology of gap junctions will be made in Sect. 2.2.4.)

2.2.1 Structure of plasmodesmata

Plasmodesmata tend not to be distinguished by their presence between cells but by their absence; that is, virtually all plant cells are found to be interconnected (see Gamalei, 1985a, for a review). However, having said

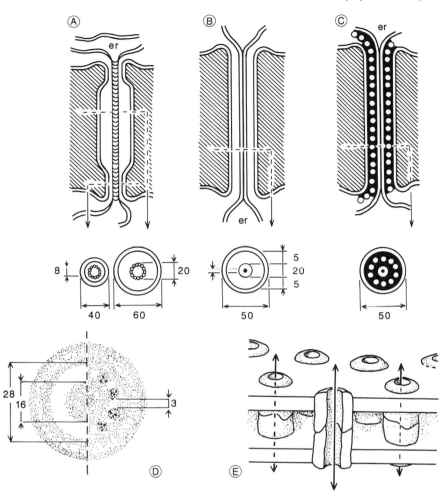

Fig. 2.1 Ultrastructural models of plasmodesmata and animal gap junctions. *A*: Robards (1971) model in which the cytoplasmic annulus or the axial component (desmotuble) is thought to function in terms of solute movement via a contiguous lumen of the endoplasmic reticulum (er). *B*: Lopez-Saez *et al.* (1966) model in which the axial component is thought to represent tightly appressed er without a central lumen. *C*: Overall *et al.* (1982) model in which the axial component is identical to the Lopez-Saez *et al.* model but in which the cytoplasmic annulus is partially occluded by protein particles. *D*: Terry and Robards (1987) representation of a plasmodesma in the stalk cell distal wall of mature trichomes from flowers of *Abutilon megapotamicum*. The cytoplasmic annulus (region not stippled on left-hand side for clarity) is shown with a width of 6.0 nm and, as in the Overall *et al.* model, particles are thought to occupy much of this space/volume. The desmotubule is depicted as a tightly rolled cylinder of membrane with a virtually closed central channel. *E*: Structure of animal cell gap junctions (from Loewenstein & Rose, 1978). (All dimensions indicated are in nanometres.)

this, we must also point out that the 'typical' structure invoked for a plasmodesma is not always seen between all cell types, and that considerable variations on a basic theme are often seen, depending on the tissue being examined. An extreme case in point in leaf tissues is exemplified by the branched plasmodesmata found connecting companion cells and sieve elements.

In higher plants, the simplest and most common type of plasmodesmatal structure appears, in TEM thin sections, to be a membrane-lined pore about 40 to 50 nm in diameter (Fig. 2.1). In some cases, the cytoplasmic pore may be constricted at the neck regions and appears to be partially occluded by proteinaceous sub-units. The central region of the pore contains an electron opaque strand, the axial component or desmotubule, which is associated with the endoplasmic reticulum. Further ultrastructural details are not clearly resolved by the electron microscope, necessitating the use of models to account for supposed structure/function relationships. The models differ principally in the role of the axial component and the number of protein sub-units within the cytoplasmic pore (Fig. 2.1A–D).

The concept of the desmotubule as the transport channel (the Robards model, Fig. 2.1A) has recently been challenged, as no opening within the tubule has been clearly discerned (Overall *et al.*, 1982; Thomson & Platt-Aloia, 1985). Indeed, in a recent study (Terry & Robards, 1987) Robards himself now concludes, on the basis of dye exclusion studies, that the effective pathway for solute movement is likely to be through the spaces between the protein sub-units which partially occlude the cytoplasmic annulus (the 'Terry and Robards' model, Fig. 2.1D). These spaces have an effective diameter of about 1.5 nm and show exclusion size limits of the same order seen for animal gap junctions (Loewenstein & Rose, 1978).

2.2.2 Indirect evidence for cell-to-cell transport in leaf tissues: plasmodesmatal frequencies

It has long been postulated that rates of solute flow through the symplasmic pathway could be dictated by the frequency and distribution of plasmodesmata between cells. In recent years, there has been a renewed interest in extensive and thorough documentation of plasmodesmatal structure and frequency in source leaf tissues, with the aim of determining the role of plasmodesmata in assimilate movement. Most of the available data now come from the extensive studies of Gamalei (e.g. see Gamalei & Pakhomova, 1981 and Gamalei, 1985a,b) and Evert and co-workers (e.g. see Evert & Mierzwa, 1986).

In most leaf tissues studied, symplasmic continuity has been demonstrated along the entire length of the pathway leading from the mesophyll to the minor veins (Table 2.1). Nonetheless, this pathway has been considered an unlikely one for assimilate transport (Evert *et al.*, 1978;

Table 2.1 Summary of frequency of plasmodesmata (average number of plasmodesmata per μm of interface) between cell types of minor veins and contiguous tissues in leaves of *Beta vulgaris* (Evert & Mierzwa, 1986), *Populus deltoides* (Russin & Evert, 1985), *Coleus blumei* (Fisher, 1986), and *Amaranthus retroflexis* (Fisher & Evert, 1982)

	Beta	*Populus*	*Coleus*	*Amaranthus*
MS/MS	0.09	0.19	0.26	—
MS/PVM		0.34		
BS/MS	0.09	0.28	0.18	1.09
BS/BS	0.09	0.33	0.17	0.18
BS/VP	0.13	0.43	0.07	0.53
BS/CC	0.06	0.73	0.01	0.11
VP/VP	0.16	0.69	0.16	0.59
VP/CC	0.12	0.40	0.10	0.25
VP/ST	0.02	0.09	0.01	0.05
CC/ST	0.14	1.71	0.21	0.13
IC/BS			1.33	
IC/ST			0.88	
IC/VP			0.27	
IC/CC			0.00	
IC/IC			0.57	

BS: bundle-sheath; CC: companion cell; IC: intermediary cell; MS: mesophyll; PVM: paraveinal mesophyll; ST: sieve tube; VP: vascular parenchyma

Fisher & Evert, 1982; Evert & Mierzwa, 1986), a conclusion based partly on the frequency of plasmodesmata found between the various leaf cells, but primarily on the apparent steep positive solute gradient usually found at the interface between the vascular parenchyma and the sieve element-companion cell complex.

In some leaf tissues, notably in members of the Cucurbitaceae such as *Cucurbita pepo*, this uphill solute gradient does not appear to exist (Richardson *et al.*, 1984). Additionally, plasmodesmatal frequencies in these species (*Cucurbita*, Turgeon *et al.*, 1975; *Cucumis*, Schmitz *et al.*, 1987) and physiological studies (*Cucurbita*, Madore & Webb, 1981; *Cucumis*, Schmitz *et al.*, 1987) would tend to support an entirely symplasmic mode of assimilate transfer to the veins of these leaves. In this context, an interesting hypothesis has been advanced by Gamalei (1985b) on the basis of his own studies of plasmodesmatal frequencies in leaves (Gamalei & Pakhomova, 1981). He has drawn an interesting correlation between the presence of abundant plasmodesmata along the symplasmic route and the synthesis of oligosaccharides of the raffinose family in source leaves. Gamalei divides plants into three groups on this basis: *Group I* consists of plants having large amounts of raffinose-type oligosaccharides in the phloem, as well as companion cells with abundant symplasmic connections to adjacent parenchyma and sieve tubes; *Group II* plants translocate mainly sucrose with only very small amounts of other oligosaccharides present, and companion cells are linked symplasmically predominantly (only) on the sieve tube side; whilst plants of *Group III* have only sucrose in the phloem (no other oligosaccharides present) and their companion cells are both modified into transfer cells and symplasmically isolated from the surrounding vascular tissue.

Gamalei envisages that plants of *Group I*, to which the Cucurbitaceae belong, probably have an entirely symplasmic mode for the transfer of raffinose-type oligosaccharides into the phloem, while sucrose may move either via the symplasm or the apoplasm. Plants in *Group II* and *Group III* are thought to move assimilates symplasmically up to a point close to the sieve element-companion cell complex, where they are then released into the apoplasm and 'loaded' directly into the phloem. (It is interesting that those plant families such as the Apiaceae (Umbelliferae) and Scrophulariaceae, which translocate mannitol in addition to sucrose, fall into this third category.)

However, on the basis of the frequency analyses provided in Table 2.1, it is more difficult to assess the validity of this arbitrary assignment of plant species to specific 'phloem loading groups'. For example, *Coleus* (Lamiaceae translocates high amounts of raffinose-family oligosaccharides and therefore falls into Gamalei's *Group I*, but it has been assigned, on the basis of plasmodesmatal frequency studies, a mixed method of loading (Fisher, 1986). Similarly, *Populus* (Salicaceae) translocates relatively lower amounts of these sugars (possibly on a seasonal basis) and therefore falls into *Group II*, but the method of loading is not clear from either ultrastructural or plasmolytic studies (Russin & Evert, 1985). In *Beta*, assigned to *Group II* as it translocates sucrose, not raffinose (although raffinose is stored in the tap root; French, 1954), plasmodesmatal frequency between *all* cell types appears to be unusually low, but the number of connections *increases* as one approaches the vascular tissues (Table 2.1). Additionally, we have found, using dye injection methods (Madore *et al.*, 1986), that the plasmodesmata between *Beta* mesophyll cells would appear to be fully functional and allow rapid cell-to-cell transport of dye. That this occurs even though the frequency of plasmodesmata between these cells is one of the lowest in the leaf (Table 2.1), suggests that the other plasmodesmata along the path may also be operational and that a symplasmic pathway for assimilate transport may also exist in the *Beta* leaf, despite its overall low numbers of intercellular connections.

There are obvious limitations to the inferences that can be made, on the basis of structural data alone, concerning the functionality of the symplasm. It is also not clear that solute distribution, as inferred from plasmolytic studies, is a legitimate method for assessment of symplasmic transport, as Russin and Evert (1985) demonstrate that in the paraveinal mesophyll of *Populus* very high concentration gradients can exist even across cells which are abundantly connected by plasmodesmata. They caution that 'overall sucrose gradients may not be the driving forces for sucrose diffusion from the photosynthetic sites toward the vein'. In other words, solute distributions determined by plasmolytic methods are probably *not* representative of *cytosolic* solute levels; that is, the solute concentrations that presumably would be responsible for driving solute flow through a plasmodesma.

In summary, it may also be noted that analysis of the frequency and distribution of plasmodesmata in leaf tissues, while providing valuable ultrastructural information, suffers from the implicit inference that all plasmodesmata can be functionally equated to holes or pores. As a result, other inferences are made, first of all that an understanding of the control of solute flow through plasmodesmatal 'pores' is currently available (which it is not), and secondly, that the dissimilarities in plasmodesmatal structure can be effectively ignored, as they are unlikely to be part of any mechanism(s) regulating solute flow through the symplasm. Gunning and Robards (1976) stressed that plasmodesmata were 'much more complicated than mere membrane-lined holes in the cells walls', a feature readily recognized from ultrastructural data, but not always appreciated in the physiological interpretations of symplasmic solute transport.

2.2.3 *Direct evidence for cell-to-cell transport: dye tracer studies*

The adaptation of dye tracer techniques used in probing animal cell gap junctions has recently provided a more direct means of establishing the functional status of plasmodesmata in plant tissues. The dyes used do not cross the plasma membrane (Stewart, 1981); therefore, intercellular movement can be monitored by conventional fluorescence microscopy following micro-injection of a single cell. From these micro-injection techniques, convincing evidence for operative symplasmic connections have now been presented for a variety of plant tissues, including the leaf of *Elodea* (Goodwin, 1983), leaf and extrastelar tissues of *Egeria* (Erwee & Goodwin, 1983, 1984, 1985), apices of *Silene* (Goodwin & Lyndon, 1983), leaf tissues of *Allium* (Palevitz & Hepler, 1985), staminal hairs of *Setcreasea* (Tucker, 1982; Tucker & Spanswick, 1985), the source leaf of *Commelina* (Erwee *et al.*, 1985), source leaves of *Beta* and *Ipomoea* (Madore *et al.*, 1986) and nectary trichomes of *Abutilon* (Terry & Robards, 1987).

In the *Elodea* symplasm, the size exclusion limit, as judged by the passage of low molecular weight probes, appears to be about 874 daltons (Da), corresponding to an *effective* pore size of less than 2 nm (Goodwin, 1983). In *Egeria* it appears to be somewhat smaller, of the order of 749 Da (Erwee & Goodwin, 1983), which is similar to that of staminal hairs of *Setcreasea* in which the exclusion limit lies somewhere between 700 and 800 Da (1.5 nm) (Tucker, 1982). In the *Abutilon* nectary, the effective pore size is somewhat larger, of the order of 1090 Da, or 2.4 nm (Terry & Robards, 1987).

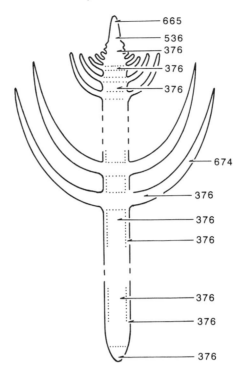

Fig. 2.2 A schematic diagram of *Egeria densa* showing the molecular exclusion limits of different tissues (in daltons) and the proposed barriers to dye movement (. . . .) (from Erwee & Goodwin, 1985).

The effective pore size may vary within the plant (Fig. 2.2). In *Egeria*, exclusion limits in the shoot and apical regions are of the order of 665 Da, while exclusion limits for stem and root epidermal tissues are much lower, of the order of 376 Da (Erwee & Goodwin, 1985). Additionally, there are apparent barriers to dye movement in certain tissues, notably between epidermal and cortical cells of roots and stems, and between root cap and other root cells (Erwee & Goodwin, 1985). On this basis, Erwee and Goodwin (1985) postulated the existence of symplasmic 'domains' in plant tissues; that is, that barriers to symplasmic movement occur in tissues which otherwise appear to be adequately connected by plasmodesmata of normal appearance. (Again, this finding should once more caution against drawing conclusions on functioning of plasmodesmata on the basis of structural evidence alone.)

Other experiments provide some insight into the possible nature of the control mechanisms that may be regulating solute flow in these otherwise 'normal' cellular channels. From ultrastructural examinations of plasmodesmata at different locations in the *Egeria* plant, Erwee & Goodwin (1985) noted a correlation between the radius of the cytoplasmic annulus at the neck region of the plasmodesmata and the size exclusion limits observed by dye injection. In an earlier study, Belitser *et al.* (1982) reported that treatment of barley root tips with calcium chloride resulted in the appearance of electron-dense particulate occlusions, possibly possessing ATPase activity, at the neck regions of plasmodesmata. Erwee & Goodwin (1983) found that in the *Egeria* leaf, in which probes of molecular weight of 665 Da or less moved freely through the symplasm, a pre-injection with calcium chloride could completely abolish this movement. Other divalent cations (magnesium, strontium) also affected movement, but monovalent cations such as potassium and sodium had no effect. The effect of calcium was transient, with full recovery of dye movement seen within 30 min of calcium injection. Treatments designed to elevate artificially cytosolic calcium levels (i.e. treatment with the calcium ionophore A23187, carbonyl cyanide *p*-trifluoromethoxyphenyl hydrazone or trifluralin), also reduced dye mobility. Calmodulin inhibitors such as trifluoperazine had no effect, suggesting that the response to elevated calcium was not mediated through a calmodulin-linked reaction. Furthermore, cytochalasin B treatment did not affect dye movement, suggesting that the effects of elevated calcium on dye movement were also not due to cessation of cytoplasmic streaming.

Interestingly, this apparent regulation by calcium could be completely overcome if the tissues were first plasmolysed and then deplasmolysed (Erwee & Goodwin, 1984). Plasmolysed tissues did not allow movement of probes but following deplasmolysis, the tissues showed a great increase in permeability, allowing the passage of probes of up to 1678 Da. Deplasmolysed tissues also allowed passage of aromatic amino acid probes, which normally are excluded from passage through the symplasm (Tucker, 1982; Erwee & Goodwin, 1984; but see Terry & Robards, 1987). The deplasmolysed tissues showed no response to calcium pre-treatment, suggesting that normal regulatory mechanisms had also been perturbed by the disruption of plasmodesmatal integrity caused by plasmolysis.

In leaf tissues, there may also be a potential for differential regulation of plasmodesmata located at different sites on the pathway leading from the mesophyll to the minor veins. In the species studied to date (*Commelina*, Erwee *et al.*, 1985; *Ipomoea*, Madore *et al.*, 1986; *Cucurbita foetidissima*, Madore, unpublished; *Coleus*, D. G. Fisher, unpublished), dye micro-injection studies indicate that a symplasmic transport pathway exists from the mesophyll to the cells of the minor veins. However, in *Ipomoea* (Madore *et al.*, 1986), we have found that barriers to symplasmic solute movement can be imposed by exposure of leaf tissues to an alkaline pH (pH 8 or 9). Following alkaline pH treatment, the movement of micro-injected dye *between* mesophyll cells is apparently unaffected (Fig. 2.3; cf. A and C). However, as illustrated in Fig. 2.3, this movement of dye to the minor veins is almost completely or entirely abolished by alkaline pH treatments (Madore *et al.*, 1986). A similar report by Lyalin *et al.* (1986) of

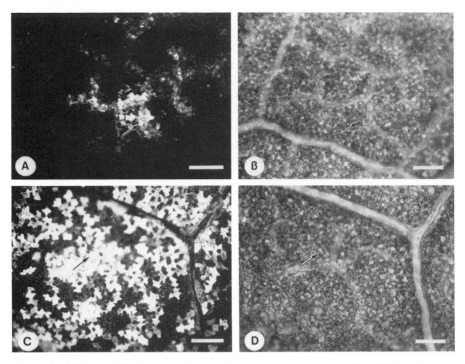

Fig. 2.3 Effect of bathing medium pH on movement of Lucifer Yellow in the source leaf symplasm of *Ipomoea tricolor*. All bars represent 200 μm ($\times$ 64). Pattern of dye movement (white areas) 60 min after injection into tissue given a pH 5.5 (A) or pH 8.0 (C) treatment. Bright field micrographs (B) and (D) depict regions illustrated in (A) and (C), respectively (from Madore *et al.*, 1986).

plasmodesmatal uncoupling by alkaline pH treatments in *Salvinia* trichomes strongly suggests that the site of action of alkaline pH in the *Ipomoea* leaf may also be plasmodesmata, but in this case only those interconnecting certain cells, possibly the bundle-sheath and vascular parenchyma. This interpretation is supported by our finding (Madore and Lucas, 1987) that the movement of recently fixed ^{14}C-labelled photo-assimilates into the minor veins is prevented when tissue is pre-treated at alkaline pH (see Fig. 2.4). The actual site of the blockage induced at alkaline pH and the mechanisms involved in inducing plasmodesmatal closure still remain to be determined.

In general, there would appear to be a valid basis for postulating a role for calcium in the control of solute movement through plasmodesmata. The nature of the calcium-regulated processes is as yet uncertain, as is the role that calcium may play in the control of photoassimilate transport between leaf cells. In recent years, however, significant advances have been made towards developing a complete understanding of calcium-regulated solute movement through animal cell gap junctions. Although

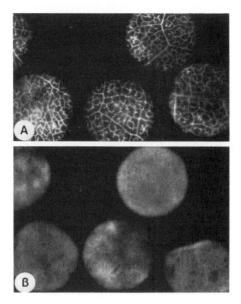

Fig. 2.4 Influence of bathing medium pH treatment on the distribution of [^{14}C] photosynthate in *Ipomoea tricolor* source leaf discs. Discs were pre-treated (4 h) in bathing medium buffered at pH 5.0 (A) or pH 9.0 (B) prior to a 5 min exposure to $^{14}CO_2$; leaf discs were then given a 2 h chase period at their pre-treatment pH, before being processed for autoradiography (from Madore & Lucas, 1987).

structurally quite dissimilar (Fig. 2.1E) the functional analogy between gap junctions and plasmodesmata no longer seems questionable and again, it appears that the knowledge gained from these animal studies may be directly applicable to cell-to-cell transport in plant tissues. With this in mind, we will give a brief review of the regulation of solute movement through gap junctions.

2.2.4 Analogies between plasmodesmata and cell-to-cell transport through animal cell gap junctions

In animal systems, the permeability of gap junctions, as determined by the passage or exclusion of micro-injected probes of different molecular weight, has been extensively studied by Loewenstein's group (Simpson *et al.*, 1977; Flagg-Newton *et al.*, 1979). It is apparent that the gap junction permeability varies for different cell types, with exclusion limits being as low as 464 Da for some mammalian cells and as high as 1830 Da for insect cells; in general, for mammalian cells, the limit falls somewhere around 700–800 Da. In addition, mammalian gap junctions appear to differ in structure from insect gap junctions, since they discriminate against probes on the basis of charge as well as size (Flagg-Newton *et al.*, 1979).

However, as is seen for plasmodesmata, the size exclusion limits of gap junctions also are not absolute. Rather, permeability appears to be under the control of cytosolic calcium levels (Loewenstein & Rose, 1978). In an elegant series of experiments using fluorescent probe pairs of different molecular weight and fluorescence characteristics, Loewenstein's group have shown that the permeability of gap junctions can be regulated in either a graded manner or an all-or-none manner by calcium ions. They postulate that in an injury situation, which induces a high and unbalanced influx of calcium into the cell due to the steep gradient across the cell membrane, gap junctions are totally sealed off, isolating the injured cell from healthy ones. In contrast to injury or death symptoms, small changes in cytosolic calcium are thought to give rise to physiological responses, and result in a graded opening or closure of the gap junctions, such that the unit behaves like a sieve with the mesh size controlled by the level of calcium in the cytosol.

The actual mechanism for calcium control is not yet fully elucidated. It was initially postulated that calcium could change the channel protein charge or configuration, thereby affecting permeability by decreased pore size or possibly by misalignment of the two sub-unit proteins which span the membranes (Fig. 2.1E). Calcium was envisaged as binding to the channel proteins themselves, or perhaps to a membrane lipid, affecting the subsequent conformation of the embedded protein (Loewenstein & Rose, 1978). However, the discovery that the product of the viral *src* gene down-regulated gap junction permeability and also promoted extensive protein and membrane lipid phosphorylations has radically changed this rather simplistic view.

It has been known for a number of years that the protein products of certain viral and cellular genes could down-regulate the permeability of vertebrate cell gap junctions. The mechanism was tied in with the peculiar protein phosphorylations that these gene products elicited, and the discovery that lipid precursors of the phosphoinositide pathway were also phosphorylated by these viral gene products (Yada *et al.*, 1985). This suggested that the phosphatidylinositol transmembrane signalling system for control of cytosolic calcium levels was involved (see review by Berridge (1987) for a complete description of this signalling system; and by Pooviah *et al.* (1987) for evidence of its operation in plant tissues).

Recently, extensive evidence has been obtained for the involvement of the diacylglycerol branch of the phosphoinositide pathway in the control of gap junction permeability (Yada *et al.*, 1985). Diacylglycerol pre-treatment of cultured epithelial cells was found to down-regulate the gap junction permeability to Lucifer Yellow CH in a manner similar to that seen with calcium. Application of the phorbol ester, 12-0-tetradecanoyl-phorbol-13-acetate, which bipasses diacylglycerol and mobilizes internal calcium, resulted in a similar response. The effect of diacylglycerol could be blocked by application of 8-*N*,*N*-(diethylamino)octyl-3,4,5-trimethoxy-benzoate (TMB-8), a blocker of internal calcium mobilization.

Temperature-sensitive mutants of Rous sarcoma virus have been isolated which show thermolabile expression of the *src* gene protein kinase activity (Rose *et al.*, 1986). Thus, a change of a few degrees, from 34°C (the 'permissive' temperature for gene expression) to 41°C (the non-permissive temperature), could be used to assess the importance of the *src* gene product on junctional permeability. As expected, junctional transfer was much greater at the non-permissive temperature of 41°C, when the *src* gene was not being expressed and phosphorylations were therefore not being activated. However, at this temperature permeability could be greatly reduced by application of diacylglycerol. Changing to the permissive temperature of 34°C reduced junctional permeability to zero, but this effect could be prevented by application of TMB-8. This effect of temperature was fully reversible, with junctional permeability being restored to 75% of the control value 12 min after return to the non-permissive temperature.

Interestingly, vanadate, an inhibitor of tyrosyl phosphatases (the *src* gene product selectively phosphorylates proteins at tyrosine residues), inhibited the recovery at the non-permissive temperature (Rose *et al.*, 1986). It was postulated that the effects of vanadate may have been due to inhibition of dephosphorylating activities required for restoration of normal permeability. Alternatively, it was also postulated that vanadate may have inhibited plasma membrane ATPase activity, allowing the maintenance of high internal cytosolic levels of calcium, which in turn maintained elevated levels of protein kinase activity.

In summary, then, in animal gap junctions, calcium seems to play a role in the control of junctional permeability. This role would appear to be tied in with the activation of calcium release through the diacylglycerol branch of the phosphatidylinositol transmembrane signalling system and may be mediated through phosphorylation of specific proteins, perhaps even the channel proteins themselves (Loewenstein, 1985). There is increasing evidence for the operation of this membrane signalling system in plant tissues (Pooviah *et al.*, 1987) and, in view of the effects of calcium seen on plasmodesmatal permeability, this may also prove to be a fruitful line of pursuit in future studies on transport of assimilates between leaf cells.

2.2.5 *Viral movement through the symplasm*

As previously mentioned, Erwee & Goodwin (1984) have demonstrated that the size exclusion characteristics (selectivity) of plasmodesmata can be overcome by plasmolysis/deplasmolysis treatments. However, it is also well known that certain types of viruses can spread through the symplasm of the plant by moving through modified plasmodesmata (De Zoeten &

Gaard, 1969; Kitajima & Lauritis, 1969; Allison & Shalla, 1974; Weintraub *et al.*, 1976). Esau (1968), for example, has shown that for plants infected by the beet yellows virus, 'both the longitudinal and the transectional views of plasmodesmata containing virus particles show no cores (desmotubules) and no endoplasmic reticulum associations'. This viral infection process seems to offer a radical way to modify the plasmodesmata, provided this can be achieved without the complications associated with viral infection *per se*. Molecular biology has aided considerably the elucidation of the nature of the viral protein(s) involved in conferring symplasmic mobility, as well as providing transgenic plants in which these specific proteins have been expressed (see Abel *et al.*, 1986; Meshi & Okada, 1986).

Working on tobacco, Nishiguchi *et al.* (1978, 1980) were able to isolate a temperature-sensitive mutant of the tobacco mosaic virus (TMV Ls1) which, under restrictive temperatures, replicated and assembled normally in both epidermal cells and protoplasts, but did not move from cell to cell in inoculated *leaves*. An analysis of two-dimensional tryptic peptide maps of proteins synthesized by *in vitro* translation of viral RNAs from the temperature-restrictive mutant and a closely related strain that could move from cell to cell, revealed a slight difference *only* in a 30 kDa protein (Leonard & Zaitlin, 1982). Leonard & Zaitlin (1982) concluded that this 30 kDa polypeptide must be involved, in some way, in mediating cell-to-cell movement of TMV (see also Deom *et al.*, 1987). A similar conclusion was recently drawn by Huisman *et al.* (1986) concerning the function of a closely homologous 30 kDa protein of alfalfa mosaic virus.

Further support for the hypothesis that this 30 kDa viral protein is involved in modifying the plasmodesmatal substructure comes from the observation that viruses *defective* in cell-to-cell movement (like TMV Ls1) can be complemented by a 'helper' virus. For example, when TMV Ls1 was *co-inoculated* with a common strain, TMV Ls1 also spread through the symplasm (Taliansky *et al.*, 1982; Meshi & Okada, 1986). This observation that distinctly different viruses can function as helpers, or can complement a *defect* in a non-spreading virus in a non-host plant, suggests that the underlying mechanism of cell-to-cell movement is common to many groups of plant viruses (Meshi & Okada, 1986).

The limitation at present is that little is known about the mode by which this 30 kDa viral protein facilitates cell-to-cell movement. It may cause changes in the substructure of the plasmodesmata, or alternatively, it may have an effect on the cell wall in the region of the plasmodesmata which could increase the diameter of the cytoplasmic annulus. Concerning the second possibility, Godefroy-Colburn *et al.* (1986) recently reported finding the 30 kDa protein of alfalfa mosaic virus in a cell wall fraction isolated from infected tobacco plants. Clearly, by developing an understanding of the molecular and biochemical regulation of the synthesis and action of this protein we will have available an exciting new way to further our investigation of plasmodesmatal function in terms of solute movement between cells (mesophyll) or tissues (mesophyll to phloem). The development of transgenic plants in which the 30 kDa viral coat protein is under promoter control (like heat shock) will greatly enhance the sophistication of these studies.

2.3 Apoplasmic transport

The hydrated cell walls of plant tissues form a continuum and, therefore, provide an alternative route to plasmodesmata for solute transport between leaf cells. This pathway is important in the retrieval of incoming solutes delivered from the roots via the xylem, but may, in some cases, also be important in the transport of photoassimilates within leaf tissues. A case has been made for a role for apoplasmic transport in phloem loading (Giaquinta, 1983) but the data are still open to question (Lucas, 1985; Lucas & Madore, 1988). In this section we will review what is known concerning transport to and retrieval from the apoplasm of leaf cells. The specific question of phloem loading is addressed in a later chapter (see Ch. 5) but will be dealt with briefly here in the context of solute retrieval mechanisms.

2.3.1 Membrane transport by leaf cells

As early as 1978, Guy *et al.* invoked, on the basis of the energy-dependent, pH-sensitive uptake of a glucose analogue (3-0-methyl glucose) by isolated pea mesophyll protoplasts, a sugar proton co-transport mechanism operating across the plasma membrane of pea mesophyll cells (Guy et al., 1978, 1981). Scorer (1984) reported the same energy dependence for assimilate retention in isolated tobacco mesophyll cells, and postulated that the 'retention' of assimilates could in fact be due to retrieval of effluxed solutes back into the cells by these proton-dependent carriers. However, another possibility has been raised, namely that the role of these carriers may be to *release* photoassimilates to the apoplasm from the mesophyll for subsequent retrieval by the minor veins (Huber & Moreland, 1980; Anderson, 1983, 1986; Secor, 1987).

For some time, though, the concept of active, proton-dependent carrier mechanisms operating on *mesophyll* cell membranes has been ignored in favour of the more intriguing possibility of similar mechanisms operating across *phloem* membranes, giving rise to the phenomenon of 'phloem loading'. Unfortunately, this focus on phloem tissues has left the impression that *only* minor veins possess the ability to take up exogenously supplied solutes or to perform a retrieval function (e.g. Turgeon, 1984; Daie & Wyse, 1985). This idea has, again unfortunately, been reinforced

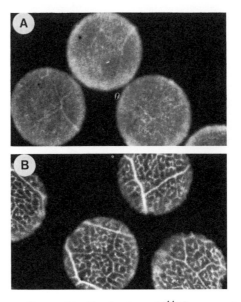

Fig. 2.5 Macro-autoradiographic distribution of [^{14}C] sucrose in peeled source leaf discs of *Beta vulgaris*. *A*: Leaves were excised at the base of the petiole while the crown of the plate was submerged under water. Petioles were then placed in distilled water and the leaves returned to the controlled environment chamber for 24 h. Uptake from 20 mM (mol m^{-3}) [^{14}C] sucrose (pH 5.0) was for 30 min. *B*: Control experiment in which discs were cut from leaves that were still attached to the plant (other details as in (A)) (from Wilson & Lucas, 1987).

by autoradiographic images which suggest that uptake of exogenously supplied [^{14}C]-sucrose into leaf discs occurs primarily into the veins (e.g. Giaquinta, 1976).

Recently, the validity of these images (van Bel *et al.*, 1986; Madore & Lucas, 1987; Wilson & Lucas, 1987), as well as the conclusions drawn from them concerning 'phloem loading' mechanisms (Maynard & Lucas, 1982; Wilson *et al.*, 1985; van Bel *et al.*, 1986; Madore & Lucas, 1987; Wilson & Lucas, 1987), have been questioned. It had been evident from early work by Fondy and Geiger (1977) that uptake of exogenous sucrose into *Beta* leaf discs was occurring, for the most part (60%), into the mesophyll compartment. In 1982, Maynard and Lucas questioned the kinetic studies of 'phloem loading', as their results with *Beta* showed conclusively that the pH-sensitive, PCMBS-sensitive saturable component of sucrose uptake, invoked as the 'phloem loading' mechanism (Giaquinta, 1980), was in fact a constituent of *all* leaf cell membranes and was not restricted to the phloem. A similar study by Wilson *et al.* (1985) showed that in the *Allium* leaf, all leaf cell types, including the phloem, possess the same saturable and linear mechanisms for uptake of exogenously supplied sucrose. More importantly, this study showed that uptake by the vascular tissues may in

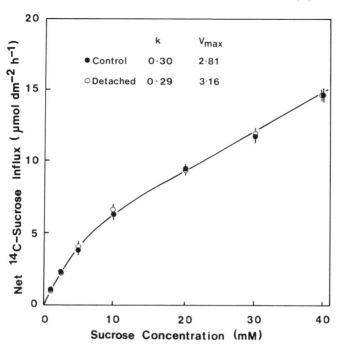

Fig. 2.6 Kinetic profiles obtained using peeled leaf discs of *Beta vulgaris* that were either cut from plant material that had been detached for 24 h (o) or had remained attached to the parent plant (●). (Macro-autoradiographs presented in Fig. 2.5 were sub-sampled from the 20 mM (mol m^{-3} [^{14}C] sucrose treatment). The kinetic constants for the saturable (V_{max}, μmol dm^{-2} h^{-1}) and first order kinetic (k, μmol dm^{-2} h^{-1} mM^{-1} sucrose) systems have been included (from Wilson & Lucas, 1987).

fact be quite small, a finding in agreement with the original data of Fondy and Gieger (1977) and also now substantiated for leaf tissues of *Commelina* (van Bel *et al.*, 1986) and *Ipomoea* (Madore & Lucas, 1987).

As a consequence, the concept of retrieval mechanisms must be reintroduced into our present-day thinking on solute transport processes in leaf tissues. van Bel *et al.* (1986) have suggested, on the basis of comparative experiments with *Commelina* leaf discs and isolated mesophyll cells, that what was originally determined from autoradiographic data to be phloem loading was in fact, to a large extent, uptake of exogenously supplied sucrose into the mesophyll followed by (symplasmic) relay to the minor veins. A similar conclusion was reached by Madore & Lucas (1987) for sucrose uptake into *Ipomoea*. Particularly convincing evidence that this may be true has recently been obtained by Wilson & Lucas (1987). As seen in Fig. 2.5A, leaf discs obtained from mature *Beta* leaves that were detached and placed in water for 24 h retrieved exogenously supplied [^{14}C]-sucrose almost exclusively into the mesophyll compartment. Unlike control leaf discs, this [^{14}C]-sucrose remained in the mesophyll rather than accumulating in the minor veins (cf. Fig. 2.5A and B). However, the kinetics of the retrieval process were identical, irrespective of whether the discs were obtained from control (0 h excised) or 24 h excised leaves (Fig. 2.6). Clearly the kinetics of sugar uptake in this case cannot be representative of sugar uptake by the phloem, but must represent the kinetics for uptake into the mesophyll.

2.3.2 Transfer cells

As indicated in the previous section, the concept of phloem loading from the apoplasm must be looked at more carefully. One case in which an apoplasmic step may be involved in phloem loading is the special situation where the companion cells or associated vascular parenchyma are modified into transfer cells. As noted by Gamalei (1985b), herbaceous species tend to be designed for apoplasmic transport, a feature also noted by Pate and Gunning (1969) in their survey of transfer cell distribution in various plant families. Pate and Gunning point out that herbaceous species have very rapid and extensive circulation of inorganic and organic solutes within and between organs. In addition, such plants tend not to have large reserves of photoassimilates compared to the demands of reproduction and storage. Pate and Gunning postulated that in these plants the nature of their growth habit demands that there be a method for rapid recycling of nutrients. Thus, solutes are constantly being delivered, via the xylem, predominantly to the mature source leaves which have the most rapid rates of transpiration. A pathway must exist for these xylem-borne nutrients to be re-channelled to the phloem for export to developing tissues.

The labyrinth of wall ingrowths present in transfer cells may provide an efficient system for taking up solutes from the apoplasm; a corollary would be that carriers for specific solutes must exist on the membranes lining these ingrowths. The origin of these solutes may be predominantly from the xylem; indeed in the Fabaceae, or other nitrogen-fixing species, it may be particularly important to retrieve nitrogenous compounds from the xylem and deliver them to the phloem. In this context, it would be particularly interesting to examine the possible correlation of the presence of transfer cells (i.e. Gamalei's *Group III* plants) with respect to the principal sites of nitrogen (Andrews, 1986) or possibly sulphur assimilation. In the case of legumes, in particular, the association of vascular transfer cells and delivery of organic fixed nitrogen products from the roots would appear to be especially clear. In this case, the delivery of nitrogenous compounds to the phloem may represent a case of apoplasmic retrieval *par excellence*.

In many leaves which possess vascular transfer cells, the onset of phloem transport corresponds with the onset of transfer cell development (Pate & Gunning, 1969; Henry & Steer, 1980). Thus, it is entirely possible that loading of the sieve elements is the primary function of vascular transfer cells, particularly those which are modified phloem parenchyma and companion cells. However, as indicated by Gunning and Pate, and as suggested by the existence of *xylem* parenchyma transfer cells, the recycling of nutrients is likely to be just as important a function of these cells. It must be remembered that the minor veins represent not only the origin of the phloem transport system, but also the terminus of the transpiration stream. Therefore, it is equally possible that modified vascular parenchyma cells evolved to deal with retrieval of extracellular solutes other than photoassimilates. Of course, they may also be responsible for 'phloem loading', although in most cases symplasmic routes for loading are also in evidence (Gunning & Pate, 1969). Clearly, dye micro-injection studies should be initiated on leaves with these types of modified parenchyma to determine if these cells represent an end-point of the symplasmic pathway into the minor veins.

2.4 Transport between specialized leaf compartments

In leaf tissues not all soluble photoassimilates are destined for immediate phloem transport. In most cases, kinetically distinct pools of photo-assimilates, predominantly sucrose but often other compounds as well, can be distinguished, representing separate storage and export pools. In other leaf types, most notably those of C_4 plants like *Zea mays*, extensive intercellular transfer of photosynthetic products must occur between specialized tissues prior to the synthesis of sucrose or other phloem-mobile metabolites. Transport of assimilates between specific leaf tissues with specialized storage or synthetic functions represents another interesting aspect of assimilate transport between leaf cells.

2.4.1 Transport between storage pools within the leaf

In *Glycine*, the existence of kinetically distinct pools of sucrose was thought initially to be a consequence of a storage pool that existed in the paraveinal mesophyll, a specialized layer of tissue found between the mesophyll and the phloem in this species (Fisher, 1970). However, distinct pools of photoassimilates have been determined even in tissues which lack this specialized layer (e.g., *Vicia*, Outlaw & Fisher, 1975; Outlaw *et al.*, 1975; Fisher & Outlaw, 1979; *Beta*, Geiger *et al.*, 1983). Hence, storage pools may be a common feature of most leaves. The site of the soluble carbohydrate storage pool is presently thought to be the vacuoles of the mesophyll cells (Fisher & Outlaw, 1979), but clearly the vacuoles of other cell types could also be involved. The important question that needs to be answered is how transport into and out of these storage pools is regulated in terms of providing carbohydrate to the phloem. Unfortunately, very few studies have focused on sugar transport across the tonoplast of mesophyll cells, and at present there are no details available on likely regulatory mechanisms. This area clearly warrants immediate attention.

Recent light, electron microscopy, physiological and molecular studies conducted by Franceschi and co-workers have greatly expanded our knowledge of the complexities in assimilate partitioning that occur within a mature soybean leaf. As mentioned above, the soybean leaf contains a unique cell type, the paraveinal mesophyll (PVM), which forms a one-cell-thick tissue that interconnects the palisade parenchyma and spongy mesophyll to the vascular bundles (see Fig. 2.7A). This position of the PVM within the leaf dictates that *all* photosynthate produced in the palisade and spongy mesophyll must pass through these cells *en route* to the phloem. Franceschi & Giaquinta (1983a) found that plasmodesmata interconnect the PVM to the palisade parenchyma, spongy mesophyll and the bundle-sheath. Thus, a symplasmic route for photoassimilates is possible at least on structural grounds.

Franceschi & Giaquinta (1983b) discovered that starch is unequally compartmented *and* degraded in the various cell types of the soybean leaf. For example, prior to or during anthesis, the greatest amount of starch (at the end of the photoperiod) is present in the second palisade layer followed by the spongy mesophyll and then the first (uppermost) palisade layer (Fig. 2.7B). By the end of the dark period, the starch reserves in the first palisade layer are *completely* degraded, whereas those in the second palisade and spongy mesophyll are *not* remobilized to any appreciable extent.

In the post-anthesis period, large amounts of starch are stored in all cell types, *except* the PVM (Fig. 2.7C). That the PVM cells contain normal (apparently) chloroplasts that do not accumulate starch to anywhere near the degree found in the nearby palisade cells (see Fig. 2.8), illustrates the type of spatial and temporal regulation that may also be present in other leaf types (see Franceschi, 1986). These extensive, often engorging, amounts of starch are not remobilized until the photosynthetic capacity of the leaves falls below the carbohydrate requirement of the maturing seeds.

Fig. 2.7 Light micrographs illustrating general anatomical features of the soybean leaf. *A*: Cross-section of a soybean leaf, stained with methylene blue, illustrating the two layers of palisade parenchyma (a and b), the large-celled paraveinal mesophyll (PVM, c) and spongy mesophyll (d). Note the vein with the bundle-sheath connecting to the paraveinal mesophyll. ($\times$ 310) *B*: Pre-anthesis leaf stained for polysaccharide (starch). Note that the PVM has little or no starch and that differential starch accumulation is noticeable in the palisade layers. ($\times$ 200) *C*: Post-anthesis leaf stained for starch. A few small starch grains can be seen in the PVM (arrows) while the lower palisade chloroplasts are engorged with starch. Following anthesis, the mature leaf has thickened due to cell expansion. ($\times$ 200) (Light micrographs kindly provided by V. R. Franceschi, Washington State University, Pullman, Washington, USA.)

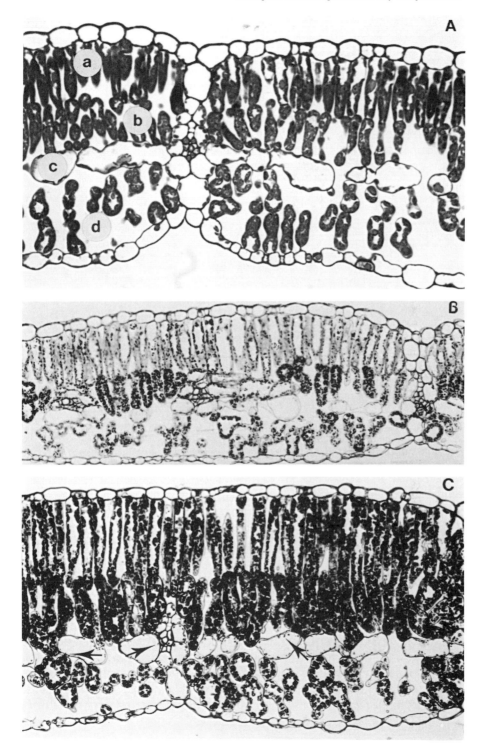

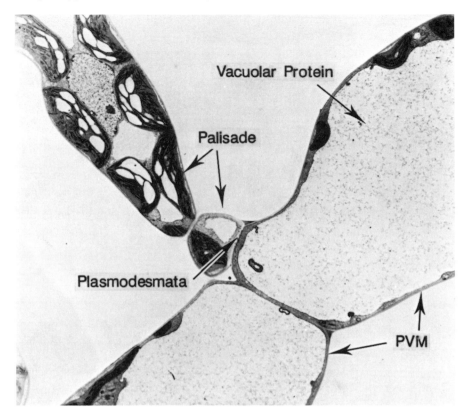

Fig. 2.8 Transmission electron micrograph of a soybean leaf showing two para-veinal mesophyll cells joined to palisade parenchyma cells. Material in the PVM vacuole has been identified as a storage glycoprotein. Note the difference in starch accumulation in the chloroplasts of the PVM and palisade parenchyma. ($\times$ 2600) (Micrograph kindly provided by V. R. Franceshi, Washington State University, Pullman, Washington, USA.)

Although little information is available on how the cells of a particular layer control the remobilization of these starch reserves within their chloro-plasts, the soybean system provides us with an excellent example of carbohydrate regulation *within* the symplasmic compartment of the leaf.

The soybean PVM not only functions as the structural linkage between the photosynthetic cells and the phloem, but it also acts as a temporary storage pool for nitrogen. Franceschi & Giaquinta (1983b) have shown that during the pre-anthesis period, the vacuole of the PVM stores protein that was synthesized by these cells. This protein appears to be a glycoprotein (Wittenbach, 1983) and may be composed of two sub-units having molecular weights of 27 and 29 kDa (Wittenbach *et al.*, 1984). Remark-ably, all of this nitrogen reserve is then remobilized and exported to the pods during the first two weeks of seed-fill.

As is the case with the remobilization of starch, little information is presently available about the mechanism by which the soybean plant controls events in the PVM to mobilize these protein reserves. Depodding experiments have confirmed that the reserves are destined for the seeds and that in the absence of this specific sink, the PVM vacuolar storage proteins remain undegraded. Although the nature of the signalling mechanism that integrates the specific sink (pod or seed) demand to the reserves (carbohydrate and nitrogen) in the source leaves remains unresolved, molecular biology may provide an important 'avenue' through which these processes can be elucidated. Franceschi and co-workers have begun this molecular characterization using isolated soybean mesophyll and PVM protoplasts (Franceschi *et al.*, 1983, 1984; Wittenbach *et al.*, 1984; Costigan *et al.*, 1987). Studies of this type are essential if we are to make progress in furthering our understanding of the regulation of phloem translocation/ resource allocation.

2.4.2 *Transport between mesophyll and bundle-sheath cells in Zea mays*

It has long been postulated that the transport of malic acid, between the mesophyll and bundle-sheath of *Zea*, probably occurs via the symplasm, as the cell walls of the bundle-sheath are suberized, but abundant plasmodesmata interconnect these two cell types (Evert *et al.*, 1977). In fact, the spatial separation of carbon fixation mechanisms in the *Zea* leaf requires a rapid shuttle of, apart from C_4 acids, certain other metabolites such as triose phosphate and 3-phosphoglyceric acid (Hatch & Osmond, 1976). As indicated in Fig. 2.9, 3PGA is produced by Calvin cycle activity in the bundle-sheath, and is then shuttled to the mesophyll, presumably again through the symplasm via the plasmodesmata that interconnect these cells. In the mesophyll the 3PGA is reduced to triose-P. It was originally believed that this triose-P had to return to the bundle-sheath to allow continuation of the Calvin cycle, as well as the synthesis of starch and sucrose (Hatch & Osmond, 1976).

Recent compartmental studies have confirmed that a concentration gradient of 3PGA does exist between bundle-sheath and mesophyll cells (Leegood, 1985; Furbank *et al.*, 1985). Although these studies add experimental support to the concept that diffusion through plasmodesmata can couple biochemical and/or physiologically separated compartments, a further complexity was added to the *Zea mays* system. Furbank *et al.* (1985) discovered that in this species the synthesis of sucrose occurs primarily in the mesophyll! A spatial separation of starch and sucrose synthesis therefore also occurs in the *Zea* leaf, in addition to the known biochemical separation of C_3 and C_4 photosynthetic reactions. (Details of the biochemistry and the regulation of these reactions are provided in Ch. 1.) Thus, at least in *Zea mays*, the C_4 biochemical pathway does not

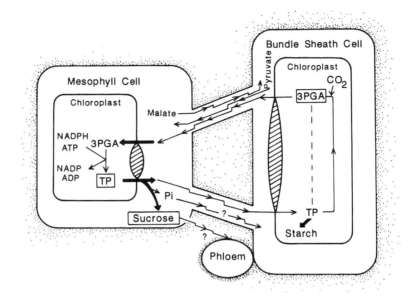

Fig. 2.9 Compartmentation of sucrose synthesis in leaves of *Zea mays*. Metabolites enclosed in boxes are thought to be at high concentrations to support *diffusion* between compartments. Proposed diffusion down respective concentration gradients has been indicated by stepped lines. The shaded regions represent the phosphate translocator. TP, triose phosphate; 3PGA, 3-phosphoglycerate. (Redrawn from Furbank *et al.*, 1985.)

give rise to the production of sucrose at a location close to the site of phloem 'loading'. Consequently, one must now add sucrose to the list of substrates that traffic through these C_4 plasmodesmata. One cannot help but be impressed by the efficacy of plasmodesmata in this situation, where the process of random thermal motion has been 'biochemically and physiologically focused' through these incredibly small 'organelles'.

2.5 Conclusions

It is now well established that photosynthetic carbon assimilation in leaf cells is indirectly controlled by levels of certain key intermediates (e.g. P_i) in the cytosol. Considerable biochemical information is now available concerning the regulation of processes involved in photoassimilate production, in particular the synthesis of sucrose, within a leaf cell (see Walker & Sivak, Ch. 1). However, it is only in recent years that the importance of cellular transport processes in establishing and regulating these cytosolic metabolite levels has been recognized. The work on compartmentation in *Zea* (Furbank *et al.*, 1985; Leegood, 1985; Stitt & Heldt, 1985) makes the importance of short-distance transport processes in leaf metabolism particularly clear. With this in mind, perhaps the biochemical 'end-products' of the photosynthetic reactions, such as starch, amino acids and sucrose, should now be looked on in a new light as the starting 'substrates' for numerous transport processes occurring both within green cells and within the complex of differentiated tissues that make up the green leaf.

References

Abel, P. P., Nelson, R. S., De, B., Hoffmann, N., Rogersa, S. G., Fraley, R. T. & Beachy, R. N. (1986). Delay of disease development in transgenic plants that express the tobacco mosaic virus coat protein gene. *Science*. **232**, 738–43.

Allison, A. V. & Shalla, T. A. (1974). The ultrastructure of local lesions induced by potato virus X: A sequence of cytological events in the course of infection. *Phytopathology* **64**, 784–93.

Anderson, J. M. (1983). Release of sucrose from *Vicia faba* L. leaf discs. *Plant Physiology* **71**, 333–40.

Anderson, J. M. (1986). Sucrose release from soybean leaf slices. *Physiologia Plantarum* **66**, 319–27.

Andrews, M. (1986). The partitioning of nitrate assimilation between root and shoot of higher plants. *Plant, Cell and Environment* **9**, 511–19.

Belitser, N. V., Zaalishuili, G. V. & Sytrianskaja, N. P. (1982). Ca^{2+} binding sites and Ca^{2+}-ATPase activity in barley root tip cells. *Protoplasma* **111**, 63–78.

Berridge, M. J. (1987). Inositol trisphosphate and diacylglycerol: Two interacting second messengers. *Annual Review of Biochemistry* **56**, 159–228.

Costigan, S. A., Franceschi, V. R. & Ku, M. S. B. (1987). Allantoinase activity and ureide content of mesophyll and paraveinal mesophyll of soybean leaves. *Plant Science* **50**, 179–87.

Daie, J. & Wyse, R. (1985). Evidence on the mechanism of enhanced sucrose uptake at low cell turgor in leaf discs of *Phaseolus coccinus*. *Physiologia Plantarum* **64**, 547–52.

Deom, C. M., Oliver, M. J. & Beachy, R. N. (1987). The 30-kilodalton gene product of tobacco mosaic virus potentiates virus movement. *Science* **237**, 389–94.

De Zoeten, G. A. & Gaard, G. (1969). Possibilities of inter- and intracellular translocation of some icosahedral plant viruses. *Journal of Cell Biology* **40**, 814–23.

Erwee, M. G. & Goodwin, P. B. (1983). Characterization of the *Egeria densa* Planch. leaf symplast. Inhibition of the intercellular movement of fluorescent probes by group II ions. *Planta* **158**, 320–8.

Erwee, M. G. & Goodwin, P. B. (1984). Characterization of the *Egeria densa* Planch. leaf symplast: Response to plasmolysis, deplasmolysis and to aromatic amino acids. *Protoplasma* **122**, 162–8.

Erwee, M. G. & Goodwin, P. B. (1985). Symplast domains in extrastelar tissues of *Egeria densa* Planch. *Planta* **163**, 9–19.

Erwee, M. G., Goodwin, P. B. & van Bel, A. J. E. (1985). Cell–cell communication in the leaves of *Commelina cyanea* and other plants. *Plant, Cell and Environment* **8**, 173–8.

Esau, K. (1968). *Viruses in Plant Hosts*, 225 pp. Madison, University of Wisconsin Press.

Evert, R. F., Eschrich, W. & Heyser, W. (1977). Distribution and structure of the plasmodesmata in mesophyll and bundle-sheath cells of *Zea mays* L. *Planta* **136**, 77–89.

Evert, R. F., Eschrich, W. & Heyser, W. (1978). Leaf structure in relation to solute transport and phloem loading in *Zea mays* L. *Planta* **138**, 279–94.

Evert, R. F. & Mierzwa, R. J. (1986). Pathway(s) of assimilate movement from mesophyll cells to sieve tubes in the *Beta vulgaris* leaf. In Cronshaw, J., Lucas, W. J. & Giaquinta, R. T. (eds.). *Phloem Transport*, pp. 419–32. New York, Alan R. Liss.

Fisher, D. B. (1970). Kinetics of C-14 translocation in soybean. II. Kinetics in the leaf. *Plant Physiology* **45**, 114–18.

Fisher, D. B. & Outlaw, W. H. (1979). Sucrose compartmentation in the palisade parenchyma of *Vicia faba* L. *Plant Physiology* **64**, 481–3.

Fisher, D. G. (1986). Ultrastructure, plasmodesmatal frequency, and solute concentration in green areas of variegated *Coleus blumei* Benth. leaves. *Planta* **169**, 141–52.

Fisher, D. G. & Evert, R. F. (1982). Studies on the leaf of *Amaranthus retroflexis* (Amaranthaceae): ultrastructure, plasmodesmatal frequency and solute concentration in relation to phloem loading. *Planta* **155**, 377–87.

Flagg-Newton, J., Simpson, I. & Loewenstein, W. R. (1979). Permeability of the cell-to-cell membrane channels in mammalian cell junction. *Science* **205**, 404–7.

Fondy, B. R. & Geiger, D. R. (1977). Sugar selectivity and other characteristics of phloem loading in *Beta vulgaris* L. *Plant Physiology* **59**, 953–60.

Francheschi, V. R. (1986). Temporary storage and its role in partitioning among sinks. In Cronshaw, J., Lucas, W. J. & Giaquinta, R. T. (eds.). *Phloem Transport*, pp. 399–409. New York, Alan R. Liss.

Francheschi, V. R. & Giaquinta, R. T. (1983a). The paraveinal mesophyll of soybean leaves in relation to assimilate transfer and compartmentation. I. Ultrastructure and histochemistry during vegetative development. *Planta* **157**, 411–21.

Francheschi, V. R. & Giaquinta, R. T. (1983b). The paraveinal mesophyll of soybean leaves in relation to assimilate transfer and compartmentation. II. Structural, metabolic and compartmental changes during reproductive growth. *Planta* **157**, 422–31.

Francheschi, V. R., Ku, M. S. B. & Wittenbach, V. A. (1984). Isolation of mesophyll and paraveinal mesophyll protoplasts from soybean leaves. *Plant Science Letters* **36**, 181–6.

Francheschi, V. R., Wittenbach, V. A. & Giaquinta, R. T. (1983). Paraveinal mesophyll of soybean leaves in relation to assimilate transfer and compartmentation. III. Immunohistochemical localization of specific glycopeptides in the vacuole after depodding. *Plant Physiology* **72**, 586–9.

French, D. F. (1954). The raffinose family of oligosaccharides. *Advances in Carbohydrate Chemistry* **9**, 149–83.

Furbank, R. T., Stitt, M. & Foyer, C. H. (1985). Intercellular compartmentation of sucrose synthesis in leaves of *Zea mays*. L. *Planta* **164**, 172–8.

Gamalei, Y. V. (1985a). Plasmodesmata: intercellular communication in plants. *Fiziologiia Rastenii* **32**, 134–48.

Gamalei, Y. V. (1985b). Characteristics of phloem loading in woody and herbaceous plants. *Fiziologiia Rastenii* **32**, 866–75.

Gamalei, Y. V. & Pakhomova, M. V. (1981). Distribution of plasmodesmata and parenchyma transport of assimilates in leaves of several dicots. *Fiziologiia Rastenii* **28**, 901–12.

Geiger, D. R., Ploeger, B. J., Fox, T. C. & Fondy, B. R. (1983). Sources of sucrose translocated from illuminated sugar beet source leaves. *Plant Physiology* **72**, 964–70.

Giaquinta, R. T. (1976). Evidence for phloem loading from the apoplast. Chemical modification of membrane sulfhydryl groups. *Plant Physiology* **57**, 872–5.

Giaquinta, R. T. (1980). Mechanism and control of phloem loading of sucrose. *Berichte Deutschen Botanischen Gesellschaft* **93**, 187–201.

Giaquinta, R. T. (1983). Phloem loading of sucrose. *Annual Review of Plant Physiology* **34**, 347–87.

Godefroy-Colburn, T., Gagey, M.-J., Berma, A. & Stussi-Garaud, C. (1986). A non-structural protein of alfalfa mosaic virus in the walls of infected tobacco cells. *Journal of General Virology* **67**, 2233–9.

Goodwin, P. B. (1983). Molecular size limit for movement in the symplast of the *Elodea* leaf. *Planta* **157**, 124–30.

Goodwin, P. B. & Lyndon, R. F. (1983). Synchronization of cell division during transition to flowering in *Silene* apices not due to increased symplast permeability. *Protoplasma* **116**, 219–22.

Gunning, B. E. S. (1976). The role of plasmodesmata in short distance transport to and from the phloem. In Gunning, B. E. S. & Robards, A. W. (eds.). *Intercellular Communication in Plants: Studies on Plasmodesmata*, pp. 203–27. Berlin, Springer-Verlag.

Gunning, B. E. S. & Pate, J. S. (1969). 'Transfer cells'. Plant cells with wall ingrowths, specialized in relation to short distance transport of solutes – their occurrence, structure and development. *Protoplasma* **68**, 107–33.

Gunning, B. E. S. & Robards, A. W. (1976). Plasmodesmata and symplastic transport. In Wardlaw, I. F. & Passioura, J. B. (eds.). *Transport and Transfer Processes in Plants*, pp. 15–41. New York, Academic Press.

Guy, M., Reinhold, L. & Laties, G. (1978). Membrane transport of sugars and amino acids in isolated protoplasts. *Plant Physiology* **61**, 593–6.

Guy, M., Reinhold, L., Rahat, M. & Seiden, A. (1981). Protonation and light synergistically convert plasmalemma sugar carrier system in mesophyll protoplasts to its fully activated form. *Plant Physiology* **67**, 1146–50.

Haberlandt, G. (1914). *Physiological Plant Anatomy*. Translation from 4th German edition. London, MacMillan.

Hatch, M. D. & Osmond, C. B. (1976). Compartmentation and transport in C_4 photosynthesis. In Stocking, C. R. & Heber, U. (eds.). *Encyclopedia of Plant Physiology*, New Series, Vol. 3. *Transport in Plants*, pp. 144–84. Berlin, Springer-Verlag.

Henry, Y. & Steer, M. W. (1980). A re-examination of the induction of phloem transfer cell development in pea leaves (*Pisum sativum*). *Plant, Cell and Environment* **3**, 377–80.

Huber, S. C. & Moreland, D. E. (1980). Efflux of sugar across the plasmalemma of mesophyll protoplasts. *Plant Physiology* **65**, 560–2.

Huisman, M. J., Sarachu, A. N., Ablas, F., Broxterman, H. J. G., Van Vloten-Doting, L. & Bol, J. F. (1986). Alfalfa mosaic virus temperature-sensitive mutants. III. Mutants with a putative defect in cell-to-cell transport. *Virology* **154**, 401–4.

Kitajima, E. W. & Lauritis, J. A. (1969). Plant virons in plasmodesmata. *Virology* **37**, 681–5.

Leegood, R. C. (1985). The intercellular compartmentation of metabolites in leaves of *Zea mays* Planta. **164**, 163–71.

Leonard, D. A. & Zaitlin, M. (1982). A temperature-sensitive strain of tobacco mosaic virus defective in cell-to-cell movement generates an altered viral-coded protein. *Virology* **117**, 416–24.

Loewenstein, W. R. (1985). Regulation of cell-to-cell communication by phosphorylation. *Biochemical Society Symposium* **50**, 43–58.

Loewenstein, W. R. & Rose, B. (1978). Calcium in (junctional) intercellular communication and a thought on its behaviour in intracellular communication. *Annals of New York Academy of Science* **307**, 285–307.

Lopez-Saez, J. F., Giminez-Martin, G. & Risueno, M. C, (1966). Fine structure of the plasmodesmata. *Protoplasma* **61**, 81–4.

Lucas, W. J. (1985). Phloem loading: a metaphysical phenomenon? In Heath, R. L. and Preiss, J. (eds.). *Regulation of Carbon Partitioning in Photosynthetic Tissue*, pp. 254–71. Rockville, MD, American Society of Plant Physiology.

Lucas, W. J. & Madore, M. A. (1988). Recent advances in sugar transport. In Preiss, J. (ed.). *The Biochemistry of Plants, a Comprehensive Treatise, Vol. 21: Carbohydrates*. New York, Academic Press (in press).

Lyalin, O. O., Ktitorova, I. N., Barmicheva, E. M. & Akhmedov, N. I. (1986). Intercellular contacts of submersed trichomes in *Salvinia*. *Fiziologiia Rastenii* **33**, 442–6.

Madore, M. A. & Lucas, W. J. (1987). Control of photoassimilate movement in source-leaf tissues of *Ipomoea tricolor* Cav. *Planta* **171**, 197–204.

Madore, M. A., Oross, J. W. & Lucas, W. J. (1986). Symplastic transport in *Ipomoea tricolor* source leaves. Demonstration of functional symplastic connections from mesophyll to minor veins by a novel dye-tracer method. *Plant Physiology* **82**, 432–42.

Madore, M. A. & Webb, J. A. (1981). Leaf free space analysis and vein loading in *Cucurbita pepo*. *Canadian Journal of Botany* **59**, 2550–7.

Maynard, J. W. & Lucas, W. J. (1982). Sucrose and glucose uptake into *Beta vulgaris* leaf tissues. A case for general (apoplastic) retrieval systems. *Plant Physiology* **70**, 1436–43.

Meshi, T. & Okada, Y. (1986). Systemic movement of viruses. In Kosuge, T. & Nester, E. W. (eds.). *Plant-Microbe Interactions: Molecular and Genetic Perspectives* Vol. 2, pp. 285–304. New York, MacMillan.

Münch, E. (1930). *Die Stoffbewegung der Pflanze*. Jena, Gustav Fisher.

Nishiguchi, M., Motoyoshi, F. & Oshima, N. (1978). Behavior of a temperature-sensitive strain of tobacco mosaic virus in tomato leaves and protoplasts. *Journal of General Virology* **39**, 53–61.

Nishiguchi, M., Motoyoshi, F. & Oshima, N. (1980). Further investigation of a temperature-sensitive strain of tobacco mosaic virus: its behavior in tomato leaf epidermis. *Journal of General Virology* **46**, 497–500.

Outlaw, W. H. & Fisher, D. B. (1975). Compartmentation in *Vicia faba* leaves. I. Kinetics of ^{14}C in the tissues following pulse labelling. *Plant Physiology* **55**, 699–703.

Outlaw, W. H., Fisher, D. B. & Christy, A. L. (1975). Compartmentation in *Vicia faba* leaves. II. Kinetics of ^{14}C-sucrose redistribution among individual tissues following pulse labelling. *Plant Physiology* **55**, 704–11.

Overall, R. L., Wolfe, J. & Gunning, B. E. S. (1982). Intercellular communication in *Azolla* roots. I. Ultrastructure of plasmodesmata. *Protoplasma* **111**, 134–50.

Palevitz, B. A. & Hepler, P. K. (1985). Changes in dye coupling of stomatal cells of *Allium* and *Commelina* demonstrated by microinjection of lucifer yellow. *Planta* **164**, 473–9.

Pate, J. S. & Gunning, B. E. S. (1969). Vascular transfer cells in angiosperm leaves. A taxonomic and morphological survey. *Protoplasma* **68**, 135–56.

Pooviah, B. W., Reddy, S. N. & McFadden, J. J. (1987). Calcium messenger system: Role of protein phosphorylation and inositol bisphospholipids. *Physiologia Plantarum* **69**, 569–73.

Richardson, P. T., Baker, D. A. & Ho. L. C. (1984). Assimilate transport in cucurbits. *Journal of Experimental Botany* **35**, 1575–81.

Robards, A. W. (1971). The ultrastructure of plasmodesmata. *Protoplasma* **72**, 315–23.

Rose, B., Yada, T. & Loewenstein, W. R. (1986). Down regulation of cell-to-cell communication by the viral *src* gene is blocked by TMB-8 and recovery of communication is blocked by vanadate. *Journal of Membrane Biology* **94**, 129–42.

Russin, W. A. & Evert, R. F. (1985). Studies on the leaf of *Populus deltoides* (Salicaceae): Ultrastructure, plasmodesmatal frequency, and solute concentrations. *American Journal of Botany* **72**, 1232–47.

Schmitz, K., Cuypers, B. & Moll, M. (1987). Pathway of assimilate transfer between mesophyll cells and minor veins in leaves of *Cucumis melo* L. *Planta* **171**, 19–29.

Scorer, K. (1984). Evidence for energy-dependent ^{14}C-photoassimilate retention in isolated tobacco mesophyll cells. *Plant Physiology* **76**, 753–8.

Secor, J. (1987). Regulation of sucrose efflux from soybean leaf discs. *Plant Physiology* **83**, 143–8.

Simpson, I., Rose, B. & Loewenstein, W.R. (1977). Size limit of molecules permeating the junctional membrane channels. *Science* **195**, 294–6.

Stewart, W. W. (1981). Lucifer dyes – highly fluorescent dyes for biological tracing. *Nature* **292**, 17–21.

Stitt, M. & Heldt, H. W. (1985). Control of photosynthetic sucrose synthesis by fructose-2,6,bisphosphate. Intercellular metabolite distribution and properties of the cytosolic fructose bisphosphatase in leaves of *Zea mays* L. *Planta* **164**, 179–88.

Taliansky, M. E., Malyshenko, S. I., Pshennikova, E. S., Kaplan, I. B., Ulanova, E. F. & Atabekov, J. G. (1982). Plant virus-specific transport function. 1. Virus genetic control required for systemic spread. *Virology* **122**, 318–26.

Terry, B. R. & Robards, A. W. (1987). Hydrodynamic radius alone governs the mobility of molecules through plasmodesmata. *Planta* **171**, 145–57.

Thomson, W. W. & Platt-Aloia, K. (1985). The ultrastructure of the plasmodesmata of the salt glands of *Tamarix* as revealed by transmission and freeze-fracture electron microscopy. *Protoplasma* **125**, 13–23.

Tucker, E. B. (1982). Translocation in the staminal hairs of *Setcreasea purpurea*. I. A study of cell ultrastructure and cell-to-cell passage of molecular probes. *Protoplasma* **113**, 193–201.

Tucker, E. B. & Spanswick, R. M. (1985). Translocation in the staminal hairs of *Setcreasea purpurea*. II. Kinetics of intercellular transport. *Protoplasma* **128**, 167–72.

Turgeon, R. (1984). Efflux of sucrose from minor veins of tobacco leaves. *Planta* **161**, 120–8.

Turgeon, R., Webb, J. A. & Evert, R. F. (1975). Ultrastructure of the minor veins in *Cucurbita pepo* leaves. *Protoplasma* **83**, 217–32.

van Bel, A. J. E., Amerlaan, A. & Blauw-Jansen, G. (1986). Preferential accumulation by mesophyll cells at low and by veins at high exogenous amino acid and sugar concentrations in *Commelina benghalensis* L. leaves. *Journal of Experimental Botany* **37**, 1899–910.

Weintraub, M., Ragetli, H. W. J. & Leung, E. (1976). Elongated virus particles in plasmodesmata. *Journal of Ultrastructural Research* **56**, 351–64.

Wilson, C. & Lucas, W. J. (1987). Influence of internal sugar levels on apoplasmic retrieval of exogenous sucrose in source leaf tissue. *Plant Physiology* **84**, 1088–95.

Wilson, C., Oross, J. W. & Lucas, W. J. (1985). Sugar uptake into *Allium cepa* leaf tissue: an integrated approach. *Planta* **164**, 227–40.

Wittenbach, V. A. (1983). Purification and characterization of a soybean leaf storage glycoprotein. *Plant Physiology* **73**, 125–9.

Wittenbach, V. A., Franceschi, V. R. & Giaquinta, R. T. (1984). Soybean leaf storage proteins. In Randall, D. D., Belvins, D. G., Larson, R. L. & Rupp, B. J. (eds.). *Current Topics in Plant Biochemistry and Physiology*, pp. 19–30. Columbia, University of Missouri Press.

Yada, T., Rose, B. & Loewenstein, W. R. (1985). Diacylglycerol down regulates junctional membrane permeability. TMB-8 blocks this effect. *Journal of Membrane Biology* **88**, 217–32.

3 Structure of the phloem

H.-D. Behnke

3.1 Introduction

The phloem is the tissue responsible for the distribution of photoassimilates within the plant and as such is also one of the major structural
features within the wide range of adaptations prerequisite for the successful
evolution of our land plants. Therefore it is obvious that among the
present day vascular plants a phylogenetic sequence can be followed up
within the phloem tissue itself. This chapter, however, is concerned
primarily with the phloem structure of the most highly evolved plants, i.e.
of the angiosperms. In addition, some peculiarities of gymnosperm phloem
and more general phylogenetic aspects of sieve elements, including structural features within bryophyte and pteridophyte phloem, are dealt with
under separate headings.

3.2 Development and organization of the phloem

3.2.1 Vascular bundles

The phloem is a complex tissue composed of several structurally and functionally different cell types (see Sect. 3.3) and almost always associated
with the xylem, the principal water-conducting system. Both phloem and
xylem co-occur in vascular bundles, discrete strands which traverse the
entire plant supplying every part of it with water and nutrients. Vascular
bundles in turn are part of the vascular tissue system which represents an
integral part of the plant body. The vascular system can be visualized as
a kind of skeleton composed of either a continuous hollow cylinder or of
separate bundles interlinked by anastomoses (cross-bridges) in the nodes
and internodes. The integration of lateral or branching systems, e.g. of

leaves and buds, into the preferentially vertical arrangement of the stem/root system is guaranteed by the insertion of their bundles between, and eventual junction with, stem bundles.

The close spatial relation between phloem and xylem is certainly an expression of their functional interdependence and has led to the evolution of different patterns within the vascular plants, of which the most conspicuous and most often recorded are recognized as bundle types. The main distinction between vascular bundles is based on the location of phloem and xylem relative to each other, the respective types being referred to as the concentric bundle, the central radial bundle (system), and the collateral bundle. An additional classification into open and closed bundles makes use of the presence or absence of a cambium and thus is helpful to accentuate their potential to develop secondary vascular tissues.

Concentric bundles have their phloem tissue either surrounding the xylem (amphicribal or periphloematic) – and are commonly found in ferns – or enclosed by the xylem (amphivasal or perixylematic) – and are occasionally found in angiosperms (dicotyledons as well as monocotyledons). *Radial* bundle systems are characteristic of the root vasculature of seed plants. Within the central cylinder of a root the xylem is present in two to five (or more) parts, which in cross-section are aligned radially (i.e. like spokes within a wheel), the respective arrangement being referred to as diarch to pentarch (or polyarch), with the phloem being located in the same multitude between the spokes.

Fig. 3.1 *A*: Transverse section through stem of *Aristolochia salpinx* with open collateral vascular bundles. C = cambium, P = phloem, S = sclerenchyma, X = xylem. *B*: Transverse section through secondary phloem of *Annona reticulata* showing hard bast (H) composed of sclerenchyma cells, alternating with soft bast (W) composed of sieve elements and parenchymatic elements. The sieve elements are almost occluded by the heavy formation of nacreous wall parts (arrows). Chlorzinc iodine staining for cellulose (violet colour). *C*: Transverse section through a closed collateral vascular bundle of stem of *Zea mays*. P = phloem, S = sclerenchyma, X = xylem. *D*: Longitudinal section through a nodal anastomosis of *Dioscorea reticulata*. Several layers of consecutively smaller sieve elements (1 to 3) connect two internodal sieve tubes. Sieve plates and sieve areas are visible by the blue staining of their callose deposits. The innermost cells (3) have sieve pores distributed all over their walls (light-blue colour). Resorcin blue staining for callose. *E*: Longitudinal section through primary stem phloem of *Austrobaileya maculata* viewed with the fluorescence microscope after staining with aniline blue and DAPI. Nuclei (N) stained with DAPI are prominent in parenchyma cells and degenerating in young sieve elements (arrows). Callose in sieve plates (SP) and lateral sieve areas (arrow heads) brightly stained with aniline blue. *F*: Longitudinal section through secondary phloem of *Tilia cordata* showing two sieve areas of a compound sieve plate. The over 1 μm wide sieve pores are narrowed down by callose deposits (blue collars). Arrows point to plasmatic connections (black dots). Resorcin blue staining. *G*: Longitudinal section through node of *Dioscorea reticulata* with parts of the xylem (X) and phloem (P) anastomoses. Internodal sieve tubes (IST) branch into successively smaller and narrower sieve elements (1 to 3). Sieve plates visible by callose deposits stained with resorcin blue.

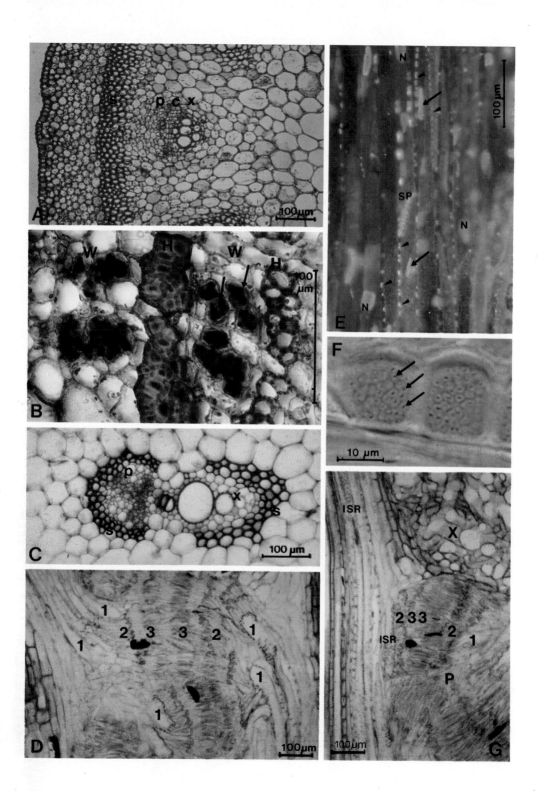

The *collateral* bundle is the specific pattern found in the aerial parts of the seed plants. It is generally characterized by the placement of phloem and xylem opposite each other and may be subdivided into three distinct sub-types: the closed collateral, open collateral and bicollateral bundles. Closed collateral bundles have their phloem and xylem surrounded by a more or less conspicuous sheath of thick-walled sclerenchyma cells, terminate their development at the primary stage and are typically present in the monocotyledons (Fig. 3.1C). Open collateral bundles with only caps of sclerenchyma near the phloem and xylem poles and the potential to develop secondary vascular tissues are the most common type in conifers and dicotyledonous angiosperms (Fig. 3.1A), while bicollateral bundles – a modification of the open collateral which has two phloem parts on each side of the xylem – is restricted to a few dicotelydon taxa (e.g. Myrtales, Cucurbitaceae, Solanaceae).

There certainly are intermediates between the established bundle types and, as with other characters, deviations from the recognized pattern (see e.g. compound bundles in several monocotyledon families; cf. French & Tomlinson, 1986). However, the introduced classification provides a most helpful tool for morphological, taxonomic, evolutionary and other primarily descriptive purposes.

3.2.2 *Differentiation of vascular tissues*

The differentiation of the vascular tissues within a bundle is intimately associated with the ontogenetic development of a plant. A major distinction is made between primary (proto- and metaphloem, proto- and metaxylem) and secondary vascular tissues.

The first vascular elements, representing the *protophloem* and protoxylem, differentiate in close proximity to the apical meristems and serve as conduits of the actively growing plant parts, i.e. for the short period prior to and at the beginning of elongation growth. They are disrupted during more rapid phases of elongation and crushed by the extension of neighbouring cells differentiating likewise from the procambium. Protophloem sieve elements do become obliterated (i.e. become completely destroyed), while phloem parenchyma cells may differentiate into sclerenchymatic fibres. The functional continuity of the vascular tissues is maintained by the concomitant replacement of protophloem and protoxylem by elements of *metaphloem* and metaxylem and their subsequent differentiation into larger and more effectively translocating cells.

Primary vascular tissues are present in all vascular bundles; the area and direction of their expansion, however, depend upon the respective bundle type. While in all collateral bundles the first protophloem and protoxylem elements are located at opposing poles and further maturation proceeds from the periphery toward the centre of a bundle (i.e. with respect to their location in the plant body the developmental directions are: centripetal for phloem and centrifugal for xylem), in radial bundle systems the differentiation of both primary phloem *and* xylem starts at the periphery of the central cylinder and progresses centripetally.

For many plants, e.g. all monocotyledons and some annual and herbaceous dicotyledons, the entire procambial tissue is exhausted in the formation of the metaphloem and metaxylem, which then represent their life-time conducting tissues (closed bundle types).

Biennial and perennial plants continue to produce vascular tissues, by either normal (conifers and most woody dicotyledons) or anomalous growth patterns. Their life-time vascular tissues are the secondary phloem and xylem; their conducting elements, however, are functional for different short periods of time only, depending upon growth activity, seasonality and other external parameters.

Normal secondary growth is always dependent upon the regular activity of a cambium developing in open bundle types. Open collateral bundles leave a few cell rows of procambial tissue undifferentiated, which may develop into a (primary) fascicular cambium and be further combined with the (secondary) interfascicular cambium (derived from reactivated parenchyma cells) to form a complete meristematic ring. In radial bundle systems of roots the entire cambium is secondary.

Based on the orientation of their cells – roughly characterized as vertical or horizontal – and dependent on two different types of cambial initials, both secondary phloem and secondary xylem are classified into two morphologically and physiologically distinct parts, the respective axial and ray systems. The axial systems are composed of prosenchymatic cells (in phloem, e.g., of sieve elements, companion cells, phloem parenchyma cells, phloem fibres) arranged parallel to the longitudinal axis of the stem, root or other organs and preferentially participating in the long-distance transport within the plant. These derive from vertically extended, long and spindle-shaped 'fusiform initials', whilst horizontally orientated, relatively short and often isodiametric 'ray initials' give rise to parenchymatic cells extended in radial directions, which facilitate translocation and exchange between bark and pith and provide additional storage capacities. The communication between the two systems (axial and ray) is realized by parenchyma cells mutually interconnected through plasmodesmata.

Whatever the origin of the vascular cambium is, it produces secondary phloem towards the periphery, secondary xylem towards the centre of the plant and ray cells in both directions. The relative proportion of vascular and parenchymatic (primary ray) tissues found during secondary growth is dependent on the arrangement of their primary tissues and the differential cambial activity, according to which three patterns are generally discerned:

1. Primary and secondary vascular tissues appear as a network of discrete strands, i.e. the interfascicular cambium produces only ray cells (this pattern is present in many dicotyledon lianas).
2. Discrete vascular strands in the primary condition are replaced by a complete hollow cylinder of secondary vascular tissues, i.e. the interfascicular cambium produces (mainly) elements of the axial and (less) of the ray system (conifers and many dicotyledons).
3. Primary and secondary vascular tissues form a continuous hollow cylinder with only narrow interfascicular (ray) portions (many woody dicotyledons).

Anomalous secondary growth is recorded for a variety of – preferentially tropical – seed plant taxa and is less clearly defined than normal growth. From the different patterns recognized three examples should suffice in this context:

1. In several dicotyledon families (mainly within the Caryophyllales) and in a few gymnosperms (e.g. *Cycas* and *Gnetum*) successive (supernumerary) cambia are formed which originate in the first (normal) cambium, but form a new outer hollow cylinder producing, as normal, xylem to the inside and phloem to the outside.
2. Some perennial monocotyledons (e.g. *Dracaena, Sansevieria, Yucca*) contain peripheral thickening meristems which build new vascular bundles towards the inside and parenchyma to the periphery, while primary thickening meristems occur in other monocotyledons.
3. *Bougainvillea* (Nyctaginaceae) combines two different patterns: the primary vascular system consists of irregularly distributed open collateral bundles, which in the process of stem thickening are supplemented by new bundles originating in a peripheral, primary thickening meristem producing parenchymatic and vascular tissues. Eventually, after closure of a first complete cambial hollow cylinder, successive cambia are produced in connection with the former, thus giving the *Bougainvillea* stem an appearance similar to that found in representatives of (1).

3.3 Composition of the phloem

The term phloem was coined by Nägeli (1858), who defined it as being composed of parenchyma, (hard) bast, and soft bast (including sieve elements). His essay entitled 'The growth of the stem and root of vascular plants and the arrangement of vascular strands in the stem' (translated from the German) was one of the early contributions to our knowledge of vascular tissue systems and the first attempt to replace the then commonly used term 'bast' (referring to bast fibres), which after Hartig's (1837) description of the sieve elements became misleading. Haberlandt (1879) later introduced the term 'leptome' as a substitute for the old term 'soft bast', and consisting of sieve elements and parenchymatic cells as opposed to the sclerenchymatic cells of the hard bast. Nägeli's phloem concept became generally accepted, restricting Haberlandt's leptome to those plants which, like bryophytes, lack bast fibres in their conducting tissue.

Thus, the phloem of seed plants as primarily covered in this chapter is defined to be composed of sieve elements, parenchymatic and sclerenchymatic elements – each of which contains several morphologically distinct cell types. However, the detailed cellular composition of the phloem is dependent upon several different parameters: (1) taxonomic position of the plant; (2) location within the plant body; and (3) developmental stage of the tissue.

1. The gymnosperm phloem is generally composed of sieve elements (type: sieve cell), Strasburger cells, phloem parenchyma cells and phloem fibres (Pinaceae lack phloem fibres), while in the angiosperms the phloem consists of sieve elements (type: sieve tube member), companion cells, phloem parenchyma cells and phloem fibres. (For distinct patterns of secondary phloem see (3).)
2. The main differences in the phloem composition of the various plant organs are in the extent (i.e. relative percentage) of the different cell types.
3. Within the primary phloem, protophloem sieve elements are comparatively narrower than their metaphloem successors, often have less distinct sieve plates and sometimes are not associated with companion cells.

   The secondary phloem is generally divided into parts belonging to the axial and ray systems. The growth pattern is responsible for the relative amounts of the two parts; in addition, the ray system may be augmented by secondary rays produced during dilatation growth.

   Preferentially in cross-sections periodic structures corresponding to yearly increments or alternating leptome and bast portions (soft bast/hard bast) may be recorded in some plant families (e.g. Annonaceae, Fig. 3.1B), and in all woody taxa complete radial files of cells derived from a fusiform initial can be traced in both phloem and xylem directions. Sometimes these radial files are composed of one cell type only, e.g. sieve elements or parenchyma cells in the secondary phloem of *Gnetum* and tracheids or parenchyma cells in the xylem of *Pinus*.

   In addition to a strict radial arrangement, in some conifer families (Taxodiaceae, Taxaceae, Cupressaceae and others) the entire axial phloem is characterized by the very constant and repeating sequence of: sieve element – fibre – sieve element – parenchyma cell – sieve element – fibre, etc. (i.e. every second derivative of a fusiform initial is a sieve element, with phloem fibres and phloem parenchyma cells regularly alternating in between). Since all fusiform initials more or less synchronously give rise to the same cell type, in these plants the secondary phloem can be said to be constructed of successive homogenous hollow cylinders ('bands' in cross-sections), each composed of one cell type and interrupted by ray parenchyma only.

Sieve elements and parenchymatic elements of phloem will be described in detail below, but because of the topic of this book sclerenchymatic elements (e.g. regularly present phloem fibres and sclereids) and secretory structures (e.g. occasionally interspersed laticifers and resin ducts) will not be covered.

3.4 Sieve elements

Sieve elements are the most highly specialized cells of phloem and the structural basis for a rapid translocation of food material. Moreover, the ontogenetically established degeneration of their nuclei, the disintegration of the central vacuole, and loss of most other organelles, which eventually turn the sieve elements into effective conduits, is a unique feature not recorded in any other type of living plant cells. First anatomically described by Hartig in 1837, but not functionally characterized as the phloem cells responsible for long-distance translocation until 1930 by Schumacher, sieve elements have over the last 150 years been the object of continuous and very intensive structural research.

3.4.1 Development, structural identification and terminology

3.4.1.1 Development

Sieve elements are procambial or cambial derivatives and develop either direct or via one to several precursors which by unequal divisions may give rise to closely related parenchyma cells. Protophloem sieve elements differentiate acropetally into a continuum of mature cells, influenced by a basipetal auxin flux and an acropetal sucrose sink. Metaphloem and secondary phloem sieve elements differentiate successively, in predetermined patterns and according to the growth requirements of the respective organs – ultimately also governed by the procambium.

The structural identification of a cambial derivative as a future sieve element is often difficult, especially in cases where simple positional criteria (see Sect. 3.3) are missing. In several seed plant taxa protein crystals developing rather early within their (P-type) plastids can be used as a reliable marker. The onset of other typical changes (e.g. first callose deposition in future pore sites) are disclosed only by the experienced phloem cytologist.

3.4.1.2 Identification

The main diagnostic character of sieve elements is the formation of pores, specialized plasmatic connections through parts of the cell walls common to two neighbouring cells. Pores always occur in clusters, the so-called sieve areas which by their resemblance to a sieve (German: *Sieb*) obviously inspired Hartig (1837) to coin the terms *Siebfaser* and *Siebröhre*. Sizes of pores, their pattern and distribution within transverse and/or longitudinal walls and consequently the arrangement of sieve areas between vertically or laterally adjoining sieve elements are taxon-specific. There are, however, a couple of common structural features: the continuity through the pores of the plasma membrane as well as some cytoplasmic contents, and the involvement of callose in the formation, obstruction or definite closure of the pores (see Sect. 3.4.2.3.2). The presence of callose in almost all preparations of phloem makes it a comparatively easy marker for the microscopic identification of a sieve element. Callose shows a light blue colour in stainings with resorcin blue (Fig. 3.1 D, F, G) and can be identified in hand or microtome sections for light microscopy, but a more sensitive stain is aniline blue applied to fresh or pre-fixed sections in the fluorescence microscope: its bright yellow fluorescence even traces small amounts of callose in primary pit fields (Fig. 3.1E). In ultrathin sections viewed with a transmission electron microscope (TEM) callose remains unstained (Figs. 3.7C, 3.8: Ca).

Another diagnostic wall feature is the formation of nacreous wall thickenings, which in many woody dicotyledons considerably reduce the transsectional area of a sieve element (Fig. 3.1B). Nacreous walls, by their light pearly lustre, facilitate the recognition of sieve elements in unstained transverse or longitudinal sections of vascular tissues. A comparably quick location of mature sieve elements in ultrathin sections viewed with the TEM is by their paucity of contents relative to the neighbouring cells; in low magnifications the sieve elements stand out as lighter cells (Figs. 3.6A, 3.9, 3.10).

3.4.1.3 Terminology

Pores and their specific data (sizes and distribution, mainly) have also served as the primary distinctive features used to differentiate between the two types of sieve elements currently recognized, namely the sieve cell and the sieve tube member. Based on observations in the monocotyledons these were defined by Cheadle & Whitford (1941) as follows:

A sieve cell is a cell, enucleate at maturity, in which the sieve areas of all walls are of the same degree of specialization; a sieve-tube element or member is a cell, enucleate at maturity, in which certain sieve areas are more highly specialized than others, the former being largely localized on end walls to form the sieve plates; a sieve tube is a series of sieve-tube members arranged end to end.

However, structural details from an increasing number of seed plants, subsequently made available through the higher resolving power of the electron microscope, combined with an intensified focusing on vascular cryptogams, not only yielded additional distinctive characters (cf. Behnke, 1986), but also blurred the limits between the two categories of sieve elements and thus not only confused the non-specialist but sometimes even intrigued the phloemologist. Therefore, throughout this chapter the collective term 'sieve element' will be used, supplemented by a taxon-specific epithet only, where necessary (cf. Sects. 3.4.3 and 3.4.4).

3.4.2 Ontogeny and mature structure of the sieve element

3.4.2.1 Sizes of sieve elements

Sizes (and shapes) of sieve elements depend upon their heritage expressed by the taxonomic position of the plant (reflecting their evolutionary level), their developmental stage (i.e. origin within primary or secondary phloem), location in the plant body, and – wherever appropriate – size and division pattern of fusiform cambial initials. Comparative measurements in the primary phloem are sparse and mostly restricted to the metaphloem of monocotyledons (Cheadle, 1948). Broader surveys exist for the secondary phloem of seed plants (Holdheide, 1951; Esau & Cheadle, 1959; Zahur, 1959). The length of sieve elements ranges between 50 and 5000 μm, their width between less than 10 and more than 100 μm. Since there is considerable size variation within one plant depending upon the part considered – in the vine of *Dioscorea* the extremes for length and width of a sieve element are 40–50 μm and 5–10 μm in the node and 3500 μm and 110 μm in the internode (Behnke, 1965) – a quotation of general averages is not possible. If, however, sizes of sieve elements within the secondary phloem of woody plants are compared, some phylogenetic trends can be visualized:

1. The length of the sieve elements decreases from a maximum of about 5000 μm in the conifers to 100–200 μm in the advanced dicotyledons – effected by a shortening of cambial initials, but in many taxa veiled by the so-called secondary partitioning, a post-cambial division of the sieve element precursors – and is positively correlated to the length and inclination of its end walls bearing sieve areas. That is,
2. Sieve element end walls concomitantly change from the very long,

almost vertical through the rather oblique, scalariform to the short, transverse or nearly transverse.

3. The gradual reduction of the number of sieve areas from well over 100 (in compound sieve plates) through several steps to the one only (in a simple sieve plate) is a consequence of (1) and (2).

4. The size of the pores in sieve areas increases, correlated to the former trends. However, quantitative calculations of the total pore area of different sieve plates had an ambiguous result: sieve elements with simple sieve plates were narrower than those with scalariform plates, and the relative total pore area (i.e. total area of a sieve plate occupied by the pores expressed as a percentage of the transverse cell area) was higher in the sieve elements with scalariform end walls.

A comparative study of the metaphloem sieve elements in the different organs of a representative number of monocotyledons revealed additionally that some of the trends listed are paralleled by the same specializations within an individual plant: the percentage of sieve elements with transverse end walls is least in roots, greater in aerial stems and rhizomes, and greatest in leaves, inflorescence axes, bulbs and corms (Cheadle, 1948).

3.4.2.2 The differentiation of the sieve element protoplast

A procambial cell or cambial derivative which is to develop into a sieve element starts its specific differentiation with a complete set of the cell organelles usually present in meristematic cells. The ontogenetic changes which culminate in the mature sieve element almost concomitantly affect the structure of all cell components – each to a different degree, but resulting in irreversible alterations. And although in almost all organelles many changes occur as interdependent developments, a stepwise description of the complex but continuous differentiation process as a whole is not practicable. Therefore, in order not to lose the general idea, the ontogenetic changes of the organelles will be dealt with as isolated but self-contained units.

3.4.2.2.1 The nucleus The degeneration of the nucleus is one of the major events during the differentiation of a sieve element, unique among living plant cells and well known already to the classical plant anatomists of the late nineteenth century (see e.g. Strasburger, 1891). Intensive studies using material selectively stained for chromatin (Feulgen reaction) enabled Salmon (1946/47) to describe different steps and to distinguish between two main modes of nuclear disintegration, which she named 'pycnosis' and 'dechromatinization'.

During *pycnosis* the sieve element nucleus elongates, often becomes deeply lobed (amitotic appearance), gradually shrinks and finally loses its envelope. These changes in the nuclear shape are paralleled by a degenerative sequence of the chromatin material, which initially evenly penetrates the entire nucleus (being responsible for the pycnotic staining behaviour), later reticulates (condensation process) and subsequently disintegrates into droplets which eventually – and concomitant with the disappearance of the nuclear envelope – dissolve (pulverization process).

In contrast to this the *chromatolysis* (as 'dechromatinization' was renamed by Resch, 1954) is characterized by a gradual loss of all stainable contents – i.e. chromatin and nucleolar material – (often accompanied by a volume increase of the nucleus), the maintenance of the nuclear envelope for some time beyond this depletion of the nucleus, and, eventually, the rupture and complete breakdown of the envelope.

Ultrastructural investigations largely corroborated these views but also contributed additional details and different aspects of nuclear degeneration. Initially, the structure of the sieve element nucleus is not much different from that of other meristematic cells, e.g. neighbouring phloem parenchyma cells (Fig. 3.3A). It is surrounded by an envelope, the perinuclear cisterna of ER, which is specialized by the presence of nuclear pores and the association of its inner membrane with chromatin. The nuclear matrix contains one to several nucleoli and condensed chromatin distributed in the form of small patches over the entire nucleus. Greater irregularities in sizes and distribution of heterochromatin and smaller nucleoli are, according to Esau & Gill (1972, 1973), first indications of a nucleus belonging to a prospective sieve element. Other changes frequently reported during the early phase of sieve element differentiation are a lobing of the nucleus and a partial or complete dilatation of the nuclear envelope.

In the chromatolytic mode the nuclear disintegration proceeds with the steady decrease of all stainable contents (the heterochromatin attached to the envelope being the last to disappear; cf. Figs. 3.2A, 3.6B, 3.7A, 3.9), until the nuclear matrix is almost translucent and often hardly distinguishable from the surrounding cytoplasm. At this stage the nuclear envelope is still delimiting the complete area of the former nucleus. When, finally, the envelope is ruptured, larger pieces may continue to be identified by their pore complexes. It has repeatedly been suggested (and partly demonstrated in wound sieve elements by Schulz, 1987) that eventually the nuclear membranes are recycled to the endoplasmic reticulum, which in the mature sieve element is present in the form of agranular stacks.

A stepwise documentation of the complete sequence of nuclear degeneration is difficult and realized in a few angiosperm species only (e.g. Esau & Gill, 1972, 1973; Thorsch & Esau, 1981c; Eleftheriou, 1986; all using root protophloem). However, the type of disintegration could be recognized by screening a few characteristic stages. Accordingly, chromatolytic nuclei could be recorded within angiosperms, dicotyledons as well as monocotyledons, whereas nuclear degeneration of the investigated gymnosperms (many species from the Pinaceae and representatives of the

Cycadales and Gnetales) differed considerably and was repeatedly referred to the pycnotic type.

In the sieve element of gymnosperms the lobed nucleus continues its degeneration with a gradual pycnosis of the chromatin material, expressed by its initial coarsening (clumping) and subsequent increasing condensation (Fig. 3.11). The nucleolus obviously dissolves during the early coarsening of the chromatin – at least it has not been seen later. Since at the same time the nuclear matrix decreases, the size of the nucleus diminishes, ending up as an irregular body composed of an electron-opaque material tightly surrounded by the nuclear envelope. Further stages of disintegration have not definitely been followed up. However, a break-up into several parts which persist for a longer time must be assumed, since smaller nuclear remnants associated with ER material have repeatedly been recorded in old sieve elements. Therefore, in gymnosperms, the last steps of Salmon's (1946/47) pycnosis – the loss of the nuclear envelope and the pulverization of the clumped chromatin – are delayed, in the extreme case to the final obliteration of the sieve element. Although in gymnosperms secondary phloem was the preferred tissue, details in nuclear degeneration did not differ when primary phloem sieve elements were studied (Neuberger & Evert, 1976).

In angiosperms pycnosis or pycnotic stages have been recorded during nuclear degeneration of some – but not all – investigated protophloem sieve elements. On the basis of a continuous series of 32 sieve elements of cotton root protophloem (some cells dissected completely by serial ultrathin sections), Thorsch & Esau (1981c) were able to follow up distinct changes in the disintegrating nucleus almost in their natural course: the observed early association of aggregated heterochromatin with the nuclear envelope (Fig. 3.3B), and the shrinkage of the nucleus combined with its constriction and fragmentation are described as steps of a pycnotic degeneration (defined by Salmon, 1946/47). The resulting nuclear fragments are dissolved completely, although considerably late. Similar views of the peripheral aggregation of heterochromatin (sometimes combined with some other, unidentified material) and of a folded and distorted nucleus have been found occasionally in primary phloem of other dicotyledons (Pacini & Cresti, 1972) and repeatedly in monocotyledon root

Fig. 3.2 Young sieve elements from the metaphloem of a dicotyledon, *Aristolochia clematitis*. *A*: Longitudinal section through phloem of young shoot showing two sieve elements (SE) accompanied by companion cells (CC) and phloem parenchyma cells (PC). Sieve elements still contain nuclei (n), vacuoles (v) and a parietal layer of cytoplasm (sp) = developing sieve plate). Companion cells are distinguished from phloem parenchyma cells by their dense protoplast and smaller vacuoles. *B*: Part of a young sieve element demonstrating some of the changes initiated during cell differentiation. er = tubular network of agranular ER, m = mitochondria, arrows point to microtubuli, parallel to the cell wall, p = amoeboid P-type sieve element plastids, PP = P-protein, t = tonoplast, v = vacuole.

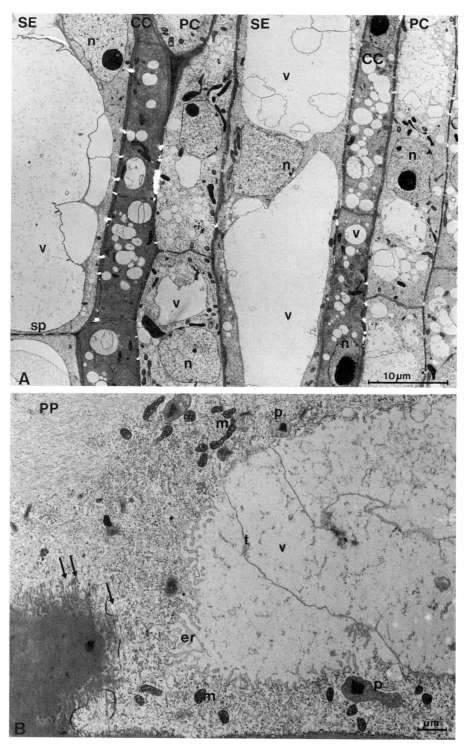

protophloem (Esau & Gill, 1973; Danilova & Telepova 1980; Eleftheriou & Tsekos, 1982c; Eleftheriou, 1986).

One of the more frequent features associated with both pycnotic and chromatolytic degeneration is the association of ER membranes with the nuclear envelope. Very early during sieve element ontogeny stacked cisternae of agranular ER are aligned parallel to the nuclear envelope and remain associated with it beyond the final breakdown of the nucleus, most likely integrating the perinuclear cisterna into the parietal ER stacks of the mature sieve element. In angiosperm sieve elements the intensity of the association between ER and the nuclear envelope varies considerably depending upon the species and developmental stage considered (e.g., protophloem versus metaphloem, cf. Thorsch & Esau, 1981b,c), while in gymnosperms a close association has not been observed, although the elaborate tubular ER of the mature sieve element is often seen in loose (secondary) contact with nuclear remnants (Behnke & Paliwal, 1973; Neuberger & Evert, 1974). However, an intimate association between ER and degenerating nuclei is also present in some fern sieve elements.

The fate of the *nucleolus* has been commented upon in many papers dealing with the sieve element differentiation. While most of the ultrastructural studies limited themselves to a statement on when the nucleolus disappeared (prior to or after pycnosis or lysis of chromatin), Thorsch & Esau (1981c) for the first time described details of its dissolution. In cotton sieve elements the initially compact granular and fibrillar nucleolar material starts to disintegrate soon after the peripheral aggregation of the chromatin, spreads within the nuclear matrix, but remains distinct until its final disappearance – at the time when the nuclear envelope breaks down.

In a few species studied with the light microscope the extrusion of the nucleolus during nuclear degeneration and its persistence in the mature sieve element have been reported (Esau, 1947; Kollmann, 1960). *Extruded nuclei* were subsequently reported within many additional taxa, and glob-

Fig. 3.3 A and B Young sieve elements from the phloem of monocotyledons. *A*: Longitudinal section through phloem of a young shoot of *Posidonia oceanica* with mature protophloem sieve element (PSE), several young metaphloem sieve elements (MSE) and companion cells (CC). Nuclei (n) contain many patches of heterochromatin, P-type sieve element plastids (p) are amoeboid and deeply stained. The young sieve elements contain parallel cisternae of granular ER. PC = phloem parenchyma cells. *B*: Part of a young sieve element of *Tulipa sylvestris* with part of lobed, heterochromatic nucleus (N), several P-type sieve element plastids (P) (including cuneate protein crystals, typical of monocotyledons), mitochondria and a dense cytoplasm. Arrow heads point to microtubuli which are in close contact with the plasmalemma. *C*: Sieve pore/plasmodesmata connections between a young sieve element (SE) and a companion cell (CC) of *Myristica fragrans* (serial section to 3.6A). The sieve pores are not yet fully developed, callose (Ca) marks their future size. Vacuole (v) and tonoplast (t) of the sieve element are still present. The plasmodesmata on the companion-cell side are branched.

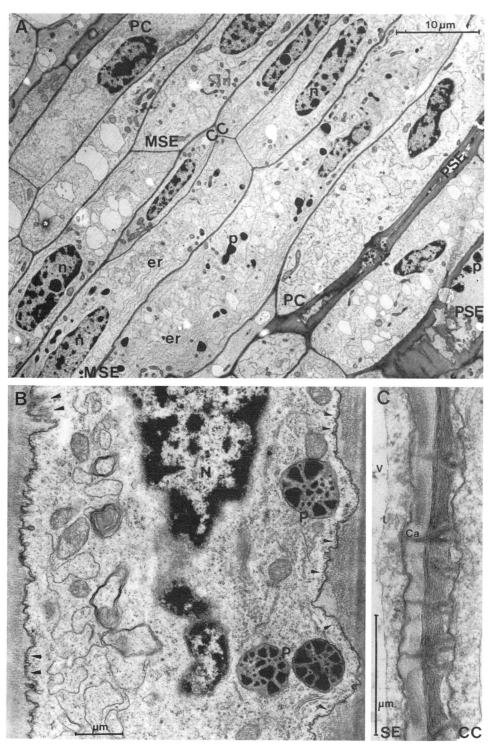

ular inclusions previously found in various species and interpreted as slime bodies were subjected to a reinterpretation (cf. Zahur, 1959). Eventually ultrastructural investigations were able to demonstrate that most of these inclusions are persistent, crystalline protein bodies which never resided within the nucleus (Deshpande & Evert, 1970; Oberhäuser & Kollmann, 1977). In only two families did the crystalline inclusions found in mature sieve elements originate in the nucleus (but not in close contact with the nucleolus), namely in the Boraginaceae (Esau & Magyarosy, 1979) and the Myristicaceae (Fig. 3.6; cf. Behnke, 1988).

The described processes of nuclear degeneration basically extend from the early recognition of a sieve element through the entire time of drastic cytoplasmic changes – in part dependent on the nucleus – to about the time when the sieve plate pores are opened. There has been considerable controversy about whether the final disintegration occurs before or after the pore differentiation. Results from more species now available clearly demonstrate that in many plants (e.g. cotton and all gymnosperms) nuclear remnants persist far beyond the opening of the pores. This brings us to the question of when a degenerating nucleus ceases to function. In the main two methods have been applied to correlate the function of a nucleus to its structure, both suggesting an early blockage of transcription. High-resolution autoradiography showed that the incorporation of tritiated thymidine and uridine into the sieve element nucleus ceases at about the time when the nucleolus disappears (Zee & Chambers, 1969) and that nuclear remnants within mature gymnosperm sieve elements equally lack any incorporation (Hébant, 1975). The staining of degenerating nuclei with the DNA-specific fluorochrome DAPI (cf. Fig. 3.1E) traced the chromatin until its accumulation towards the nuclear envelope, but not beyond (Schulz, 1987).

3.4.2.2.2 The endomembrane system After the nucleus, the endomembrane system as a whole undergoes the second series of drastic changes within the differentiating sieve element. Its components, however – the plasmalemma, dictyosomes (Golgi apparatus), endoplasmic reticulum (ER), vacuole and tonoplast, which are roughly characterized as the one-membrane-bound compartments of the cell and thought to be functionally interdependent (the basis of a membrane flow) – display individual extents of disintegration.

In the young sieve element functional and structural relations between granular ER, dictyosomes and plasmalemma are clearly illustrated during the formation of specific wall parts. Later, the involvement of all parts of the endomembrane system in the rapid structural changes of the maturing sieve element obscures the interconnections between the different organelles.

The *plasmalemma* is present until the final obliteration of a sieve element and is responsible for the maintenance of its functional integrity.

Except for the incorporation of vesicle membranes during the granulocrine excretion of cell wall material it seems to be the most stable part of the sieve element (cf. Fig. 3.5C, 3.8C: Plm). Beyond these visible alterations, however, conformational changes at the molecular level are necessary to enable the plasmalemma to adapt to continuously changing transport situations within the differentiating sieve element, e.g. to the obligate reversion of the transport polarity in the primary phloem when leaves switch from import to export.

Freeze-fracture analysis of the plasma membranes in tissue cultures revealed that intramembranous particles of the sieve element plasmalemma are smaller and – based on overall area countings – slightly more frequent than those of their parenchymatic counterparts. If, however, their distribution within the two membrane faces is considered, the discrimination between the two cell types is obvious (PF/EF ratio 1.2:1 in sieve element plasmalemma against 9.6:1 in parenchyma). Sjolund & Shih (1983b) conclude from their observations that the high particle density of the sieve element E-face is a morphological expression of an intensive active transport across the plasmalemma.

Nucleoside phosphatase activity has been detected by histochemical staining techniques in the plasmalemma of both young and mature sieve elements (for details see Cronshaw, 1981, but cf. Katz *et al.*, 1988).

Dictyosomes are abundant in the young sieve element, but missing in mature sieve elements. Initially their structure does not differ from that in other plant cells, i.e. they are composed of, on average, five to seven stacked cisternae which show structural and functional polarity (Figs. 3.5A, 3.7A). At the forming face (secretion side) numerous smooth-surfaced Golgi vesicles are formed by budding off from the edges of the cisternae or by an entire disintegration of a (distal) cisterna; whilst at the maturing face (regeneration side) new (proximal) cisternae arise by fusion of primary vesicles derived from the ER. In addition, coated vesicles are seen to protrude from smooth vesicles or distal cisternae (Fig. 3.5A). When during sieve element ontogeny the synthetic capacities of the ER system are lost and therefore a regeneration of Golgi cisternae is made impossible, dictyosomes gradually disappear, probably depleting themselves in the formation of vesicles.

Dictyosomes are active in the secretion of cell wall material. Within sieve elements this has repeatedly been demonstrated during the formation of the nacreous lateral walls by using high-resolution autoradiography and cytochemical staining, both confirming that precursor material is secreted from dictyosomes through Golgi vesicles to the plasmalemma (Arsanto & Coulon, 1974).

The *endoplasmic reticulum (ER)*, an elaborate system of membrane-bound cavities suitable for intracellular transport, may be organized as a network of flat cisternae or in the form of small tubules – and both are able to partly disintegrate into, or at least generate, vesicles. The attachment of polysomes at the cytoplasmic face of the ER membranes makes the granular (rough) ER distinct from the ribosome-free, agranular (smooth) ER. All forms of ER do occur during sieve element ontogeny, often at the same time. Some additional complex formations (e.g. the convoluted ER) and specific locations (e.g. the parietal ER) have contributed to its accentuation and led to the collective term 'sieve element reticulum' (SER). In the following, however, the descriptive terms will be used.

Young sieve elements, at the beginning of their differentiation, contain granular cisternal ER (Fig. 3.3A) connected directly to the nuclear envelope and via vesicles to dictyosomes and (in some plants) to the vacuole. Small cisternae are also found parallel to the plasmalemma and limited to the pore sites of future sieve plates (see Sect. 3.4.2.3.2).

Concomitant with the first changes observed in the nucleus a reorganization of the ER is initiated, affecting both its conformation and distribution. Although absolute quantities of ER and relative proportions of the different structural configurations may be dependent on developmental (proto- versus meta- or secondary sieve elements), taxonomic or other criteria, principally two different sequences of structural changes can be distinguished: the one involving an early formation of cisternal stacks and the other characterized by the preferential transformation of granular cisternae into smooth tubules. Both, however, end in the formation of a parietal anastomosing network of agranular ER located in the vicinity of the plasmalemma and supplemented in part by cisternal stacks, loose tubular associations or by more dense convoluted or paracrystalline complexes.

In metaphloem sieve elements of cotton leaves and stem the first changes in the ER observed by Thorsch and Esau (1981a) were an increase in its total amount, combined with a loose parallel assembling of a few granular cisternae. The subsequent formation of real stacks and further condensation into convoluted aggregations is paralleled by the separation of ribosomes from their membranes. In almost mature sieve elements stacked and convoluted ER are found preferentially alongside the longitudinal walls, while during all stages of nuclear disintegration stacks of ER are also associated with the nuclear envelope.

A local swelling of granular cisternae starts the structural changes of the ER in the sieve element of *Smilax* (Behnke, 1973). It is followed by the transformation of straight cisternae into twisted tubuli and the concomitant gradual shedding of ribosomes from the ER membranes. Still in the nucleate condition of the sieve element, ER tubuli first assemble to form loose associations and are successively condensed into convoluted aggregates. The complex convoluted ER is also present in mature sieve elements and occurs close to their wall-lining plasmalemma. Similar transformations of cisternae into tubules were reported in a few other angiosperms (cf. Figs. 3.2B, 3.5A).

The parietal anastomosing ER present in the majority of the investigated mature sieve elements consists of at least one agranular cisterna being tightly attached to the plasmalemma and in many species continuous with the stacked and convoluted ER (Figs. 3.5C, 3.8C). Sjolund & Shih (1983a), applying freeze-fracture techniques to sieve elements of tissue cultures, were able to confirm results obtained with thin-sectioning techniques about the close spatial relationships between plasmalemma and ER and to demonstrate that the parietal cisterna is fenestrated. They suggested that the function of the ER-plasmalemma association is to provide a micro-environment for proton-efflux and proton-sucrose co-transport through the plasmalemma, spatially separated from but not without communication (function of the fenestrated regions) to the rapid translocation stream in the sieve element lumen.

The endoplasmic reticulum also associates with organelles other than the plasmalemma and nuclear envelope. In young as well as mature sieve elements several ER cisternae may be found surrounding plastids and/or mitochondria, often separating them from the cytoplasm by a multiple membranous sheath. The function of such associations is unknown.

Histochemical localization of acid phosphatases, nucleoside phosphatases and peroxidases in the ER of young and mature sieve elements (see Cronshaw, 1981) led to the suggestion that this organelle may play an important role in the autophagic processes of the cell. On the other hand, and with reference to the prolamellar bodies of etioplasts, the convoluted and paracrystalline forms in particular have been discussed as a sequestration of membranes in an inactive form and at a minute space.

During sieve element ontogeny, *vacuoles* share the fate of other one-compartment organelles – they disintegrate completely. The loss of the vacuole and the breakdown of the *tonoplast* as its membranous boundary towards the protoplast was first concluded from the changing response of differentiating sieve elements to neutral-red staining (Crafts, 1932). EM studies confirmed the absence of a vacuole in mature sieve elements.

Young sieve elements either contain several small vacuoles – which subsequently enlarge and in some plants may fuse to form a single one – or already start their development with a central vacuole, depending upon whether they are derivatives of the procambium or cambium. Discrete vacuoles are depicted throughout most of the differentiation time of a sieve element (Figs. 3.2A, 3.4A); their disintegration seems to be a rather sudden event at about the time when the nuclear envelope is ruptured. So far, however, neither exact sequences of the morphological changes implied are known nor their basic physiological events understood.

One of the few structural details observed in many plants just prior to the disappearance of the vacuole is a partial withdrawal of the tonoplast from the parietal protoplast (Figs. 3.2B, 3.7A). Although repeatedly interpreted as an artefact this lability of the tonoplast is likely to be an expression of naturally occurring changes in membrane permeability, or of a pressure reversal between vacuole and protoplast, or of both. The influx of osmotic substances into the maturing sieve element, incipient in translocation, could serve as a simple explanation for a gradual shrinkage of the vacuole. A flat cisterna-like compartment, probably fragmented into several parts, would eventually associate with other remnants of the endomembrane system. Narrow elongate protrusions from disintegrating vacuoles, and the fact that a final dissolution of the tonoplast membrane was often discussed but never shown, support this idea.

3.4.2.2.3 Cytosol, ribosomes, microfilaments, microtubules

3.4.2.2.3 Cytosol, ribosomes, microfilaments, microtubules The *cytosol* as the cytoplasmic matrix in which all organelles are embedded certainly undergoes drastic changes during the influx of sugars and the disintegration of the vacuole – the most obvious being the incorporation of the former vacuolar area into cytosolic control. Buvat (1963) interpreted the changes involved during the expansion of the cytoplasmic compartment as a dilution of the cytosol, visualized by a gradual depletion of its structural components. Nevertheless, the preferential location of organelles like mitochondria and plastids within a parietal zone of the mature sieve element casts doubt on the homogeneity of the cytoplasmic matrix throughout the entire sieve element lumen.

Both cytoplasmic (free) and ER-bound *ribosomes* (Fig. 3.8B) are numerous in the young sieve element and easily identified if organized into polysomes (a helical array discriminating cytoplasmic ribosomes form their ER-bound counterparts showing spiral patterns). Cytoplasmic ribosomes are the logical organelles responsible for the synthesis of P-protein, and the commonly found close spatial relationships between ribosomes and so-called precursor filaments of P-protein lend support to this theory, yet the ultimate proof for this statement is still missing. All ribosomes disappear as a consequence of nuclear degeneration; their disintegration is sometimes delayed until the opening of the sieve plate pores.

Actin-like *microfilaments* are among the normal constituents of plant cells, including sieve elements, but largely due to their small size and lability they are not always encountered. Parthasarathy & Pesacreta (1980) found bundles of microfilaments in young sieve elements of several vascular plants. Their preferential occurrence in the peripheral parts and orientation parallel to the longitudinal axis of the cell has been put forward as an indication of their involvement in the cytoplasmic streaming observed in young sieve elements. The absence of microfilaments in mature sieve elements (also lacking cytoplasmic streaming) clearly negates any connection with either the mechanism of translocation or (as has been occasionally suggested) anchoring of the larger organelles in the parietal cytoplasm.

Cortical *microtubules (MTs)* are present in young sieve elements from the very first stage of development, i.e. immediately following the last (mostly unequal) division of a phloem mother cell, and are regularly depicted in differentiating sieve elements of many angiosperms (Figs. 3.2B,

3.3B, 3.5A). A few detailed studies (e.g. Eleftheriou, 1987) have demonstrated that the density of MTs (number per wall length) is positively correlated with the formation of cell wall thickenings in sieve elements. MTs overlie only those wall parts that are being thickened, and are absent from areas to be penetrated as lateral sieve areas or by sieve pore/plasmodesmata connections. Thus, in longitudinal sections their distribution anticipates the pattern of the differentiated lateral walls composed of alternating broad ridges and narrow depressions. The orientation of the MTs is perpendicular to the longitudinal axis of the cell and parallel to the cellulose microfibrils within the growing cell wall. This, in combination with micrographs showing numerous Golgi vesicles caged between MTs (which implicates a tracking of the vesicles), would further support the theory of the direct involvement of MTs in the orientation of cellulose microfibrils and moreover in cell wall development. After completion of the cell wall thickenings the number of MTs rapidly declines to zero. Occasionally MTs are found trapped between stacked or convoluted ER being associated with the nuclear envelope or the plasmalemma of differentiating sieve elements (Thorsch & Esau, 1981b). Their disintegration will be delayed for some time.

Thorsch & Esau (1982) also observed MTs overlying end walls which differentiate into sieve plates. The orientation of these MTs is not yet disclosed, – probably random, but at least different from the one described for the lateral wall MTs.

3.4.2.2.4 *Phloem proteins*

Sieve elements of the great majority of angiosperms (and in addition phloem parenchyma cells or companion cells of a few species) synthesize large amounts of protein material, which was collectively named slime (and slime bodies) whenever it became visible with the light microscope.

Early electron microscopy studies were concerned with the structural composition of the 'slime', its origin (cytoplasmic or vacuolar), and – as a result of these two things – its proper renaming. The basic ultrastructural components of 'slime' were first disclosed by Kollmann (1960) in *Passiflora* sieve elements and described as protoplasmic threads (*Plasmafäden*) with a diameter of 10 to 15 nm. Identical structures were subsequently revealed in other dicotyledons but named 'fibrils of slime'. In an attempt to clarify terminology Behnke & Dörr (1967) defined the basic units as 'plasmatic filaments' (*Plasmafilamente*), referring to both their origin within the protoplast and their threadlike structure (and confining the term 'fibril' to an aggregation of many filaments). Almost simultaneously Esau & Cronshaw (1967) replaced the misnomer slime by the term 'P-protein' and first applied it to the ontogeny of sieve elements in *Nicotiana* (Cronshaw & Esau, 1967). Tubular components (with a diameter of 23 nm and similar to those first described in *Primula* by Tamulevich and Evert, 1966) present within young sieve elements were shown to be reorganized into smaller (diameter 15 nm) 'fibrils' at about the time of nuclear and vacuolar disintegration. Eventually, Parthasarathy & Mühlethaler (1969) presented a first model of the P-protein tubule composed of helically arranged subunits and its transition into stretched and loosened forms, the 'fibrillar' P-protein.

In the many ultrastructural investigations which followed the P-protein concept was generally accepted, but over time the terms 'filament' or 'filamentous' replaced 'fibril' and 'fibrillar'. The focusing on the ontogeny of P-protein in an ever-increasing number of angiosperms, however, also revealed that there are at least two different types of protein inclusions, distinguished by their origin in the cytoplasm or in the nucleus.

The term P-protein was redefined by Cronshaw (1975a) as 'the protein-aceous material in the phloem which is sufficiently characteristic when observed with the electron microscope to warrant a special term'. It literally comprises all protein inclusions within phloem cells, but in actual use it only refers to proteins which reside in the cytosol and covers neither nuclear proteins (e.g. nuclear crystals), nor proteins synthesized by the granular ER (e.g. within dilated cisternae), nor those deposited in P-type plastids. Despite these inconsistencies P-protein is a characteristic feature of angiosperm sieve elements, it is encountered in practically all investigated dicotyledons and the great majority of monocotyledons, but it is absent from the grass family, from some (but not all) palm species and from a few other monocotyledons. No P-protein has been found in the phloem of gymnosperms or vascular cryptogams.

Within the young, nucleate sieve element of an average angiosperm the first inclusions definitely regarded as P-protein are small aggregates of tubular or filamentous components discernible in the (parietal) protoplast (Figs. 3.4D, 3.5A, 3.8A) at a time when other organelles are not yet visibly implicated in degenerative processes. As these aggregates increase in size they are identified at the light microscopic level as morphologically distinct bodies giving protein-positive staining reactions. Depending upon the species investigated, one to many of these P-protein bodies are recorded within one sieve element (Fig. 3.4A). During further maturation of the sieve element the P-protein bodies begin to disaggregate, and by about the time that nucleus and vacuole disappear the P-protein gradually disperses over the lumen of the sieve element, whereby tubular

Fig. 3.4 Dispersive P-protein. *A*: Young metaphloem sieve elements (SE) with their companion cells (CC) and phloem parenchyma cells (PC) of *Verbena utricifolia* with nucleus (n), P-protein bodies (PB), vacuoles (v) and dense cytoplasm. sp = developing sieve plate. *B*: Detail of P-protein body within a sieve element of serial section to *A*. The P-protein is tubular and partly aggregates into hexagonal order. *C*: High magnification of hexagonally arranged P-protein tubules. *D*: Small P-protein body of sieve element of *Aristolochia clematitis* showing parallel arrangement of P-protein tubules.

100

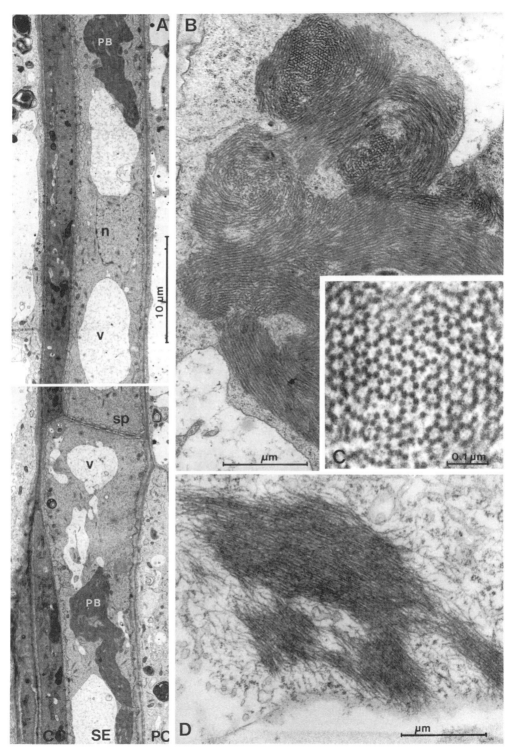

components are changed into structures of filamentous appearance. P-protein filaments are the morphological form generally present in mature sieve elements (see Sect. 3.4.2.4 for controversial views on their natural distribution).

Following the detailed description in *Nicotiana*, the presence of P-protein tubules and their conformational change into filaments have been verified in many dicotyledons using a step-by-step ontogenetic analysis (cf. Fig. 3.4). However, there is some evidence that the formation of P-protein tubules is not obligate to all angiosperms. In quite a few species – several investigated ontogenetically – P-protein of filamentous appearance was recorded in both young and mature sieve elements, but tubules were never found. On the other hand, in *Nelumbo* P-protein filaments were missing and tubules were also found in mature sieve elements (Esau, 1975). Other species may contain additional morphological forms of P-protein, e.g. granular in *Cucurbita* and crystalline in the legumes.

Tailed or tailless, spindle-shaped crystalline P-protein bodies are a characteristic feature of the Fabaceae (and of taxonomic value for the legume family; Behnke & Pop, 1981); they have been well known from LM studies since Strasburger's (1891) description. Crystalline P-protein bodies appear very early during sieve element ontogeny and seem to be related to co-appearing other forms of P-protein. Although most of the spindle-shaped bodies do persist in mature sieve elements (Lawton, 1978), some show a considerable loosening of their hexagonally packed sub-units combined with a transformation into striated P-protein. It is therefore suggested that the crystalline P-protein of the Fabaceae may represent an intermediate (with extended life-span) between the tubular and filamentous forms (see LaFlèche, 1966; Wergin & Newcomb, 1970; Palevitz & Newcomb, 1971).

Many other dicotyledons also contain non-dispersive, persistent P-protein bodies in their sieve elements (Fig. 3.5); these had been seen with the light microscope but were often mistaken for so-called extruded nucleoli. Ultrastructural studies revealed that all of them are proteinaceous and almost exclusively originate in the cytoplasm (Deshpande & Evert, 1970; Oberhäuser & Kollmann, 1977), some being apparently independent and in addition to dispersive P-protein. Presence of non-dispersive P-protein bodies and their specific morphology (e.g. globular, stellate,

Fig. 3.5 Non-dispersive P-protein bodies. *A*: Young sieve element of *Drymis winteri* showing dispersive P-protein (PP), a non-dispersive P-protein body (PB), mitochondria (M), tubular ER, dictyosomes (D) with smooth (sv) and coated vesicles (cv), small vacuoles (V), and microtubuli (arrow heads) adjacent to plasmalemma. W = cell wall. *B*: Young sieve element of *Fagus sylvatica* with nucleus (N), vacuole (V) and non-dispersive P-protein body (PB) in parietal cytoplasm. P-protein tubules (pt) surround the protein body and seem to protrude from it (arrows). M = mitochondria, D = dictyosomes. *C*: Stacked parietal agranular ER adjacent to the plasmalemma (Plm) in a mature sieve element of *Fagus sylvatica*. (Fig. 3.5B,C from Schulz & Behnke, 1987)

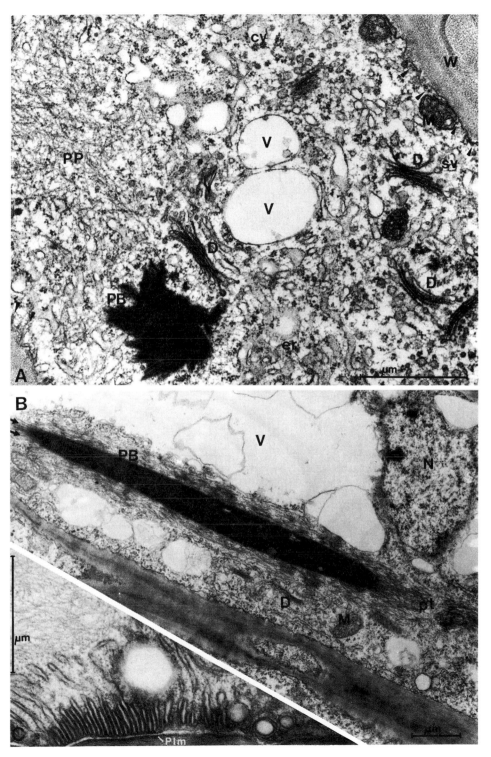

compound-spherical) are largely taxon-specific (cf. Behnke, 1981). Their structural composition is often (para-)crystalline with hexagonally arranged tubular units of a substructure comparable to that of P-protein tubules (Nehls *et al.*, 1978).

The origin of P-protein is not definitely known as yet. Ribosomes – most likely free cytoplasmic ones, but membrane-associated ones have been suggested as well – are the logical sites of protein synthesis, and the exact timing of P-protein formation and of nuclear and ribosomal disintegration favours such a hypothesis. In some plants, precursor (granular, nascent) P-protein has been shown prior to the formation of tubules or filaments (e.g. Wergin & Newcomb, 1970; Esau, 1978), and, repeatedly, free (helical) ribosomes were seen almost veiling P-protein components (e.g. Buvat, 1963; Wergin & Newcomb, 1970; Behnke, 1974a). Spiny vesicles occasionally found in association with protein bodies are also thought to participate in the formation of P-protein (Newcomb, 1967), e.g. in the transfer of precursor protein to the sites of protein body development (Arsanto, 1982). However, ultrastructural investigations are not adequate to answer this question, especially if it is considered that the structural units of P-protein may be the result of a self-assembly of precursor proteins.

A superhelical model attempts to explain the substructure of P-protein tubules and their interconversion into filaments. Based on the results of freeze-etched, negatively stained and ultrathin-sectioned P-protein of four different dicotyledons (including a legume with crystalline P-protein) viewed with a high resolution TEM, Lawton & Johnson (1976) and Arsanto (1982) proposed that a 20–25 nm tubule is composed of two 6–9 nm filamentous strands wound around a central lumen and forming a 'super-double helix'. Each strand is made up of two 4 nm elementary filaments likewise arranged in a (minor) double helix. Elementary filaments are composed of 3 nm sub-units, of which six are found per turn of the minor double helix. Such extremely thin filaments have previously

Fig. 3.6 Nacreous walls and nuclear crystals in sieve elements of the dicotyledon family Myristicaceae. *A*: Cross-section through primary phloem of *Myristica fragrans* with sieve elements (SE), companion cells (CC) and phloem parenchyma cells (PC). The cell lumen of the sieve elements is narrowed by the formation of a nacreous wall part (na). Sieve element plastids (p) and protein crystals (nc) are discerned in young and mature sieve elements. The nacreous wall shows a parallel arrangement of fibrillar material (arrows). sp = sieve plate, n = nucleus. Rectangle marks a sieve area/plasmodesmata connection (see Fig. 3.3C). *B*: Nucleus with protein crystal in an almost mature sieve element of *Knema*. na = nacreous wall part. *C*: Longitudinal section through young and mature sieve elements of *Myristica sylvestris*, the contents of which are dominated by S-type sieve element plastids (s). Mature sieve elements are narrowed down by broad nacreous wall parts (na). sp = compound sieve plates. *D*: Mature sieve elements (SE) of *Knema* containing S-type sieve element plastids (s) and a protein crystal (nc) derived from the degenerated nucleus. na = nacreous wall. sp = sieve plate.

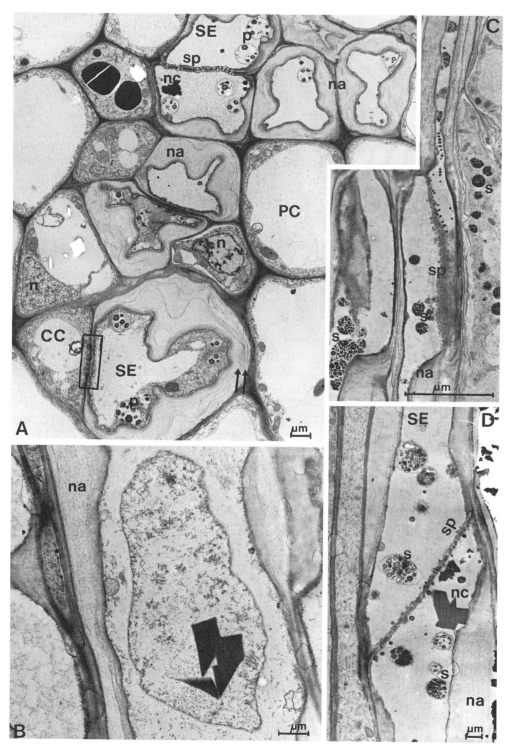

been depicted in isolated and negatively stained P-protein of *Nicotiana* (Kollmann *et al.*, 1970). Within a P-protein body tubules are interconnected by filamentous side arms, thus anchoring them together. Later, during the disaggregation of the P-protein bodies the super double helix of the tubules will be stretched and partially loosened to result eventually in the striated appearance of P-protein filaments.

The results of biochemical analyses undertaken on the phloem or sieve element exudate of a few dicotyledons indicate that, depending on the species considered, phloem proteins are acid or (more often) basic proteins with molecular weights from less than 30 000 to more than 100 000. In *Cucurbita maxima* two distinct fractions each representing up to 40% of the total phloem protein were further characterized: one (116 000 MW) readily reaggregates into filaments and represents the P-protein seen in ultrathin-sectioned phloem tissue (Kollmann *et al.*, 1970; Beyenbach *et al.*, 1974), while the other (a dimer of 48 000 MW) is identified as lectin (Sabnis & Hart, 1978; Read & Northcote, 1983).

In addition, critically performed tests demonstrated that P-proteins are neither actin- nor tubulin-like, i.e. they do not bind with heavy mero-myosin and are likewise insensitive to specific drugs (see literature in Koll-mann, 1980).

From all of the foregoing it is concluded that P-protein seems to be a new, independent, structural protein restricted to some cell types only. Whilst morphological characters as well as absolute and relative (dispersive or non-dispersive) contents suggest a species-specific composition, biochemical features such as the more generally found gelling properties and lectin content have supported the idea that P-protein serves as a general sealing substance for wounded sieve elements and that lectin through its binding capacities would anchor P-protein to the plasmalemma (Smith *et al.*, 1987) or by immobilizing bacteria and fungi would sterilize the wound (Read & Northcote, 1983).

Nuclear protein crystals found in young sieve elements of the dicotyledon families Boraginaceae (Esau & Magyarosy, 1979) and Myristicaceae (Fig. 3.6) survive nuclear degeneration and are released into the sieve element lumen where they may have the same morphological characteristics as persistent P-protein bodies of other taxa. The broad definition of P-protein would encompass these protein bodies, too, but at the same time cause problems. Nuclear crystals are found in parenchyma cells of vascular and other tissues quite often amongst dicotyledons and are regarded as a taxon-specific character; they are also reported in ferns, occurring in parenchyma cells as well as in sieve elements (Evert & Eichhorn, 1974), which definitely lack P-protein. Not least, the presence of P-protein-like tubules in the nuclei of young sieve elements of *Tilia* and their transformation into filamentous structures, both indistinguishable from co-existent P-protein originating in the cytoplasm (Evert & Deshpande, 1970), should be mentioned as another example of the inconsistency of the present terminology, which largely relies on structural data.

3.4.2.2.5 Mitochondria It has been known since the application of specific staining methods in LM sections and the early EM investigation that sieve elements contain mitochondria during all stages of their ontogeny, including the mature state. Young sieve elements contain the same type of elongate mitochondria as do their neighbouring parenchymatic elements (Fig. 3.2A,B). During sieve element maturation a change into spherical shapes and a gradual depletion of matrix contents are observed. Cytochrome oxidase has been localized in sieve element mitochondria of a few species (see Cronshaw, 1981).

3.4.2.2.6 Plastids Plastids are among the few organelles that are present until the final obliteration of a sieve element. Careful preparations will always find morphologically intact sieve element plastids, i.e. provided with a complete double envelope and their protein or starch contents not released.

The young prospective sieve element contains many oblong proplastids which, like those of other meristematic cells, are characterized by a densely granular matrix, a few non-stacked thylakoids and some plastoglobuli. In general, the first small deposition of sieve element starch or protein inside the matrix demonstrates that the differentiation of a proplastid into a specific sieve element plastid has been initiated. The appearance of tiny protein crystals, in particular, has even been used as a marker for the early identification of a sieve element. As the inclusions enlarge the sieve element plastids gradually change into spherical shapes, often passing through amoeboid forms if starch grains or protein crystals are deposited eccentrically and grow more rapidly than the matrix and envelope changes would anticipate (Fig. 3.2B). Co-ordinated with the accumulation of substances and the rounding of the plastids is the depletion of their matrix, until it looks completely electron translucent. Mature sieve element plastids are characterized by their specific contents and their spherical shape confined by the double envelope (Figs. 3.7B, 3.10). There is an obvious increase in volume during plastid differentiation. However, its actual amount has never been critically measured and would be difficult to find out, since preparational or fixational influences may cause the plastids to swell.

The sizes of sieve element plastids range from 0.6 μm to over 3 μm and are dependent on the species considered and to a minor degree, the location within the plant body. The type and quantity of their contents are genetically fixed. The ability to accumulate protein (crystals and/or filaments) serves as a basic feature to distinguish between P-type and S-type plastids – the latter containing starch only – and is helpful in the systematics of higher angiosperm taxa. The monocotyledons, centrospermae, legumes and a few other groups are each characterized by specific forms of P-type plastids – discriminated by protein crystals differing in number and shapes – and are clearly separated from the great number of S-type-containing dicotyledons (cf. Figs. 3.3B, 3.8B and Behnke, 1981). Moreover, in a few dicotyledon families sieve element plastids accumulate neither protein nor starch, their form S-type plastids are very small, often mixed up with vesicles or not observable, which was the reason for early reports of sieve elements being devoid of plastids (cf. Esau & Cronshaw, 1968; Behnke, 1981, 1988).

In EM micrographs it is sometimes not easy to distinguish between small starch and protein inclusions, due to staining abilities which are often similar, but sharp edges or crystal-like substructure helps to identify the protein. Digestion with specific enzymes has been of additional help (Palevitz & Newcomb, 1970; Behnke, 1975a). Protein crystals of different origin have been digested in ultrathin sections by several proteases. The sieve element starch, known since Briosi (1873) to give a reddish staining reaction after treatment with iodine, has been demonstrated by sequential digestion with pullulanase and amylases to be especially rich in $\alpha(1 \rightarrow 6)$ branched amylopectin molecules. An especially high content in amylopectin – which is approaching the composition of glycogen – might be responsible for the often observed particulation of sieve element starch into tiny, deeply contrasted globular units (Fig. 3.7B).

3.4.2.3 Differentiation of the sieve element walls

At the beginning, structure and composition of the sieve element walls do not differ from those of other living cells. A middle lamella derived from the protopectin containing Golgi vesicles during cytokinesis separates two contiguous cells and is followed by a primary wall. The distinctive features of the sieve element wall, accentuating their specific position amongst the other plant cell walls, are stamped during differentiation growth, as are:

1. the deposition at the lateral walls of inner cellulose-rich layers, named nacreous wall parts, which are responsible for the distinctive refractive properties of the sieve elements when viewed with the light microscope and which may range from almost indistinct to very prominent thickenings (in some species nearly occluding the sieve element lumina);

Fig. 3.7 *A*: Part of young sieve element of *Lindera praecox* with lobed nucleus (N), P-type sieve element plastids (P) containing starch (st) and protein (arrow), mitochondria (M), tubular ER, dictyosomes (D) and vacuoles (v). The tonoplast (t) is disrupted, the remains forming cisternal or vesicular structures. *B*: Longitudinal section through mature sieve element (SE) and adjacent companion cell (CC) in *Hieracium staticifolium*. The companion cell is modified into the transfer cell condition (shown by the many wall ingrowths (arrows). S = S-type sieve element plastids containing many tiny starch granules, M = mitochondia. *C*: Sieve area/plasmodesmata connection in *Fagus sylvatica*. The two sieve pores are surrounded by callose deposits (ca) and connect to branched parts of plasmodesmata penetrating the companion-cell half of the wall. *D*: Longitudinal section of a transfer cell (companion cell) of *Hieracium staticifolium* showing trabeculae alongside all walls, a dense cytoplasm and a nucleus (N).

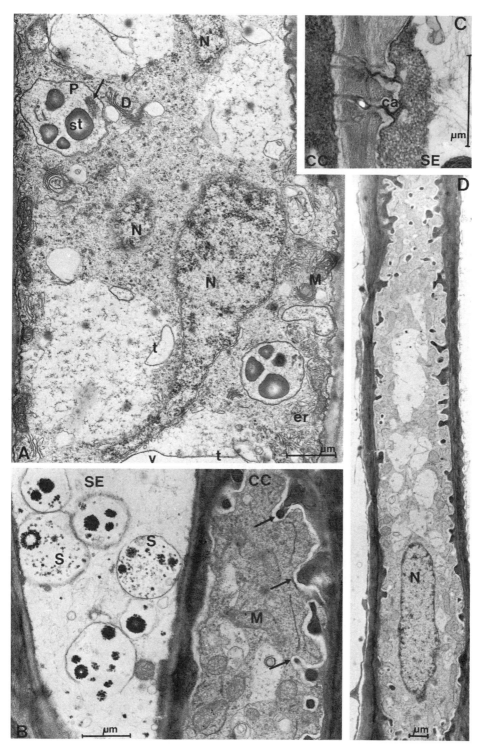

2. the formation of sieve pores, sieve areas and sieve plates in both lateral and end walls as well as of specific connections to neighbouring parenchyma cells via modified branched plasmodesmata; and
3. the involvement of callose in the development and function of sieve pores.

3.4.2.3.1 Nacreous wall parts The cell walls of the sieve element are considered primary, with positive staining reactions for pectin and cellulose only. Wall parts not covered by sieve areas, i.e. mainly lateral walls, very often are characterized by a special wall thickening, called nacreous wall, which is responsible for the extreme glistening of sieve elements when viewed with the light microscope and distinguishing them from surrounding parenchyma cell walls. The glistening is due to a specific birefringence detectable with the polarized microscope and based on a high content of ordered cellulose molecules. Nacreous walls may appear in primary and secondary sieve elements, are very conspicuous in some taxa – and often attain their maximum thickness (covering nearly 90% of the transsectional area of the cell, cf. Fig. 3.1B) when the sieve element is mature – but may be rather indistinct in others (cf. Esau & Cheadle, 1958).

In ultrathin sections viewed with the TEM the primary wall of many sieve elements consists of two distinct layers, a comparatively thin outer layer (adjacent to the middle lamella) and a thicker inner layer (facing the plasmalemma) referred to as a nacreous layer (Figs. 3.6, 3.9). With the ability to detect even small inner layers ultrastructural studies led to the recognition of a considerably greater number of species containing nacreous walls.

While histochemical and enzymatic extraction tests performed with several species yielded the uniform picture of a pectin-rich and cellulosic outer layer and a pectin-poor and (probably less?) cellulosic inner (nacreous) layer, the results of ultrastructural studies concerned with the microfibril orientation within the nacreous layer were contradictory. Cronshaw (1975b) describes a distinct striate pattern, Deshpande (1976) shows polylamellate structures in *Cucurbita*, whereas in *Pinus* Chafe and Doohan (1972) detected distinctly cross-polylamellate secondary walls. The orientation of the cellulose microfibrils in the nacreous wall is preferentially perpendicular to the long axis of the sieve element – probably forming dense spirals responsible for the birefringence in polarized light (cf. Behnke, 1971) – and parallel to adjacent cortical microtubules which are specifically restricted to wall parts undergoing nacreous thickening (Eleftheriou, 1987).

The function of the nacreous wall is not known as yet. For the thickened wall parts of root protophloem sieve elements of grasses, it has been suggested that they serve as stored wall material to be used during elongation growth of the tissue, i.e. at a time when the enucleate cell is no longer capable of synthesizing wall material, and thus enable the sieve element to escape a precocious obliteration (Danilova & Telepova, 1980; Eleftheriou & Tsekos, 1982a).

It is evident from the foregoing that not all wall parts designated as nacreous are alike. Amongst the angiosperms the most conspicuous inner layers (i.e. those occupying up to 90% of the transsectional cell area) are found in preferentially primitive woody taxa only.

3.4.2.3.2 Sieve pores In general, common wall parts between contiguous sieve elements (end walls as well as lateral walls) are perforated by sieve pores grouped into distinct sieve areas, which in angiosperms are in turn preferentially associated with larger sieve plates (cf. Fig. 3.9). In addition, lateral walls adjoining companion cells may contain specific sieve pore/plasmodesmata connections (see Sect. 3.5.2).

Each sieve pore develops from a single plasmodesma through a gradual widening of its transsectional area specifically involving a replacement of primary wall parts by callose and the subsequent dissolution of the latter (i.e. for the over 2 μm wide sieve plate pore of *Aristolochia* in Fig. 3.8C this area increase is close to 5000-fold). Plasmodesmata originate during cytokinesis. When in the course of cell plate formation an ER tubule oriented perpendicular to the new cell wall obstructs the fusion of protopectin-containing Golgi vesicles, a plasmodesma is created. Consequently, plasmodesmata are present in end and lateral walls between both contiguous sieve elements and ontogenetically related phloem parenchyma cells, e.g. companion cells, from the very beginning. They are typically composed of a narrow ER tubule (desmotubule) and a membranous hollow cylinder derived from the fusing Golgi vesicles and at the same time delimiting them against the cell wall as well as forming connections to the plasmalemma of the neighbouring cells.

Early indicators of the incipient differentiation of sieve pores are the occurrence of small ER cisternae connected to both ends of the desmotubule and restricted to a small area surrounding the plasmodesmata, and the incipient deposition of callose.

The cisternal membranes of the ER facing the cell wall are agranular and closely parallel to the plasmalemma; those facing the cytoplasm are granular (Fig. 3.8B). The cisterna itself may be flat or blown up, and at this early stage its lateral extent roughly marks the final size of the mature sieve pore opening (Fig. 3.8A). Later during sieve element ontogeny these specialized cisternae disappear, sharing the fate of the other ER conformations. Several proposals have been made as to their role in pore development, e.g: localized inhibition of wall thickening (by deviation of Golgi vesicles to neighbouring wall areas which often show a considerable thickening growth); synthesis of wall-degrading enzymes concomitant with callose deposition; or supply of membrane material for the elongating desmotubule during the extreme thickening of some sieve plates. Although all these putative roles seem convincing (either each in itself or all combined), the ultimate function of the sieve-pore-associated cisternae remains obscure.

The first visible callose is deposited as small collar-like platelets surrounding the two orifices of a plasmodesma. The callose platelets subsequently expand laterally and towards the middle of the cell wall while gradually replacing primary wall material. The size of the future sieve pore can be determined at maximum development of callose, i.e. by the extent of the callose cylinders which completely surround the plasmodesmata and penetrate the wall of the developing sieve areas excluding only the middle lamella (Fig. 3.8A). The formation of the sieve pore opening may start with a widening of the plasmalemma jacket at the level of the middle lamella; it certainly involves the loss (degradation or disruption?) of the desmotubule in conjunction with the loss of the ER cisternae; and proceeds with the complete removal of the callose cylinders.

3.4.2.3.3 Callose The role of the callose described here as part of the normal pore development is, however, not undisputed. Repeatedly, views have been expressed that all callose found in sieve elements is deposited either in response to mechanical, chemical or other manipulation of the plant during preparation for microscopy, or is due to ageing of the sieve element (dormancy and definitive callose). Reports on absence of callose from uninjured sieve elements (plantlets of *Lemna* fixed *in toto*, Walsh & Melaragno, 1976) are supportive, but the presence of small amounts of callose in sieve elements of phloem of tissue cultures fixed intact (Sjolund *et al.*, 1983) is contradictory.

Callose, a β-1,3 glucan forming helical molecules (as opposed to the straight molecule of cellulose), is used in various parts of the plant (e.g. pollen tetrades, pollen tubes), and prominently in the phloem as a sealing substance of most advantageous properties: it may be rapidly deposited and rather easily removed. Indisputably, callose seals the pores if a sieve element ceases to function – either definitely (followed by death of the cell)

Fig. 3.8 Sieve plate pores of dicotyledons. *A*: Longitudinal section of two young sieve elements of *Melocactus depressus* connected by a developing sieve plate. Two future pores can be distinguished by their callose deposits and by small ER cisternae adjacent to them (arrows). Microtubules (arrow heads) line the lateral walls. The dense cytoplasm contains P-type sieve element plastids with protein filaments (p), a small body composed of tubular P-protein (pp). dictyosomes (d) and granular ER. v = vacuole. *B*: Young sieve elements of *Achlys triphylla*. The future size of the two developing sieve pores is marked by local ER cisternae (arrows) and callose platelets (ca). Within one pore a plasmodesma is partially cut. The cytoplasm of the sieve elements contains tubular ER and precursor material of P-protein (*). r = ribosomes. *C*: Open sieve pore of a mature sieve element of *Aristolochia gigantea*. The pore is partly narrowed down by callose deposits (ca), lined by plasmalemma (plm) and penetrated by longitudinally arranged P-protein filaments. er = parietal ER cisternae, partly stacked.

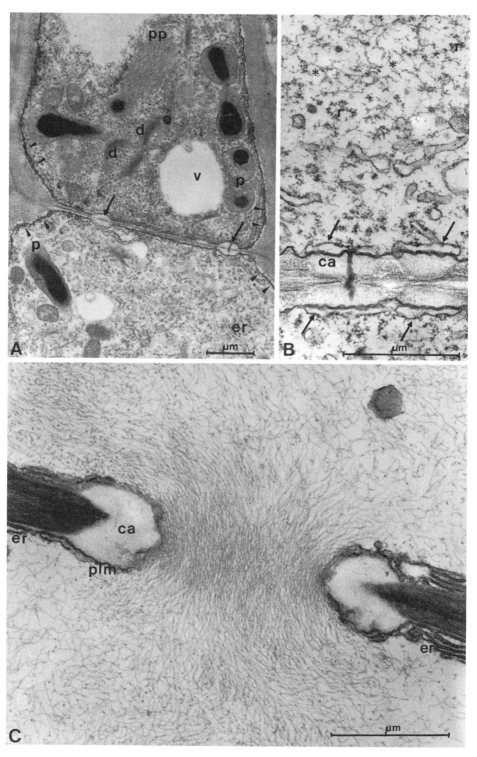

or in some plant species transitorily (at the end of a growing season). In both cases the 'senescence' callose can be removed, the definitive callose during obliteration of the entire cell and the dormancy callose in early spring of the following year. It is also beyond doubt that callose is formed as a response to injury (wound callose, Figs. 3.1F,G, 3.8C) and likewise easily removed. This does not, of course, imply that all non-senescence callose is wound callose. Certainly, the callose controversy is not likely to be solved with our microscopical methods. However, further discussions should consider that: (a) angiosperm sieve elements not only display a considerable variation in pore sizes but also in callose contents; and (b) the sealing and easy removal properties might be positively required during the ontogeny of a sieve pore.

Its specific staining response to resorcin blue in transparent light microscopy, to aniline blue in fluorescence microscopy, and its general lack of contrast in transmission electron microscopy makes callose an easy marker for sieve elements.

3.4.2.4 The mature sieve element

Sieve elements are designated mature after the protoplast has completed its selected autophagy and the sieve pores have been opened and before the final autolysis and obliteration are intitated, i.e. during the time of active translocation.

The mature sieve element is extremely sensitive to any manipulation, answering with surging effects to the slightest pressure release; it is therefore not likely ever to be adequately translated into a series of micrographs. However, counter-check using several methods and techniques and the results obtained from sieve elements fixed intact or injured intentionally before fixation, led to the following critical appraisal of what kind of structure is close to natural and which is the most likely sequence of artifactual events.

Mature sieve elements are live cells demonstrating their structural and physiological identity by the presence of a plasmalemma, which serves as a complete delimitation against the apoplast, is continuous with neighbouring cells through sieve pores and plasmodesmata, shows the common three-layered substructure, and actively responds in plasmolysis tests (cf. first report by Schumacher, 1939).

The mature sieve element is enucleate (some taxa retain nuclear remnants) and evacuolate and also devoid of or impoverished in most of its cytoplasmic contents (Fig. 3.10). The only intact organelles residing inside the plasmalemma are plastids and mitochondria. The similarity in composition between their outer envelope and the plasmalemma might protect them against lytic attacks and might thus be the only reason for their continued existence. A possible active participation in controlling translocation within sieve elements (e.g. via generation of ATP or immobilization or remobilization of sugars) has never been proven. In mature sieve elements plastids and mitochondria are found in a parietal position, have their envelopes intact and contain all components described in Sect. 3.4.2.2.5 and 3.4.2.2.6, and do not release their contents nor degenerate before the final autolysis of the cell.

A modified agranular ER resides as a parietal anastomosing system, in some plants occasionally surrounds plastids and mitochondria, and may be continuous with stacked (Fig. 3.5C), convoluted, lattice-like or other peculiar membrane condensations. Several functions, ranging from active participation in the uptake of translocates to autophagic processes and simple storage of membrane material, have been suggested but none has been experimentally verified (see Sect. 3.4.2.2.2).

In gymnosperm sieve elements an extensive network of agranular tubular ER forms a continuous system and is specifically associated with the sieve areas (see Sect. 3.4.3).

In most angiosperms P-protein is the characteristic component of the mature sieve element. With a few exceptions where P-protein tubuli were found, it is generally present as P-protein filaments, in some taxa supplemented by non-dispersive P-protein bodies. There are, however, most controversial views about how the P-protein is distributed in an actively translocating cell and what preparation or fixation yields the closest-to-nature results. Of all the structural components of the mature sieve element, P-protein has proved to be the most sensitive to even the slightest surging effect. Even if influences during fixation were disregarded, sieve elements fixed intact (cf. Sect. 3.4.2.3.3) unfortunately did not contribute to a solution, since *Lemna* did not contain P-protein and sieve elements from tissue cultures do not translocate. Thus, only indirect evidence may support the one view or the other.

Following the disintegration of the tonoplast in the sieve element the P-protein bodies continue what was initiated already within the still-tonoplast delimited cytoplasm: the dispersion of their filaments. There is debate whether this results in a parietal or more even (network-like) distribution of P-protein throughout the entire lumen of the sieve element (cf. Cronshaw, 1975a; Evert, 1984). Without unequivocal criteria by which to judge the natural distribution, a decision on which of the two interpretations is correct, or whether both are, is presently impossible. As long as their anchoring to membranes or mutual interconnection remains an unproven proposal, the idea of a random distribution of P-protein filaments combines what we know about P-protein development and observations with our various microscopical techniques (Fig. 3.10). A random distribution is in agreement with the non-oriented release of filaments from both dispersing P-protein bodies and disaggregating crystalline P-protein (of Fabaceae), is influenced by natural flow rates as well as by any pressure release answering to a severance of the phloem, may keep the main part of

115

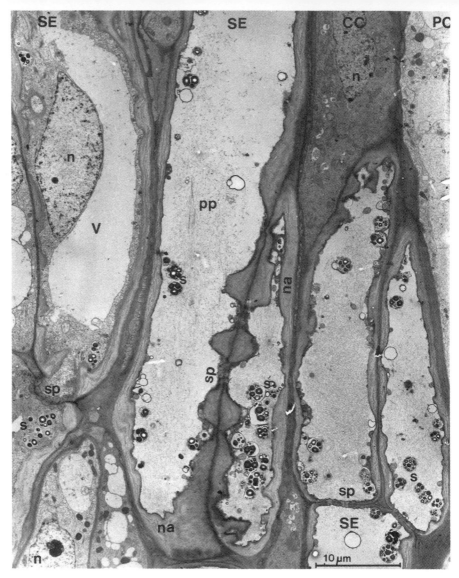

Fig. 3.9 Longitudinal section through the phloem of *Liriodendron chinense* with young nucleate and mature enucleate sieve elements (SE), companion cells (CC) and phloem parenchyma cells (PC). The young sieve elements contain a central vacuole (V) and a parietal protoplast with nucleus (n), sieve-element plastids (s) and other organelles. The mature sieve elements are almost electron-translucent, their lumen is dominated by the plastids (s) and P-protein filaments (pp). All sieve elements contain a nacreous wall (na). Many sieve elements are connected by simple or compound sieve plates (sp). Companion cells can be distinguished from phloem parenchyma cells by their dense protoplast.

Fig. 3.10 Longitudinal section through phloem of *Cheiranthus cheiri* showing two mature sieve elements (SE) connected by a sieve plate with open pores (sp), companion cells (CC) and phloem parenchyma cells (PC). The sieve elements contain S-type sieve element plastids (S) and mitochondria (M) and randomly distributed P-protein filaments (pp). Companion cells and phloem parenchma cells contain large vacuoles (V), and as prominent organelles of their parietal protoplasts nuclei (N) and chloroplasts (c). (From Behnke, 1981)

116

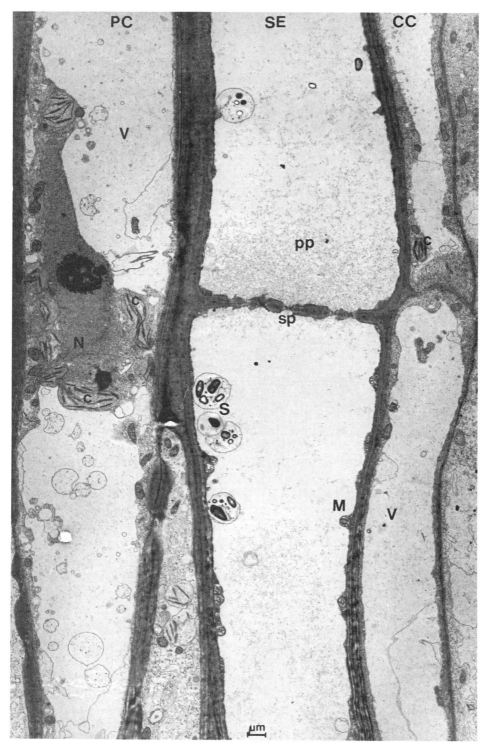

P-protein in a marginal position and only occasionally dislocate single P-protein filaments or minor aggregates through sieve element lumina and sieve plate pores, eventually relocating them to the periphery of the flowing stream. Only a pressure release dislocates massive portions of P-protein against the sieve plates, and even at a greater distance from the injury (counting in tens of sieve elements) is traceable by plugging of sieve pores, formation of coarse strands, moderately dense but orderly penetration of sieve pores, longitudinal orientation and parallel association of small groups of filaments (order of listing reflects declining artifactual influences).

Deriving from controversial views about the driving force of phloem translocation, there has been a long-lasting dispute on what morphological situation reflects the normal distribution of P-protein within the pores of actively translocating angiosperm sieve elements, i.e. whether the sieve plate pores are occluded by, loosely filled with, or completely devoid of P-protein (cf. Weatherley, 1975; Spanner, 1978; Cronshaw, 1981). Presently, the evidence from various experiments aimed at either reducing or intentionally raising the pressure release seems to favour unoccluded open pores, lined with plasmalemma, traversed by only a few P-protein filaments and devoid of any callose. Practical electron microscopy has only rarely produced micrographs which come close to this ideal (cf. Figs. 3.8C, 3.10).

Sieve pores of gymnosperms and vascular cryptogams are only marginally touched by the controversy about their content, partly due to their different ontogeny (see Sects. 3.4.3 and 3.4.4).

As a general conclusion it can be said that mature angiosperm sieve elements provide an adequate conduit for symplastic long-distance transport, which remains flexible to respond to external irritations with an occlusion of the sieve pores by P-protein and their sealing by callose.

This sequence of reactions (slight P-protein aggregation followed by massive P-protein accumulations combined with heavy callose deposition) is also recorded in cut and fixed sieve elements if one proceeds from the least affected sieve element in the middle of a specimen outward to the cut surface. Moreover, the use of non-physiological preparation methods and non-adjusted fixation media results in additional artefacts like plasmolysis of the sieve elements and/or bursting of their plastids.

3.4.2.5 Longevity of sieve elements

It is well known that enucleate sieve elements have a rather short lifetime – measured in days or weeks for protophloem and in months for metaphloem and secondary phloem elements – after which the conduction of assimilates ceases, definite callose is deposited, plasmalemma and cytoplasmic contents are destroyed, and often the entire cell is crushed. In general, the short life of sieve elements causes trouble to neither annual nor the great majority of perennial plants, since a sufficient number of cells are successively differentiated from metaphloem or through cambial activity. If necessary, e.g. in plants resuming their cambial activity rather late, immature or even mature sieve elements may survive the dormancy period (sealed by dormancy callose) and start or resume respectively their conductivity early during the next growing season. However, plants which perennate without secondary vascular tissues – like several perennial monocotyledons and vascular cryptogams – must have acquired the ability to prevent the obliteration, i.e. to change short-lived into long-lived sieve elements. Some monocotyledonous lianas solve the problem by retarding the differentiation of most vascular tissues and maturing only an appropriate amount necessary for translocation.

The longevity of sieve elements has been prominently studied in palms, where sieve elements of the basal parts may remain functional for more than a century. During their ontogeny palm sieve elements undergo the same structural changes in about the same time as do their short-lived counterparts of other angiosperms. In ultrathin sections of enucleate sieve elements estimated to be over 20 years old an intact plasmalemma, ER, plastids, and sieve pores lined with plasmalemma and some callose but otherwise unoccluded by P-protein were found (see Parthasarathy, 1980, for further details).

Longevity of sieve elements is also known within secondary phloem of dicotyledons and gymnosperms. *Vitis* and *Tilia* are prominent examples for sieve elements that survive at least 2 and a maximum of 10 years.

3.4.3 The specific structure of gymnosperm sieve elements

Sieve elements of gymnosperms were generally named 'sieve cells', a term which they share with the sieve elements of the vascular cryptogams and which accentuates differences to the 'sieve tube members' of angiosperms. Since, however, the basic structure of all so-called sieve cells is not the same, the name 'sieve elements of gymnosperms' is given preference (see also Sect. 3.4.1.3 and Behnke, 1986).

Gymnosperm sieve elements develop directly from procambial or cambial initials – i.e. without further divisions involving a phloem mother cell – and therefore are not accompanied by ontogenetically related phloem parenchyma cells. However, most of them, especially among conifers, are structurally and physiologically associated with equally highly specialized parenchyma cells, the so-called Strasburger cells (see Sect. 3.5.2).

The structural differentiation of gymnosperm sieve elements matches that of their angiosperm counterparts to a great extent, Nuclear degeneration is of the pycnotic mode (cf. Fig. 3.11) discussed in Sect. 3.4.2.2.1. Dictyosomes, ribosomes, microtubuli and vacuoles disappear as described

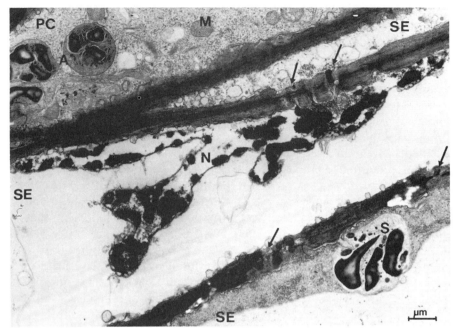

Fig. 3.11 Longitudinal section through the phloem of *Cycas revoluta* with young and mature sieve elements (SE) and a parenchyma cell (PC). The young sieve element contains a parietal protoplast and a starch-containing plastid (S). A mature sieve element contains a degenerating, deeply lobed nucleus (N) characterized by condensation of heterochromatic material. The parenchyma cell is distinguished by a dense protoplast with amyloplasts (A) and mitochondria (M). Arrows point to sieve areas between sieve elements. (From Behnke, 1986)

Fig. 3.12 *A*: Two sieve elements of *Picea abies* are connected via sieve areas composed of several branched plasmodesmata and median cavities (mc). The cytoplasm contains massive aggregations of tubular ER and several P-type sieve element plastids (p). *B*: Sieve area/primary pit field connection between sieve element (SE) and Strasburger cell (SC) in *Picea abies*. The Strasburger cell contains a dense protoplast which is rich in ribosomes. Arrows point to plasmodesmatal branches. SP = sieve pore, m = mitochondria, er = aggregates of tubular ER. (From Schulz & Behnke, 1987)

in angiosperm sieve elements, mitochondria and plastids persist, P-protein is not formed, and the ER undergoes very specific differentiations.

Sieve element plastids are of S-type with the exception of the Pinaceae which have P-type plastids. The latter specifically contain protein crystals, protein filaments and starch grains. It is the protein filaments which, if released from plastids burst during harsh manipulation, have confused many authors and led to discussion of the presence of P-protein. Careful preparations clearly demonstrate intact plastids and the absence of any protein filaments in the lumina of differentiated sieve elements (Behnke, 1974b; Neuberger & Evert, 1974).

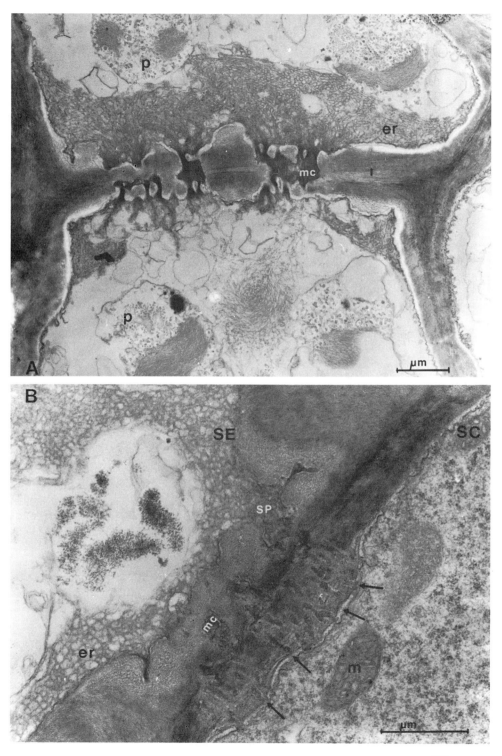

The most typical content of mature sieve elements is an elaborate system of agranular, tubular endoplasmic reticulum found at the periphery of the cell, aggregated in part (e.g. in association with nuclear remnants), and preferentially deposited at the sieve areas. This specific ER forms early during sieve element ontogeny and eventually involves a transformation of all the originally rough ER cisternae into a complex tubular network (Kollmann & Schumacher, 1964; Neuberger & Evert, 1974).

Nacreous wall parts are likewise present in gymnosperm sieve elements, with the heavy, thickened walls of Pinaceae sometimes designated as secondary. In *Pinus strobus* the composition of the sieve element wall has been described as polylamellate (Chafe & Doohan, 1972).

The structure and organization of the sieve areas are amongst the most conspicuous features distinguishing gymnosperm sieve elements from both angiosperm and lower vascular plant sieve elements. In gymnosperms sieve areas develop from (sometimes branched) plasmodesmata arranged in primary pit fields between contiguous sieve elements of almost the same age. In a very comparable way to the development of angiosperm sieve pores – although only rarely studied – ER and callose are involved in the widening up of a single plasmodesma (see Schulz & Behnke, 1987). Unlike in angiosperm sieve elements, however, during further development of the sieve area a median cavity is formed, which, in the middle of the cell wall, connects all plasmodesmata or plasmodesmatal branches (cf. Kollmann & Schumacher, 1963).

Mature sieve areas of gymnosperms are bordered on both sides by massive aggregates of the tubular ER system. Single profiles of ER are obviously continuous through the sieve pores and meet in the median cavities where they merge together. Sieve pores and median cavities are always lined by plasmalemma (Fig. 3.12A). The size of the pores rarely exceeds 0.2 μm and usually runs to two- to fourfold the diameter of a plasmodesma. Sieve elements are connected to Strasburger cells by a combination of a sieve area and a primary pit field (see Sect. 3.5.2).

Although more often investigated in the conifers, the basic structural features (e.g. pycnotic nuclear degeneration, tubular ER system, median cavities in sieve area) also apply to other taxa, like Gnetales (Behnke & Paliwal, 1973; Evert *et al.*, 1973) or Cycadales (Behnke, 1986). Consequently, any discussion of long-distance tranlocation in gymnosperms needs to consider the special impact of the tubular ER on transport rates and mechanisms.

3.4.4 Phylogenetic aspects

Below the organizational level of the seed plants sieve elements are present in various, but not all, parts of many bryophytes and regularly occur in all groups of pteridophytes. In general, sieve elements of mosses and seed-less vascular plants are extremely long and small, contain uniformly sized sieve pores on both end and lateral walls (i.e. are 'sieve cells' in the sense of Cheadle & Whitford, 1941), and are not associated with specialized phloem parenchyma cells. A great many electron microscopic studies covering all higher taxa have discussed concordance and discrepancies between the structural differentiation of the sieve elements of seed plants and those of these two groups (see Hébant, 177, for bryophytes and Warmbrodt, 1980, for pteridophytes). Some of these features are considered below.

In *bryophytes* a wide range of cell types intermediate between parenchyma cells and sieve elements ('leptoids s. str.') are encountered in the different taxonomic groups. Sieve-element-like leptoids (named sieve elements of bryophytes) are present only in the gametophyte of some Polytrichales, the most highly differentiated mosses.

Sieve elements of bryophytes are elongate cells connected in their end walls via numerous plasmodesmata, only rarely widened up to small pore-like openings (with diameters of less than 0.2 μm) and only occasionally narrowed down by callose deposits. Their lateral walls may contain plasmatic connections to neighbouring cells, sieve elements as well as parenchyma cells, and in some species develop heavy (nacreous) thickenings. The differentiation of the protoplast of bryophyte sieve elements is much like that of the seed plant sieve elements and involves the degeneration of nuclei, dictyosomes, ribosomes and vacuoles. Nuclei may show aspects of chromatolytic or pycnotic degeneration and their presence be prolonged into the mature state of the cells. Plastids contain remnants of thylakoids, partly in the form of vesicles, and plastoglobuli only; neither starch nor protein inclusions have been recorded. A remarkable feature is the extensive agranular ER found in mature sieve elements of bryophytes, which may be associated with nuclei and plastids, is in connection with the plasmodesmata, and forms stacked or convoluted membranous aggregates (see Hébant, 1977, for more details).

The sieve elements of the *pteridophytes* undergo all the ontogenetically related changes that their counterparts in the seed plants do. Mature sieve elements are plasmalemma-lined, enucleate, devoid of most of the organelles, but contain plastids, mitochondria, a parietal network of agranular ER, and – as a specific feature – the so-called refractive spherules. Sieve areas or sieve pore/plasmodesma connections penetrate the common walls to neighbouring sieve elements and parenchyma cells, respectively. Lateral walls are often heavily thickened, some displaying all the characteristics of typical nacreous walls.

The refractive spherules found, with the exception of the lycopods, in all taxa (but also in some of the polytrichaceous mosses) can be regarded as the distinctive characteristic of the pteridophyte sieve elements. They derive their name from a high refraction observed in unstained sections viewed with the light microscope. They appear early during the ontogeny of sieve elements and thus have served as a positive marker. Electron microscopy has disclosed their structure as being composed of an electron-dense globule surrounded by a single membrane and their origin as derived from the endomembrane system – most often from ER, but sometimes also in conjunction with, or simply from, dictyosomes. Although it is known from staining reactions that their contents are of proteinaceous nature, neither exact composition nor function of the refractive spherules are known.

Numerous divergent ultrastructural details found in some but not all sieve elements reflect the heterogeneity of the taxa classed under the pteridophytic organization. Nuclei may degenerate according to pycnotic, chromatolytic or 'intermediate' modes, and even change from protophloem to metaphloem sieve elements or alternate within different parts of a species. Plastids contain internal (thylakoid-derived) membranes, plastoglobuli and other spherical inclusions (rarely starch or protein) – all to a variable extent – but do not provide obvious taxonomic characters. Lateral wall thickenings are found throughout most of the taxa investigated but likewise display considerable variations, both in morphological and chemical composition of nacreous as well as other thickened wall parts (see Warmbrodt, 1980, and Evert, 1984, for details).

Of phylogenetic interest are sizes, distribution and mode of formation of the sieve pores. Sieve pores may occur as single connections to neighbouring cells in both lateral or end walls or clustered into sieve areas preferentially found in end walls. The size of the pores ranges from less than 0.1 μm to over 2 μm, and in most species there is no size difference between lateral and end wall pores. However, in some parts of *Isoetes*, *Psilotum*, and *Equisetum* sieve elements were found which had extremely wide pores specifically restricted to end walls; these sieve elements would therefore conform to the definition of a sieve tube member. The development of a sieve pore in a pteridophyte sieve element starts with a single plasmodesma which is gradually widened into an open connection. Callose is only rarely found in developing pores (e.g. in *Polypodium*, Liberman-Maxe, 1978), but wound callose is recorded in mature sieve elements of several taxa. Tubular or other ER is regularly associated with the pore sites and in most species also fills the widened sieve area pores. Dependent upon the pressure release prior to fixation, sieve elements of *Selaginella* and *Isoetes* may show sieve pores occluded by a filamentous protein derived from crystals synthesized in the ER and not comparable to P-protein, while other sieve elements of *Isoetes* contain completely unoccluded pores (Burr & Evert, 1973; Kruatrachue & Evert, 1974, 1977).

In concluding the sections on sieve elements it may be speculated that – in addition to the phylogenetically well-documented changes in cell sizes and end wall inclination (cf. Sect. 3.4.2.1) and the generally observed depletion of the protoplast – certain differential trends towards the evolution of the most highly specialized sieve elements of seed plants are evident in pteridophyte sieve elements, but are missing from those of gymnosperms. Judged from their translocation potentials, pore development by widening of individual plasmodesmata combined with the possibility of wound-induced occlusion of wide pores by protein material as encountered in some pteridophytes, is probably advantageous compared with the elaborate plasmodesma-like pores connected via ER-filled median cavities of gymnosperms. From this point of view the specialization of gymnosperm sieve elements would certainly represent an evolutionary sidetrack (cf. Behnke, 1986).

3.5 Parenchymatic elements

Parenchymatic elements typically co-occur with all sieve elements: in the phloem of vascular plants (complying with Nägeli's 1858 definition), as well as in the leptome of bryophytes and irrespective of the developmental stage of the tissue. They thus conform to the idea that translocation in an enucleate sieve element may only be maintained with the support of a nucleate cell. The term 'parenchymatic element' comprises a wide variety of nucleate phloem cells of clearly graduated relationships to sieve elements. Their distribution amongst the different plant taxa (with the most phylogenetically advanced sieve elements being associated with the most specialized, ontogenetically related parenchyma cell) again corroborates the concept of a functional relationship. Although in some plants it is often difficult to assign a parenchymatic element of phloem to a specific category, three types are commonly classified, namely: the phloem parenchyma cell, the companion cell, and the Strasburger cell.

3.5.1 Phloem parenchyma cells

Phloem parenchyma cells are generally held to be the least specialized but also the least homogeneous among the three types of parenchymatic elements classified. They comprise a wide range of cell types, distinguished: (a) by their origin and hence relation to the sieve elements; (b) in secondary phloem by their origin and position within the axial or ray system; and (c) by their ergastic content, i.e. storage of starch or accumulation of tannins, oils, and (oxalate) crystals. It is, therefore, sometimes difficult to portray all of them through their common features.

Phloem parenchyma cells are interconnected to each other and to companion cells or Strasburger cells by plasmodesmata arranged in primary pit fields or distributed all over the wall area. They are nucleate cells which usually contain large vacuoles and a protoplast not visibly specialized other than for storage or accumulation of different material (Figs. 3.2A, 3.4A).

Unrelated parenchyma cells derive from a procambial or cambial (fusiform or ray) initial different from that of an adjacent sieve element, to which they are connected only via companion cells or Strasburger cells. Since storage of assimilates retrieved from the sieve elements is their major function, chloroplasts and amyloplasts are commonly found.

Related parenchyma cells differ from the unrelated by their structural, functional, and often (if not always) also ontogenetic relationships to the sieve elements. They are connected to the sieve elements by plasmodesmata, which within the sieve element side of the common wall are widened into sieve pores (see Sect. 3.5.2). The functional difference from unrelated cells may be expressed by the absence of starch in their plastids (loss of storage function) and often also by their total collapse at the time when the associated sieve element ceases to function. Ontogenetically related parenchyma cells derive from the same precursor (phloem mother cell) but prior to the formation of companion cells.

It is obvious from the foregoing that an exact assignment of a parenchyma cell to one of the variations defined by its degree of relationship to a sieve element can only be made by an analysis of ontogenetic relationships, e.g. through a comparison of cellular development in serial sections. Ultrastructural studies are inadequate to detect more than structural relations.

Phloem parenchyma cells of some angiosperms may contain P-protein, tubular as well as filamentous. It has been suggested (but never proven) that its synthesis is dependent on a specific population of vesicles – named spiny vesicles and different from coated vesicles – found at the same time as the P-protein is being synthesized (e.g., Newcomb, 1967; Cronshaw & Esau, 1968; Steer & Newcomb, 1969).

In most families of the dicotyledon order Capparales phloem parenchyma cells contain dilated cisternae of granular ER which are filled with tubular or filamentous proteins (see Jørgensen, 1981).

3.5.2 Companion cells

Of all parenchymatic elements companion cells have the most intimate structural and functional and the closest ontogenetic relations to the sieve element. Typically, the last (unequal) division of a procambial or cambial derivative gives rise to a larger sieve element and a narrow companion cell. The companion cell may subsequently divide by anticlinal divisions to give rise to a companion cell strand or it may accompany the sieve element undivided. Relative to the longitudinal extension of its sister sieve element a companion cell (or strand) may be very short (e.g. if the companion cell is cut out at one edge of the common mother cell) or of almost the same length. However, irrespective of how extended the contact between the two cell types may be, their functional interdependence is always documented by the fact that both cease to function and die at the same time.

The functional relationships are supported by very intensive and specific plasmatic connections developing from a plasmodesma which has a single branch in the sieve-element side of the common wall and several branches in the companion-cell side (Fig. 3.3C). The single branch widens into a sieve pore and connects to the branches on the companion-cell side via a median cavity. While the desmotubule in the developing sieve pore disintegrates, it has been suggested that the desmotubules of the other branches – disconnected from the sieve pore part – apparently fuse with the widened ER-tubule of the median cavity (Fig. 3.7C). Companion cells are interconnected to each other and with phloem parenchyma cells by normal (usually unbranched) plasmodesmata.

Companion cells typically contain rather large nuclei, often occupying comparatively great portions of the cell volume, numerous elongated mitochondria and plastids and an abundance of free ribosomes, which together are responsible for the extraordinary density of their protoplasts, distinguishing them not only from the associated sieve elements but also from co-occurring phloem parenchyma cells (Figs. 3.2A, 3.4A, 3.7B,D, 3.9, 3.10). Another distinctive feature is the absence of starch in their plastids. In general, companion cell plastids are poorly differentiated plastids which, like sieve element plastids, lack chlorophylls and hence may be filed among photosynthetically inactive leucoplasts (see Behnke, 1975b, for more details).

Histochemical tests performed at the light microscopy level indicated a very strong acid phosphatase activity in the companion cells. EM histochemistry, in addition to verifying the presence of an acid phosphatase, revealed peroxidase and ATPase activities. The latter was preferentially localized in the plasmalemma and supports the idea of an ATP-driven proton-sucrose co-transport through the plasmalemma (cf. Cronshaw, 1981; Sjolund & Shih, 1983a; but see Katz *et al.* 1988).

The occurrence of companion cells is restricted to the phloem of angiosperms and is therefore often used as a second criterion for the characterization of their sieve elements (namely, sieve tube members). Although generally present in both primary and secondary phloem, companion cells have been reported absent from protophloem of several species. An anomalous pattern has also been observed in the root protophloem of several grasses, where the first protophloem sieve element may be associated with two (companion?) cells derived from the same precursor as the sieve element, while another (or two) sieve elements developing next to it will differentiate without further division and hence lack any associate cells. In *Aegilops comosa* Eleftheriou and Tsekos (1982b) were able to trace the development of the first root protophloem sieve elements and predict the position of the new cell walls by the localization of pre-prophase bands of cortical microtubules.

3.5.3 *Strasburger cells*

Strasburger cells – also called albuminous cells (translated from 'Eiweiss-zellen', the name given by Strasburger, 1891, when he first described these cells) – are the functional equivalents of companion cells in the gymnosperm phloem and, although not usually ontogenetically related, like companion cells they cease to function and die concomitant with the associate sieve element. Within the different gymnosperm taxa Strasburger cells may occur in either the axial or ray system or in both: in the axial system as part of declining tiers and in the ray system as 'erect cells' at the margins of rays.

Like companion cells Strasburger cells contain a densely staining protoplast and high acid phosphatase activity. Large nuclei, numerous mitochondria, a rather high ribosome population and a dense granular material of unknown origin account for their overall electron-dense structure in EM micrographs. Two criteria, however, are helpful in identifying Strasburger cells and separating them from other parenchyma cells, namely: the intimate contact with sieve elements via plasmodesmata/sieve area connections, and the absence of starch in their poorly differentiated leucoplasts.

The plasmatic connections between Strasburger cells and sieve elements derive from branched plasmodesmata which (unlike the situation in a companion cell/sieve element connection) penetrate *both* halves of the common cell wall. The plasmodesmatal branches on the sieve element side develop into sieve pores, which by their grouping typically resemble one half of a gymnosperm sieve area. The sieve pores are connected to the plasmodesmatal branches via a median cavity. ER tubules are continuous from the sieve element through the sieve pores and median cavity and (via desmotubules?) plasmodesmatal branches to the preferentially cisternal ER of the Strasburger cell (Fig. 3.12B; for more details see Sauter, 1980; Schulz & Behnke, 1987).

3.5.4 *The transfer cell condition*

Transfer cells, a term that was introduced by Gunning *et al.* (1968) in the description of some minor vein phloem parenchyma cells, now more generally applies to cells which, by structural modifications, are specialized for short-distance translocation; they are most effectively adapted for the exchange of translocates between the apoplast and the symplast.

The general structure of transfer cells is dominated by cell wall ingrowths (also called trabeculae, cf. Fig. 3.7B,D), which are visible in routine preparations for light microscopy. The application of histochemistry and various EM techniques reveals that wall ingrowths are rich in cellulose and pectin, provide ample space for the penetration of solutes, between their microfibrillar substructure, but differ in size, number and distribution amongst different cell types. Wall ingrowths multiply the surface area of the plasmalemma considerably (an amplification of up to 20-fold compared with equivalent walls without ingrowths has been estimated), and thus considerably enhance transmembrane transport capacities.

Transfer cells occur in a great variety of situations throughout the plant body and mediate extensive solute exchanges between cells and tissues which on physiological, developmental or genetic reasons are not able to communicate via the symplast. Amongst vascular transfer cells those present in minor veins of angiosperm leaves have been studied most extensively. A systematic survey of minor vein transfer cells lists 22 (prefentially herbaceous) dicotyledon families and one monocotyledon family (from almost 200 investigated), which were proven to have at least one species with at least one type of transfer cell (Pate & Gunning, 1969). Two parenchymatic elements of minor vein phloem may be modulated to the transfer cell condition and hence attributed specific functions, namely, the companion cell (Fig. 3.7B) and the phloem parenchyma cell (see Sect. 3.6.1).

3.6 Special phloem structures

3.6.1 Minor veins

The phloem of leaf minor veins is distinct from that of other plant parts by an inversion of the size relationships between sieve elements and associated parenchymatic elements, i.e. in minor veins sieve elements are considerably smaller than companion cells. It has been suggested that both companion cells and phloem parenchyma cells of minor veins have specific functions in the loading of photoassimilates and the retrieval of solutes from the apoplast.

The structural differences between parenchymatic elements of phloem, usually illustrated by their protoplasmic contents (see Sect. 3.5), are intensified by a specific distribution of trabeculae in plants developing different types of transfer cell. In minor vein phloem two types of transfer cell were identified, and named A-type and B-type, by Pate & Gunning (1969).

A-type transfer cells are modified companion cells, easily distinguishable from other types by their extremely dense protoplast, typical plasmatic connections with the sieve element, and the arrangement of trabeculae, which are almost evenly distributed over the entire cell wall. A-type transfer cells contain chloroplasts with starch grains, uncommon in other companion cells of other plants parts but also found in minor vein companion cells not transformed into the transfer cell condition.

B-type transfer cells, on the other hand, are looked upon as modified phloem parenchyma cells. They contain a less dense protoplast and do not elaborate plasmatic connections to sieve elements. In addition, the obvious polarity in the formation of wall ingrowths, documented by their restriction to wall parts abutting on sieve elements as well as to those shared with sieve-element-associated A-type cells, is a major morphological character of all B-type transfer cells.

3.6.2 Nodal phloem

In the nodes of stems, phloem and xylem of leaf traces are connected to the stem vascular system. While the number of leaf vascular bundles entering a stem and the number of internodes passed before they find contact are taxon-specific, the vascular elements of the leaf trace are almost always directly integrated or connected with their counterparts of the stem bundles. An interesting exception, implying anatomical and physiological consequences, has become known for the nodes of the monocotyledon family Dioscoreaceae.

In the nodes of *Dioscorea* each sieve tube of a leaf trace is connected to a sieve tube of the vine via specific anastomosing sieve elements. Such an anastomosis (also called phloem glomerulus) is composed of at least six symmetrically ordered layers of sieve elements interpolated between arriving and departing sieve tubes. From the periphery to the centre of an anastomosis the number of sieve elements in the different layers consecutively increases, whilst their sizes diminish. The innermost two layers consist of 5–10 μm wide, spindle-shaped nodal sieve elements characterized by thin walls which are penetrated all over by plasmodesma-like small sieve pores (Fig. 3.1.D,G). In their mature state, all sieve elements within the nodal anastomoses are enucleate, and in addition to a continuous plasmalemma (also lining all sieve pores) contain mitochondria, specific P-type sieve element plastids, a modified ER, and P-protein (cf. Behnke, 1965).

A nodal anastomosis contains different categories of sieve elements connected in series and so far unique in the plant kingdom: 'sieve tube members' associated with companion cells at the periphery, 'sieve cell'-like elements in the centre which are not accompanied by parenchymatic elements, and sieve elements interpolated between the two and showing intermediate characters.

Certainly, whatever the mechanism of phloem translocation and its structural limitations might be, it should be considered that the vigorously translocating lianas in the Dioscoreaceae have overcome any obstacle – e.g. imposed by the extremely small sieve pores – in their nodal anastomoses.

3.6.3 Wound phloem

The formation of wound phloem can be induced in many plants by experimentally interrupting different vascular bundles or even the entire vascular system. The complete severance of the latter has above all been used to monitor the structural and functional differentiation of new sieve elements and thus shape a model system for the timing of the different ontogenetic steps leading to an actively translocating sieve element.

Vascular stumps of a completely interrupted vascular system of pea roots are reconnected by wound sieve elements differentiated from procambial, endodermal and cortex cells which resume or regain mitotic activities. In general, after one to several preliminary divisions cutting off phloem parenchyma cells, a sieve element mother cell is formed, which after an unequal division gives rise to a wound sieve element and its companion cell. Although arising from a completely different situation, the differentiation of wound sieve elements follows exactly that of normal sieve elements, but includes some unusual features like the formation of P-type sieve element plastids from starch-storing amyloplasts and derivation of sieve pores from branched plasmodesmata combined via a median cavity to a primary pit field.

The development of a complete bridge of wound sieve elements starts by adding on new sieve elements to both ends of the interrupted vascular system and radiating into the cortex, i.e. at least for the apical root part a reversal of the direction of translocation is required. About 60 h after wounding a first continuous series of wound sieve elements bridging the interruption can be visualized using callose-specific aniline blue fluorescence for detection of sieve plates. About 10 h later – obviously the time required for the maturation of the last formed wound sieve elements – the fluorescent dye fluoresceine is first visualized to be translocated through the newly formed bridge (for details and literature see Schulz, 1987).

Acknowledgements

The micrographs used in this chapter and many unpublished details derived from the author's electron microscopy survey of the phloem of seed plants supported by grants from the Deutsche Forschungsgemeinschaft Bonn (FRG) and the Projekt Europäisches Forschungszentrum für Massnahmen der Luftreinhaltung (PEF), Karlsruhe (FRG).

References

(For more information and other aspects of phloem anatomy, see Esau, 1969.)

Arsanto, J.-P. (1982). Observations on P-protein in dicotyledons. Substructural and developmental features. *American Journal of Botany* **69**, 1200–12.
Arsanto, J.-P. & Coulon, J. (1974). Détections radio-autographique et cytochimique des sites d'élaboration ou du transit des précurseurs polysacchari-

diques parieteaux dans les cellules criblées en cours de différenciation du metaphloème caulinaire de deux Cucurbitacées voisins (*Cucurbita pepo* L. et *Ecballium elaterium* R.). *Compte rendu hebdomadaire des séances de l'Académie des sciences* **D278**, 2775–8.

Behnke, H.-D. (1965). Über das Phloem der Dioscoreaceaen unter besonderer Berücksichtigung ihrer Phloembecken. *Zeitschrift für Pflanzenphysiologie* **53**, 97–125, 214–44.

Behnke, H.-D. (1971). Über den Feinbau verdickter (nacré) Wände und der Plastiden in den Siebröhren von *Annona* und *Myristica*. *Protoplasma* **72**, 69–78.

Behnke, H.-D. (1973). Strukturänderungen des Endoplasmatischen Reticulums und Auftreten von Proteinfilamenten während der Siebröhrendifferenzierung bei *Smilax excelsa*. *Protoplasma* **77**, 279–89.

Behnke, H.-D. (1974a). Comparative ultrastructural investigations of angiosperm sieve elements: aspects of the origin and early development of P-protein. *Zeitschrift für Pflanzenphysiologie* **74**, 22–34.

Behnke, H.-D. (1974b). Sieve-element plastids of Gymnospermae: Their ultrastructure in relation to systematics. *Plant Systematics and Evolution* **123**, 1–12.

Behnke, H.-D. (1975a). P-type sieve-element plastids: A correlative ultrastructural and ultrahistochemical study on the diversity and uniformity of a new reliable character in seed plant systematics. *Protoplasma* **83**, 91–101.

Behnke, H.-D. (1975b). Companion cells and transfer cells. In Aronoff, S., Dainty, J., Gorham, P. R., Srivastava, L. M. & Swanson, C. A. (eds.). *Phloem Transport*, pp. 153–75. New York, Plenum Press.

Behnke, H.-D. (1981). Sieve-element characters. *Nordic Journal of Botany* **1**, 381–400.

Behnke, H.-D., (1986). Sieve element characters and the systematic position of *Austrobaileya* – with comments to the distinction and definition of sieve cells and sieve-tube members. *Plant Systematics and Evolution* **152**, 101–21.

Behnke, H.-D. (1988). Sieve-element plastids, phloem proteins, and the evolution of flowering plants: III. Magnoliidae. *Taxon* **37**, 699–732.

Behnke, H.-D. & Dörr, I. (1967). Zur Herkunft und Struktur der Plasmafilamente in Assimilatleitbahnen. *Planta* **74**, 18–44.

Behnke, H.-D. & Paliwal, G. S. (1973). Ultrastructure of phloem and its development in *Gnetum gnemon*, with some observations on *Ephedra campylopoda*. *Protoplasma* **78**, 305–19.

Behnke, H.-D. & Pop, L. (1981). Sieve-element plastids and crystalline P(phloem)-protein in Leguminosae: micromorphological characters as an aid to the circumscription of the family and subfamilies. In Polhill, R. M. & Raven, P. H. (eds.). *Advances in Legume Sytematics*, pp. 707–15. London, Academic Press.

Beyenbach, J., Weber, C. & Kleining, H. (1974). Sieve-tube proteins from *Cucurbita maxima*. *Planta* **119**, 113–24.

Briosi, G. (1873). Über allgemeines Vorkommen von Stärke in den Siebröhren. *Botanische Zeitung* **31**, 306–14, 322–34, 338–44.

Burr, F. A. & Evert, R. F. (1973). Some aspects of sieve-element structure and development in *Selaginella kraussiana*. *Protoplasma* **78**, 81–97.

Buvat, R. (1963). Les infrastructures et la différenciation des cellules criblées de *Cucurbita pepo*. *Portugaliae Acta Biologica* 7, 249–99.

Chafe, S. C. & Doohan, M. E. (1972). Observations on the ultrastructure of the thickened sieve cell wall in *Pinus strobus* L. *Protoplasma* **75**, 67–78.

Cheadle, V. I. (1948). Observations on the phloem in the Monocotyledoneae. II. Additional data on the occurrence and phylogenetic specialization in structure of the sieve tubes in the metaphloem. *American Journal of Botany* **35**, 129–31.

Cheadle, V. I. & Whitford, N. B. (1941). Observations on the phloem in the Monocotyledoneae. I. The occurrence and types and phylogenetic specialization in structure of the sieve tubes in the metaphloem. *American Journal of Botany* **28**, 623–7.

Crafts, A. S. (1932). Phloem anatomy, exudation, and transport of organic nutrients in cucurbits. *Plant Physiology* **7**, 183–225.

Cronshaw, J. (1975a). P-proteins. In Aronoff, S., Dainty, J., Gorham, P. R., Srivastava, L. M. & Swanson, C. A. (eds.). *Phloem Transport*, pp. 79–115. New York, Plenum Press.

Cronshaw, J. (1975b). Sieve element cell walls. In Aronoff, S., Dainty, J., Gorham, P. R., Srivastava, L. M. & Swanson, C. A. (eds.). *Phloem Transport*, pp. 129–47. New York, Plenum Press.

Cronshaw, J. (1981). Phloem structure and function. *Annual Review of Plant Physiology* **32**, 465–84.

Cronshaw, J. & Esau, K. (1967). Tubular and fibrillar components of mature and differentiating sieve elements. *Journal of Cell Biology* **34**, 801–15.

Cronshaw, J. & Esau, K. (1968). P-protein in the phloem of Cucurbita. I. The development of P-protein bodies. *Journal of Cell Biology* **38**, 25–39.

Danilova, M. F. & Telepova, M. N. (1980). Distinctive features of differentiation in proto- and metaphloem sieve elements of barley root. *Phytomorphology* **30**, 380–7. [Issued 1982]

Deshpande, B. P. (1976). Observations on the fine structure of plant cell walls. III. The sieve tube wall in *Cucurbita. Annals of Botany* **40**, 443–6.

Deshpande, B. P. & Evert, R. F. (1970). A re-evaluation of extruded nucleoli in sieve elements. *Journal of Ultrastructure Research* **33**, 483–94.

Eleftheriou, E. P. (1986). Ultrastructural studies on protophloem sieve elements in *Triticum aestivum* L. Nuclear degeneration. *Journal of Ultrastructure Research* **95**, 47–60.

Eleftheriou, E. P. (1987). Microtubules and cell wall development in differentiating protophloem sieve elements of *Triticum aestivum* L. *Journal of Cell Science* **87**, 595–607.

Eleftheriou, E. P. & Tsekos, I. (1982a). Developmental features of cell wall formation in sieve elements of the grass *Aegilops comosa* var. thessalica. *Annals of Botany* **50**, 519–29.

Eleftheriou, E. P. & Tsekos, I. (1982b). Development of protophloem in roots of *Aegilops comosa* var. thessalica. I. Differential divisions and pre-prophase bands of microtubules. *Protoplasma* **113**, 110–19.

Eleftheriou, E. P. & Tsekos, I. (1982c). Development of protophloem in roots of *Aegilops comosa* var. thessalica. II. Sieve-element differentiation. *Protoplasma* **113**, 221–33.

Esau, K. (1947). A study of some sieve-tube inclusions. *American Journal of Botany* **34**, 224–33.

Esau, K. (1969). The Phloem. In Zimmerman, W. *et al.* (eds.). *Encyclopedia of Plant Anatomy* Vol. 5, part 2, 505 pp. Berlin, Gebrüder Borntraeger.

Esau, K. (1975). The phloem of *Nelumbo nucifera* Gaertn. *Annals of Botany* **39**, 901–13.

Esau, K. (1978). Developmental features of the primary phloem in *Phaseolus vulgaris* L. *Annals of Botany* **34**, 1–13.

Esau, K. & Cheadle, V. I. (1958). Wall thickening in sieve elements. *Proceedings of the National Academy of Sciences, USA* **44**, 546–53.

Esau, K. & Cheadle, V. I. (1959). Size of pores and their contents in sieve elements of dicotyledons. *Proceedings of the National Academy of Sciences, USA* **45**, 156–62.

Esau, K. & Cronshaw, J. (1967). Tubular components in cells of healthy and tobacco mosaic virus-infected *Nicotiana*. *Virology* **33**, 26–35.

Esau, K. & Cronshaw, J. (1968). Plastids and mitochondria in the phloem of *Cucurbita*. *Canadian Journal of Botany* **46**, 877–80.

Esau, K. & Gill, R. H. (1972). Nucleus and endoplasmic reticulum in differentiating root protophloem of *Nicotiana tabacum*. *Journal of Ultrastructure Research* **41**, 160–75.

Esau, K. & Gill, R. H. (1973). Correlations in differentiation of protophloem sieve elements of *Allium cepa* root. *Journal of Ultrastructure Research* **44**, 310–28.

Esau, K. & Magyarosy, A. G. (1979). A crystalline inclusion in sieve element nuclei of *Amsinckia*. II. The inclusion in maturing cells. *Journal of Cell Science* **38**, 11–22.

Evert, R. F. (1984). Comparative structure of phloem. In White, R. A. & Dickison, W. C. (eds.). *Contemporary Problems in Plant Anatomy*, pp. 145–234. New York, Academic Press.

Evert, R. F., Bornman, C. H., Butler, V. & Gilliland, M. G. (1973). Structure and development of the sieve cell protoplast in leaf veins of *Welwitschia*. *Protoplasma* **76**, 1–21.

Evert, R. F. & Deshpande, B. P. (1970). Nuclear P-protein in sieve elements of *Tilia americana*. *Journal of Cell Biology* **44**,, 462–6.

Evert, R. F. & Eichhorn, S. E. (1974). Sieve-element ultrastructure in *Platycerium bifurcatum* and some other polypodiaceous ferns: The nucleus. *Planta* **119**, 301–18.

French, J. C. & Tomlinson, P. B. (1986). Compound vascular bundles in monocotyledonous stems: construction and significance. *Kew Bulletin* **41**, 561–74.

Gunning, B. E. S., Pate, J. S. & Briarty, L. G. (1968). Specialized 'transfer cells' in minor veins of leaves and their possible significance in phloem translocation. *Journal of Cell Biology* **37**, C7–C12.

Haberlandt, G. F. J. (1879). *Entwicklungsgeschichte des mechanischen Gewebesystems der Pflanzen*. Leipzig, Engelmann.

Hartig, T. (1837). Vergleichende Untersuchungen über die Organisation des Stammes der einheimischen Waldbäume. *Jahresbericht Forstwirtschaftliche und Forstliche Naturkunde* **1**, 125–68.

Hébant, C. (1975). Lack of incorporation of tritiated uridine by nuclei of mature sieve elements in *Metasequoia glyptostroboides* and *Sequoiadendron giganteum*. *Planta* **126**, 161–3.

Hébant, C. (1977). *The Conducting Tissues of Bryophytes*. Vaduz, Cramer.

Holdheide, W. (1951). Anatomie mitteleuropäischer Gehölzrinden. In Freund, H. (ed.). *Handbuch der Mikroskopie in der Technik* Vol. 5, Part 1, pp. 193–367. Frankfurt, Umschau-Verlag.

Jørgensen, L. B. (1981). Myrosin cells and dilated cisternae of the endoplasmic reticulum in the order Capparales. *Nordic Journal of Botany* **1**, 433–45.

Katz, D. B., Sussman, M. R., Mierzwa, R. J. & Evert, R. F. (1988). Cytochemical localization of ATPase activity in oat roots localizes a plasma membrane-associated soluble phosphatase, not the proton pump. *Plant Physiology* **86**, 841–47.

Kollmann, R. (1960). Untersuchungen über das Protoplasma der Siebröhren von *Passiflora coerulea*. II. Mitteilung: Elektronenoptische Untersuchung. *Planta* **55**, 67–107.

Kollmann, R. (1980). Fine structural and biochemical characterization of phloem proteins. *Canadian Journal of Botany* **58**, 802–6.

Kollmann, R., Dörr, I. & Kleinig, H. (1970). Protein filaments – Structural components of the phloem exudate. I. Observations with *Cucurbita* and *Nicotiana*. *Planta* **95**, 86–94.

Kollmann, R. & Schumacher, W. (1963). Über die Feinstruktur des Phloems von *Metasequoia glyptostroboides* und siene jahreszeitlichen Veränderungen. IV. Weitere Beobachtungen zum Feinbau der Plasmabrücken in den Sieb-zellen. *Planta* **60**, 360–89.

Kollmann, R. & Schumacher, W. (1964). Über die Feinstruktur des Phloems von *Metasequoia glyptostroboides* und seine jahreszeitlichen Veränderungen. V. Die Differenzierung der Siebzellen in Verlaufe einer Vegetationsperiode. *Planta* **63**, 155–90.

Kruatrachue, M. & Evert, R. F. (1974). Structure and development of sieve elements in the leaf of *Isoetes muricata*. *American Journal of Botany* **61**, 253–66.

Kruatrachue, M. & Evert, R. F. (1977). The lateral meristem and its derivatives in the corm of *Isoetes muricata*. *American Journal of Botany* **64**, 310–25.

LaFlèche, D. (1966). Ultrastructure et cytochimie des inclusions flagellées des cell-ules criblées de *Phaseolus vulgaris*. *Journal de Microscopie* **5**, 523–56.

Lawton, D. M. (1978). P-protein crystals do not disperse in uninjured sieve elements in roots of runner bean (*Phaseolus multiflorus*) fixed with glutar-aldehyde. *Annals of Botany* **42**, 353–61.

Lawton, D. M. & Johnson, R. P. C. (1976). A superhelical model for the ultra-structure of 'P-protein tubules' in sieve elements of *Nymphoides peltata*. *Cytobiologie* **14**, 1–17.

Liberman-Maxe, M. (1978). La paroi des cellules criblées dans le phloème d'une Fougère, le Polypode. *Biologie Cellulaire* **31**, 201–10.

Nägeli, C. (1858). Das Wachstum des Stammes und der Wurzel bei den Gefässpflanzen und die Anordnung der Gefässtrange im Stengel. *Beiträge zur Wissenschaftliche Botanik* **1**, 1–156.

Nehls, R., Schaffner, G. & Kollmann, R. (1978). Feinstruktur des Protein-Einschlusses in den Siebelementen von *Salix sachalinensis* Fr. Schmidt. *Zeitschrift für Pflanzenphysiologie* **87**, 113–27.

Neuberger, D. S. & Evert, R. F. (1974). Structure and development of the sieve element protoplast in the hypocotyl of *Pinus resinosa*. *American Journal of Botany* **61**, 360–74.

Neuberger, D. S. & Evert, R. F. (1976). Structure and development of sieve cells in the primary phloem of *Pinus resinosa*. *Protoplasma* **87**, 27–37.

Newcomb, E. H. (1967). A spiny vesicle in slime-producing cells of the bean root. *Journal of Cell Biology* **35**, C17–C22.

Oberhäuser, R. & Kollmann, R. (1977). Cytochemische Charakterisierung des sogenannten Freien Nucleolus als Proteinkörper in den Siebelementen von *Passiflora coerulea*. *Zeitschrift für Pflanzenphysiologie* **84**, 61–75.

Pacini, E. & Cresti, M. (1972). Dégénérescence nucléaire dans les elements criblés de l'ovule d'*Eranthis hiemalis* (L.) Salisb. *Compte rendu hebdomadaire des séances de l'Académie des Sciences* **274**, 859–61.

Palevitz, B. A. & Newcomb, E. H. (1970). A study of sieve element starch using sequential enzymatic digestion and electron microscopy. *Journal of Cell Biology* **45**, 383–98.

Palevitz, B. A. & Newcomb, E. H. (1971). The ultrastructure and development of tubular and crystalline P-protein in the sieve elements of certain papi-lionaceous legumes. *Protoplasma* **72**, 399–426.

Parthasarathy, M. V. (1980). Mature phloem of perennial monocotyledons. *Berichte der Deutschen botanischen Gesellschaft* **93**, 57–70.

Parthasarathy, M. V. & Miihlethaler, K. (1969). Ultrastructure of protein tubules in differentiating sieve elements. *Cytobiologie* **1**, 17–36.

Parthasarathy, M. V. & Pesacreta, T. C. (1980). Microfilaments in plant vascular cells. *Canadian Journal of Botany* **58**, 807–15.

Pate, J. S. & Gunning, B. E. S. (1969). Vascular transfer cells in angiosperm leaves. A taxonomic and morphological survey. *Protoplasma* **68**, 135–56.

Read, S. M. & Northcote, D. H. (1983). Subunit structure and interactions of the phloem proteins of *Cucurbita maxima* (Pumpkin). *European Journal of Biochemistry* **134**, 561–9.

Resch, A. (1954). Beiträge zur Cytologie des Phloems: Entwicklungsgeschichte der Siebröhrenglieder und Geleitzellen bei *Vicia faba* L. *Planta* **44**, 75–89.

Sabnis, D. D. & Hart, J. W. (1978). The isolation of some properties of a lectin (haemagglutinin) from *Cucurbita* phloem exudate. *Planta* **142**, 97–101.

Salmon, J. (1946/47). Différenciation des tubes criblés chez les Angiospermes: Recherches cytologiques. *Revue Cytologie et Cytophysiologie Végétale* **9**, 55–168.

Sauter, J. J. (1980). The Strasburger cells – equivalents of companion cells. *Berichte der Deutschen botanischen Gesellschaft* **93**, 29–42.

Schulz, A. (1987). Sieve-element differentiation and fluoresceine translocation in wound-phloem of pea roots after complete severance of the stele. *Planta* **170**, 289–99.

Schulz, A. & Behnke, H.-D. (1987). Feinbau und Differenzierung des Phloems von Buchen, Fichten und Tannen aus Waldschadensgebieten KfK-PEF 16 (II + 95 pp.). Karlsruhe, Kernforschungszentrum.

Schumacher, W. (1930). Untersuchungen über die Lokalisation der Stoffwanderung in den Leitbündeln höherer Pflanzen. *Jahrbuch für wissenschaftliche Botanik* **73**, 770–823.

Schumacher, W. (1939). Über die Plasmolysierbarkeit der Siebröhren. *Jahrbuch für wissenschaftliche Botanik* **88**, 545–53.

Sjolund, R. D. & Shih, Y. C. (1983a). Freeze-fracture analysis of phloem structure in plant tissue cultures. I. The sieve element reticulum. *Journal of Ultrastructure Research* **82**, 111–21.

Sjolund, R. D. & Shih, Y. C (1983b). Freeze-fracture analysis of phloem structure in plant tissue cultures. II. The sieve element plasma membrane. *Journal of Ultrastructure Research* **82**, 189–97.

Sjolund, R. D., Shih, Y. C. & Jensen, K. G. (1983). Freeze-fracture analysis of phloem structure in plant tissue cultures. III. P-protein, sieve area pores and wounding. *Journal of Ultrastructure Research* **82**, 198–211.

Smith, L. M., Sabnis, D. D. & Johnson, R. P. C. (1987). Immunocytochemical localization of phloem lectin from *Cucurbita maxima* using peroxidase and colloidal-gold labels. *Planta* **170**, 461–70.

Spanner, D. C. (1978). Sieve-plate pores, open or occluded? A critical review. *Plant, Cell and Environment* **1**, 7–20.

Steer, M. W. & Newcomb, E. H. (1969). Development and dispersal of P-protein in the phloem of *Coleus blumei* Benth. *Journal of Cell Science* **4**, 155–69.

Strasburger, E. (1891). Über den Bau und die Verrichtungen der Leitungsbahnen in den Pflanzen. *Histologische Beiträge* 3, xxxii + 1000 pp. Jena, Fischer.

Tamulevich, S. R. & Evert, R. F. (1966). Aspects of sieve element ultrastructure in *Primula obconica*. *Planta* **69**, 319–37.

Thorsch, J. & Esau, K. (1981a). Changes in the endoplasmic reticulum during differentiation of a sieve element in *Gossypium hirsutum*. *Journal of Ultrastructure Research* **74**, 183–94.

Thorsch, J. & Esau, K. (1981b). Nuclear degeneration and the association of endoplasmic reticulum with the nuclear envelope and microtubules in maturing sieve elements of *Gossypium hirsutum*. *Journal of Ultrastructure Research* **74**, 195–204.

Thorsch, J. & Esau, K. (1981c). Ultrastructural studies of protophloem sieve elements in *Gossypium hirsutum*. *Journal of Ultrastructure Research* **75**, 339–51.

Thorsch, J. & Esau, K. (1982). Microtubules in differentiating sieve elements of *Gossypium hirsutum*. *Journal of Ultrastructure Research* **78**, 73–83.

Walsh, M. A. & Melaragno, J. E. (1976). Ultrastructural features of developing sieve elements in *Lemna minor* L. – Sieve plate and lateral sieve areas. *American Journal of Botany* **63**, 1174–83.

Warmbrodt, R. D. (1980). Characteristics of structure and differentiation in the sieve element of lower vascular plants. *Berichte der Deutschen botanischen Gesellschaft* **93**, 11–28.

Weatherley, P. E. (1975). Some aspects of the Münch hypohesis. In Aronoff, S., Dainty, J., Gorham, P. R., Srivastava, L. M. & Swanson, C. A. (eds.). *Phloem Transport*, pp. 535–55. New York, Plenum Press.

Wergin, W. P. & Newcomb, E. H. (1970). Formation and dispersal of crystalline P-protein in sieve elements of soybean (*Glycine max. L.*). *Protoplasma* **71**, 365–88.

Zahur, M. S. (1959). Comparative study of secondary phloem of 423 species of woody dicotyledons belonging to 85 families. *Cornell University Agricultural Experimental Station Memoirs* 358.

Zee, S-Y. & Chambers, T. C. (1969). Development of the secondary phloem of the primary root of *Pisum*. *Australian Journal of Botany* **17**, 119–214.

4 Origin, destination and fate of phloem solutes in relation to organ and whole plant functioning

J. S. Pate

4.1 Introduction

To develop a proper understanding of the translocatory process in plants one requires access to a number of sources of information. Firstly, inventories are required of the types and amounts of solutes moving within specific segments of the phloem network. Secondly, directionalities and magnitudes of flow between elements of the system have to be prescribed. Thirdly, identification has to be made of the sites and mechanisms of origin and consumption of specific groups of phloem solutes, and, finally, an overall appreciation needs to be gained of the impact of such traffic on the nutritional interrelationships between relevant groups of donor and receptor organs.

When attempting to achieve these objectives it is highly desirable that the translocating tissues of the species under study can be induced to yield sizeable amounts of phloem exudate, for one can then conduct definitive experiments involving application of appropriate tracer substrates to a presumed source region of a plant and subsequent recovery of labelled phloem contents at selected points along the translocation pathway towards a presumptive sink. If combined with examination of any enzymatic transformations of a labelled substrate which are occurring prior to or after phloem transport, a particularly clear picture is then obtained of the particular transfer processes unde investigation and their relevance to the growth and functioning of the parent plant.

Assuming that the above approaches are feasible, a valuable extra dimension to understanding is opened up if the transport phenomena under study can be examined quantitatively in terms of bulk exchanges of major nutrient elements within the system. This may prove difficult, say, in large woody plants in which rates of production, transport and utilization of commodities such as carbon and nitrogen are difficult to measure, if only because of the sheer size and complexity of the plants in question and seasonal asynchronies in their patterns of photosynthesis, root uptake, storage and growth. In short-cycle herbaceous species, however, assimilation and growth are likely to be tightly coupled on a daily or even hourly basis, so that models of flow of major nutrients can be constructed more confidently, using measured rates of uptake and consumption by component organs, sequential assays of transport fluids, and ^{14}C feeding studies to define source:sink linkages at different stages of plant development.

The primary objective of this chapter is to describe and discuss the application of the above-mentioned approaches and experimental strategies to study of the origin and fate of phloem solutes. The sections which follow will present a series of case studies, each highlighting specific principles in the application of techniques and interpretation of data. To minimize complications, only two plant species will be considered – white lupin (*Lupinus albus* L.) and cowpea (*Vigna unguiculata* L. Walp.). These large-seeded annual grain legumes have been investigated in some detail over a number of years in the author's laboratory. White lupin, a most prolific bleeder from cut phloem, produces harvestable drops (0.1–5 μl) of phloem sap from shallow incisions in petioles, inflorescence stalks, fruit stalks and tips, and from virtually any location on the main or lateral shoot axes (see Pate *et al.*, 1974, 1979). Cowpea, by contrast, bleeds consistently from phloem tissue which has been punctured by insertion of a needle cooled in liquid nitrogen – the so-called technique of cryopuncturing (see Pate *et al.*, 1984b). The species performs in this manner only from fruit stalks or from the placental suture of near-mature fruits, so use of this approach for definitive studies of phloem transport has been confined almost entirely to solute mobilization to fruit and seed of reproducing plants.

The cases to be considered in the sections which follow are restricted mainly to transport of C- and N-containing solutes, but translocation of certain minerals will also be described. Both vegetative and fruiting stages of plant development will be studied, in some instances dealing with coupled transport activity in xylem and phloem at whole plant level, in other cases with more specific events such as the cycling of solutes through the leaf, solute transfer from nurse leaf to developing fruit, and the penultimate events whereby phloem-borne solutes are unloaded in the seed coat and then move via the endosperm to the growing embryo. Wherever possible, information from isotopic labelling of phloem donor and receptor tissues is related to parallel measurements of source and sink activity, as monitored by rates of production, throughput and consumption of the solutes or elements in question.

4.2 Case study I: Phloem transport and the partitioning of C and N in the whole non-reproductive plant

The principal objective of this section is to demonstrate how one might proceed to model empirically the differential partitioning of newly assimilated C and N within a plant, and to picture how fluxes of these elements in xylem and phloem relate to the growth and functioning of the plant's currently active donor and receptor organs. The study species for such investigations has been *Lupinus albus*, and the data to be examined in detail here refer to events happening during a specific week of growth immediately following anthesis. Full accounts of the experimental and analytical techniques used in a modelling exercise of this nature are to be found in Pate & Layzell *et al.* (1981) and Pate (1986).

Certain classes of information proved essential for the proper construction of a model depicting phloem exchanges of C and N. Firstly, net changes of C and N in dry matter of plant parts needed to be determined for the interval of study. In the example selected the nodulated root, inflorescence plus young upper lateral shoots were harvested separately from the remaining main shoot, which in turn was split into four separate strata A to D (Fig. 4.1). Each of these strata was then separated into its four adjacent sets of leaflets (L1–L4 Fig. 4.1) and the stem plus petiole tissue (SP1–SP4 Fig. 4.1) relating to these leaves. Secondly, net gaseous losses of carbon by the nodulated root and the above-mentioned shoot fractions were determined – the below-ground parts on a continuous basis, shoot parts only in terms of night respiration. Thirdly, using either a short-term $^{14}CO_2$ feeding technique involving whole shoots of intact plants, or measurements of photosynthetic (CO_2) exchanges of single leaves on the main stem, an indication was obtained of the relative and absolute rates of daytime net photosynthesis by the four strata of foliar surfaces on the shoot. Fourthly, translocatory destinations for photosynthate were determined based on a series of $^{14}CO_2$ feeding experiments involving single leaves at different positions on the shoot. This essentially prescribed which leaves fed the roots, which the shoot apex, and which nourished both of these sink regions. Finally, xylem sap from the cut stumps of nodulated

Fig. 4.1 Data utilized for construction of models of C and N partitioning in 51–58 d symbiotically dependent plants of white lupin (*Lupinus albus*). *A*: Scheme showing how plants were harvested for studying distribution of increments of C and N in dry matter, and where xylem and phloem sap samples were collected during the study interval. *B*: Diagrammatic representation of ratios by weight of carbon to nitrogen fixed by the plants and incorporated into dry matter of plant organs. The arrows relating to phloem transport depict translocatory destinations for exported photosynthate from the different strata of leaves, as revealed by $^{14}CO_2$ feeding experiments. *C*: Ratios by weight of carbon to nitrogen to translocatory streams as determined by analyses of phloem and xylem sap collected daily from a range of sampling points over the study period. (Data from Pate & Layzell 1981; Layzell *et al.*, 1981.)

MODELLING C AND N PARTITIONING IN LUPINUS ALBUS

A.

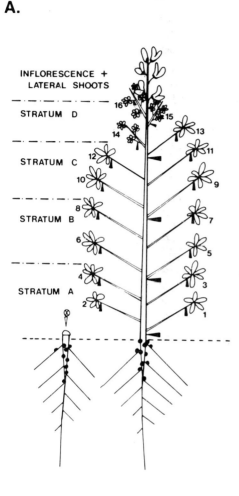

INFLORESCENCE +
LATERAL SHOOTS

STRATUM D

STRATUM C

STRATUM B

STRATUM A

KEY:

COLLECTION SITES

➤ STEM PHLOEM SAP

▲ PETIOLE PHLOEM SAP

⊗
▽ XYLEM SAP

SP-PETIOLES + STEM

L - LEAFLETS

NR - NODULATED
ROOT

B.

C:N RATIOS UTILIZATION OF FIXED C AND N

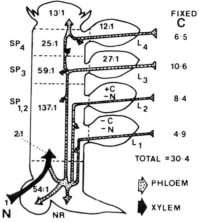

FIXED
C

13:1

12:1   6·5
L_4

SP_4 25:1

27:1   10·6
SP_3 59:1 L_3

+C   8·4
−N
$SP_{1,2}$ 137:1 L_2

−C
−N   4·9
L_1

2:1

TOTAL = 30·4

54:1   PHLOEM

N   NR   XYLEM

C.

C:N RATIOS TRANSPORT FLUIDS

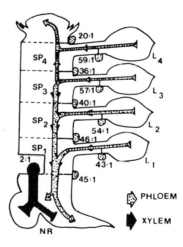

20:1

SP_4 59:1 L_4

36:1

SP_3 57:1 L_3

40:1

SP_2 54:1 L_2

46:1

SP_1 43:1 L_1

2:1

45:1

PHLOEM

NR   XYLEM

CARBON AND NITROGEN FLOW – SYMBIOTICALLY DEPENDENT LUPINUS ALBUS (51–58 DAYS AFTER SOWING)

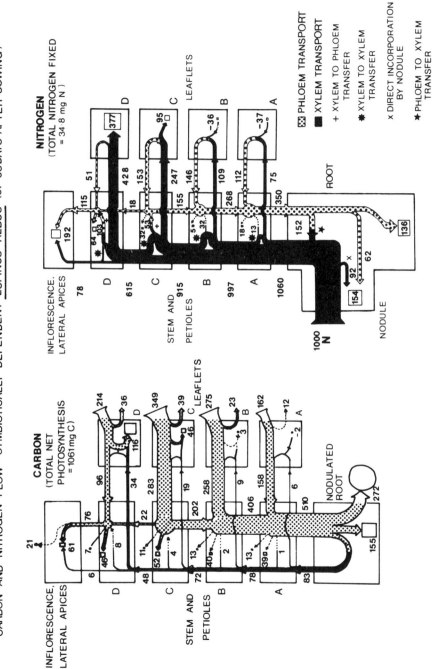

root, and phloem exudates from a range of strategic sites on the shoot system (Fig. 4.1) were collected, and the C:N weight ratio of each sample then determined.

Primary data used to construct the model were essentially as collated in Fig. 4.1. Net carbon and nitrogen production or consumption by plant parts was expressed in terms of ratios (weight basis) of C:N involved per organ for the week of the study (Fig. 4.1B). Translocatory affinities for [14C] photosynthate were as indicated by arrows (Figs. 4.1B and C) and analyses of the sap samples gave values for C:N weight ratios as shown in Fig. 4.1C.

Model construction from the data assumed that mass flow occurred in both xylem and phloem and that the relative amounts of C and N moving in these channels were as indicated by sap analysis. It then proved possible to compute progressively for each reference point within the system the exchanges of C and N through xylem and phloem required to meet the measured rates of consumption or production of C and N by the organs or sets of organs involved. Thus, at each point of interchange in the resulting models (Fig. 4.2) the ratio of C and N in each of the indicated transport segments matched precisely that determined experimentally at the nearest collection point for xylem and phloem exudate. This equivalence applied regardless of whether the organ being studied had been a net importer through both xylem and phloem (e.g. inflorescence and lateral apices), a net importer only through xylem (mature leaves) or through phloem (root), or had engaged in similarly or dissimilarly oriented long-distance xylem and phloem exchanges (segments of main stem).

Certain important principles emerge from the model. Basically, partitioning of nitrogen and carbon involves the co-operative activity of phloem and xylem and is motivated primarily by the physiological activity of leaves. Transpirational loss of water promotes an upward flow of water and solutes from roots through the xylem, while photosynthetic production of sugars provides the driving force for outward multi-directional transport from source leaves to sink organs via the phloem. Coupled to these major longitudinal flow pathways are a number of radially oriented transport components, each involving short-distance exchanges of specified amounts of carbon and nitrogen between adjacent transport channels. Four such components are indicated in the model for N (Fig. 4.2), namely: (1) xylem to xylem transfer within lower nodes of the stem; (2) xylem to phloem

Fig. 4.2 Empirically based models for uptake, flow and utilization of C and N in symbiotically dependent 51–58 d white lupin plants. The models show co-operative involvement of long- and short-distance transport in xylem and phloem, and the series of short-distance cross-transfers involving xylem and/or phloem in the partitioning of both C and N. Further data relating to the study are found in Fig. 4.1. (Data from Pate & Layzell, 1981; Layzell *et al.*, 1981.)

transfer in mature leaves; (3) xylem to phloem transfer in the upper stem; and (4) cycling of shoot-derived C and N from root back to shoot by phloem to xylem transfer.

The models (Fig. 4.2) present in quantitative terms the relative importance of these long- and short-distance pathways in circulation of both C and N. For instance, over half the carbon of net photosynthate produced by a white lupin plant in the period 51–58 d after sowing was shown to pass initially from shoot to root by phloem transport, mostly to be consumed in growth and respiration. However, some carbon, the equivalent of 8% of the plant's total net photosynthate, cycled back to the shoot attached to organic acids and newly formed nitrogenous solutes. Turning to N partitioning, transpiring leaves were found to attract 85% of the N exported from the root. Young leaves at the top of the shoot retained some of this N, but most was transferred within the leaf from xylem to phloem and thus became available to any organs in receipt of photosynthate from that leaf. For instance, four-fifths of the N moving to the root from the shoot consisted of N cycled through older leaves, while the developing inflorescence and lateral shoots developing immediately below this inflorescence gained over one-third of their N requirement cycled through upper leaves in this manner. Xylem to xylem transfer of N at lower nodes and xylem to phloem N transfer in upper reaches of the stem (Fig. 4.2) were seen to supplement significantly any N acquired by apical parts of the shoot, either on the basis of direct phloem translocation from leaves or through the relatively poor ability of these young apical parts to compete for xylem-borne N on the basis of transpirational loss.

The balance between these interacting components was shown to be perturbed in a number of experimental treatments in which inputs of C or N to the system had been artificially altered. For example, in certain experiments N_2 fixation was almost totally inhibited by exposing the enclosed nodulated roots of water-cultured plants to argon:oxygen (80%:20% v/v) (Pate *et al.*, 1984b). Such treatment caused levels of N in the xylem to decline to 10% of their former value within a day of transfer to $Ar:O_2$ (Pate *et al.*, 1984b), and repercussions within phloem transport pathways over the next few days involved a steep rise in C:N ratios of phloem sap of stems and leaves, presumably due to restricted supply of xylem N for phloem loading at the various interchange points within the system. Eventually the C:N balance of phloem was partly restored due to

Fig. 4.3 Models depicting uptake, transport and accumulation of major nutrient cations in nodulated 39–49 d white lupin. Arrow thicknesses are drawn proportional to relative rates of flow within the system and absolute rates of uptake, flow and incorporation are given as μmol gFW^{-1} h^{-1}. Numbered circles refer to: 1 – intake by root; 2 – accumulation by root; 3 – direct incorporation by root after uptake from external medium; 4 – xylem transport; 5 – phloem transport; 6 – accumulation by shoot; 7 – circulation through root by phloem to xylem transfer. (Reproduced with modifications from Jeschke *et al.*, 1985.)

CATION PARTITIONING IN LUPINUS ALBUS

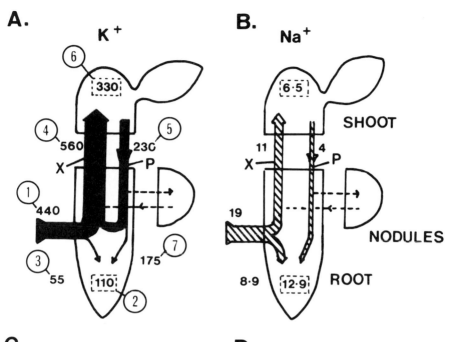

A. K⁺

⑥ ┆330┆

④ 560 ⑤ 230

X P

① 440

③ 55 ┆110┆ ②

⑦ 175

B. Na⁺

┆6·5┆

SHOOT

11 4

X P

19

NODULES

8·9 ┆12·9┆ ROOT

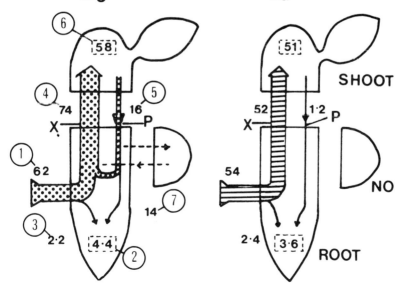

C. Mg⁺⁺

⑥ ┆58┆

④ 74 ⑤ 16

X P

① 62

③ 2·2 ┆4·4┆ ②

⑦ 14

D. Ca⁺⁺

┆51┆

SHOOT

52 1·2

X P

54

NODULES

2·4 ┆3·6┆ ROOT

X – UPWARD XYLEM TRANSPORT
P – DOWNWARD PHLOEM TRANSPORT

induction of premature senescence in lower leaves, and to the availability thereby for phloem translocation of amino compounds released from protein breakdown.

4.3 Case study II: Phloem transport and the circulation of mineral cations in the vegetative plant

The purpose of this series of studies (Jeschke *et al.*, 1985, 1986, 1987) was to extend the empirical modelling technique for studying C and N partitioning in *Lupinus albus* to follow concomitant exchanges of mineral cations within this species, and to examine quantitatively how such behaviour was affected by mineral supply, especially under conditions of salinity stress. The procedure adopted was first to assemble data on C and N increments in dry matter, on C:N ratios of xylem and phloem sap, and on CO_2 exchanges of plant parts. From these data a model for C and N intake, transport and utilization was constructed, essentially as described in the previous case study. Concentrations of mineral cations in dry matter and transport fluids were then determined and, by a modelling technique based on cation:C ratios in xylem and phloem streams serving different parts of the system, estimates were made of the rates at which each cation species was accumulating and circulating within the system. In a first generation of models (Jeschke *et al.*, 1985) attention was restricted to the cycling of minerals between shoot and nodulated root of low NaCl plants (see Fig. 4.3). The data suggested that downward phloem transport from shoot to root, followed by a subsequent return of most of this mobile component back to the shoot in the xylem, represented a major component of the flow profiles for both K^+ and Mg^{2+}, but that such circulation failed to occur or was of insignificant proportions in the case of Ca^{2+} and Na^+ (see Fig. 4.3 C,D).

A second series of studies (Jeschke *et al.*, 1986) addressed the problem of how the transport system of *Lupinus albus* coped with increasing levels (1–40 mol m^{-3}) of NaCl. Concentrations of Na^+ and Cl^- in root xylem and shoot organs were found to increase more or less in proportion to levels of applied NaCl, suggesting little ability on the parts of roots to exclude salt. However, under mild stress lower mature stems and petioles appeared to be capable of selectively abstracting Na^+ from the ascending xylem stream, thus leading to higher K:Na ratios in leaflets and apical shoot parts than in the main stem axis. Non-uniformity in distribution of Na^+ and K^+ was virtually lost at high salt levels (25 and 40 mol m^{-3} NaCl). This suggested that the salt-retaining ability of mature tissues was limited, resulting in severe reductions in shoot growth and evidence of NaCl toxicity once Na^+ and Cl^- broke through in significant amounts into phloem supplying younger parts of the shoot. One interesting outcome of

Table 4.1 Concentrations of major inorganic and organic solutes in root bleeding (xylem) sap and leaf petiole phloem sap of white lupin (*Lupinus albus*) grown at low and high levels of NaCl salinity*

Solutes (mol m^{-2})	Xylem sap[†]		Phloem sap[†]	
	external NaCl (mol m^{-3})		external NaCl (mol m^{-3})	
	1	40	1	40
K$^+$	9.0	10.8	66.9	52.6
Na$^+$	2.4	37.8	8.1	92.6
Ca^{2+}	0.43	1.65	1.5	0.91
Mg^{2+}	0.67	1.9	3.4	2.7
Cl$^-$	1.8	33.2	7.9	68.0
SO$_4^{2-}$	0.29	1.6	4.3	5.5
H$_2$PO$_4^-$	1.25	4	10.0	12.6
Sucrose	—	—	652	601
Total amino acids	7.2	30.6	41.0	110.0
Asparagine	4.4	15.6	9.3	29.0
Aspartate	1.0	8.0	3.9	6.8
Glutamine	1.4	1.6	5.1	6.3
Glutamate	0.19	1.3	2.4	7.5
Serine	—	—	2.5	20.3
Lysine	—	—	2.5	6.8
Valine	0.19	1.0	6.3	8.8
Proline	—	—	(n.d.)	2.6
Oxalate	—	—	0.04	0.12
Pyruvate	—	—	1.7	0.6
Malate	6.0	1.2	5.0	4.5
Malonate	—	—	6.4	6.6
Succinate	3.4	4.5	42.0	35.0
Fumarate	—	—	0.08	0.16
α ketoglutarate	0.25	0.29	2.2	3.6
Citrate	0.18	0.29	1.5	2.5
Tartarate	0.47	0.56	1.7	3.1

* Plants effectively nodulated and grown in minus nitrogen sand culture (see Jeschke *et al.*, 1986).
† Collected at 2 d intervals over the period 62–71 d after sowing. Phloem sap obtained from petioles of all expanded leaves.

the study was to provide virtually complete inventories of the major organic solute and inorganic ionic components of xylem and phloem sap of plants experiencing different levels of stress (see Table 4.1). The data demonstrated marked increases in Na$^+$, Cl$^-$ and certain amino compounds in xylem and phloem and in xylem sulphate and phosphate with increased exposure to NaCl, but little other evidence of the solute balance of xylem or phloem being significantly affected.

A third series of more sophisticated models of cation partitioning (Jeschke *et al.*, 1987) studied white lupin plants during an 8 d period after

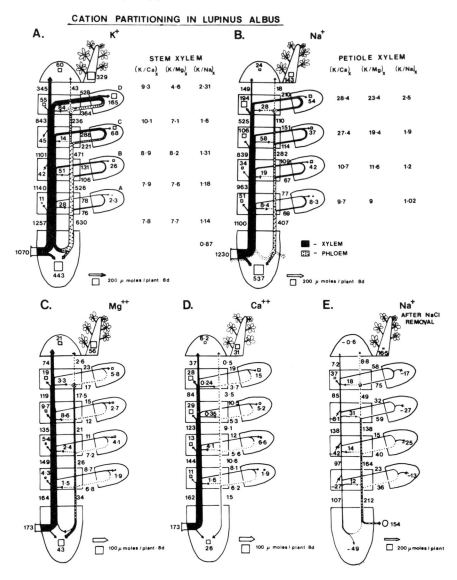

Fig. 4.4 Profiles for uptake, transport and utilization of major nutrient cations in nodulated plants of white lupin over the period 55–63 d after sowing. Rates of ion uptake, flow and incorporation are expressed in μmol plant^{-1} for the study interval. Cation ratios in stem and petiole xylem sap are estimated from the data for the four strata of shoot segments. A, B, C, and D plants remaining on full culture solution with 10 mol m^{-3} NaCl; E plants on same culture but with NaCl removed at 55 d. (Data redrawn from Jeschke *et al.*, 1987.)

flowering, when grown continuously in mild (10 mol m^{-3}) salinity or where this salinity stress had been removed at the beginning of the experimental period. The analysis involved study of cation exchanges between four strata of main shoot segments, root system, inflorescence developing lateral shoots, and collection of root-bleeding xylem sap and petiole and stem phloem sap at the sites depicted in Fig. 4.1A. Data for K$^+$, Na$^+$, Mg^{2+} and Ca^{2+} distribution in the continuous salt stress treatment, and for Na$^+$ redistribution in the plants whose NaCl supply had been removed, are illustrated in Fig. 4.4 (A–E respectively). Included are values predicted from the model for K:Ca, and K:Mg and K:Na molar ratios in the xylem streams of stem and petiole for each of the four strata of the main shoot. Data were recorded in this manner to emphasize the extent to which partitioning of K+ had differed from that of the three other cations. The study reinforced earlier work in demonstrating high rates of transport of K$^+$ and a lesser mobility of Ca^{2+}. K$^+$ was found to be transported preferentially in phloem towards the shoot apex, and Na$^+$ deposited mainly in stems, partly in exchange for K$^+$. Preferential transport of Na$^+$ to roots occurred, apparently due to selective xylem to phloem transfer within stem tissue. Plants whose external source of NaCl had been removed mobilized previously accumulated Na$^+$ to the roots in the phloem, and the model (Fig. 4.4D) predicted that this led to significant Na$^+$ losses to the medium.

Just as with the previously mentioned studies on C and N partitioning, these models of cation partitioning highlight the highly integrated nature of the long- and short-distance transport processes of the white lupin plant, and the marked selectivity that exists between one solute and another with regard to the relative extents of access to each of a number of interchange points within the system. Moreover, as shown in the studies of varying salinity (Fig. 4.3) or of relief from salt stress (Fig. 4.4E) profiles of the transport of individual solutes in the phloem can be markedly affected by the presence or absence of other solutes such as Na$^+$ and Cl$^-$, and, supposedly, also by the nutritional status of the plant at the time of examination. It is to be hoped that continuing experimentation along these lines will allow one to understand how solute exchanges at specific points in the system are regulated, and thus help to explain how plant behaviour may become so drastically modified under certain mineral regimes.

4.4 Case study III: Phloem transport in relation to the C:N relationships and amino acid balance of a single leaf

From the time of its inception through to the time when it first fixes carbon photosynthetically at rates faster than are required for its growth and respiration, the developing leaf is a net importer of C and N through both phloem and xylem. This point of graduation to self-sufficiency for C typically occurs when the leaf is one-third to one-half its full size. It is often marked by a brief period of bidirectional phloem transport during which apical more mature regions may be starting to export photosynthate through certain phloem channels, whilst proximal younger regions of the leaf are still importing assimilates from the rest of the plant (see Turgeon & Webb, 1976; Larson *et al.*, 1980). After this transitional phase, export of photosynthate rises rapidly to a maximum at or shortly after full leaf expansion, followed by declining photosynthetic activity and associated pre-senescence losses of chlorophyll, total N and dry matter.

The approach which has been used to study the phloem and xylem exchanges of C and N between a single main stem leaf and parent plant of *Lupinus albus* (Pate & Atkins, 1983; Pate, 1986) is essentially similar to that already described for the whole plant of this species in Sect. 4.3. Day by day exchanges of C as CO_2 were followed between leaf and atmosphere, alongside progressive estimates of C:N weight ratios of xylem sap and petiolar phloem sap and net changes of C and N in leaf dry matter. Leaf functioning was then assessed for a sequence of phases in leaf development in terms of net exchanges of C and N with the rest of the plant through xylem and phloem, thus making it possible to determine the extent to which newly fixed C and newly arrived xylem N had contributed to dry matter gain as opposed to phloem export during the various stages in the growth of the leaf. The complex interrelationships involved are summarized in Fig. 4.5 in terms of plots of phloem and xylem net fluxes of C and N, and in Fig. 4.6 as C:N flow profiles for the early importing phase, the principal growth phase and the mature phase of leaf development.

The first phase of leaf development of *L. albus* (0–11 d; Figs. 4.5, 4.6A) showed continued net import through xylem and phloem, with phloem supplying almost 78% of the C but only 19% of the imported N. The C:N ratio of phloem translocate was relatively low (36:1) at this stage.

The second phase of leaf development (11–20 d; Figs. 4.5, 4.6B) marked the period of most rapid leaf growth and dry matter accumulation. Export of C in the phloem rose steeply as photosynthetic rate increased and requirements of carbon for leaf growth declined. Little N was exported in the phloem (mean C:N ratio of phloem sap 104:1) since most of the N entering through the xylem was sequestered for leaf growth.

The third phase of leaf development, spanning the remainder of the leaf's life, witnessed continuously declining rates of photosynthesis and losses of C and N from dry matter, but continued high rates of cycling of N through the leaf by xylem to phloem transfer. Export of C relative to N declined accordingly, with a mean C:N ratio of phloem sap for the phase of 52:1 (Figs. 4.5, 4.6C).

By incorporating information on relative amounts of different nitrogenous solutes in the xylem and phloem streams serving the leaf, it was possible to use the data of Figs. 4.5 and 4.6 to show that xylem delivery of asparagine, aspartate and γ-amino butyric acid was at all times in excess of the leaf's requirements for these compounds, whether for direct incorporation as such into protein and/or soluble pools of N within the leaf, or for phloem re-export of the compound in an unmetabolized state. Conversely, synthesis within the leaf was deemed responsible for fulfilling all or part of the leaf's requirements for other amino acids. An exception to this was glutamine, which arrived in excess until full leaf expansion, but was later synthesized in large amounts by the leaf, especially in association with protein breakdown at the approach of leaf senescence (Atkins *et al.*, 1983).

The net balance of the leaf in respect of asparagine proved especially interesting, since it regularly comprised over 60% of the N supplied to the leaf via the xylem. Accordingly, all the N relationships of the leaf hinged upon the relative extents to which this amide was used as a N source for synthesis of other amino acids within the leaf, was incorporated directly into protein, and was cycled through the leaf unmetabolized after xylem to phloem transfer. Most rapid breakdown of asparagine was recorded during the main period of leaf growth, when asparaginase levels were highest. The leaf then entered a phase when incoming amide became increasingly sequestered for phloem export. During this phase asparaginase decreased, but asparagine aminotransferase remained relatively high (Atkins *et al.*, 1983).

A major problem with the above approach to studying net fluxes of amino acids into and out of a leaf by xylem and phloem is that there is no indication whatever as to how a particular phloem solute originated. The current balance might suggest, for example, that asparagine was arriving in vast excess in the xylem but it would not be possible to determine whether the asparagine molecules intercepted in the outflowing phloem stream were indeed those which had entered a short time before in the xylem, or whether a significant proportion had arisen secondarily from asparagine-synthesizing systems within the leaf. The converse would apply, of course, when considering the fate of a solute grossly undersupplied through the xylem. In this case the question would be whether any of the compound present in the phloem had been available directly by xylem to phloem transfer.

In an attempt to resolve the above problems labelling studies were undertaken involving the feeding singly through the xylem of ^{14}C- or ^{15}N-labelled amino compounds to detached leafy shoots of *Lupinus albus*, and subsequent assay of petiolar and fruit phloem sap for the labelled compound and for any other labelled products which had formed prior to passage between xylem and phloem (Sharkey & Pate, 1975; McNeil *et al.*, 1979; Atkins *et al.*, 1980; Pate, 1986). The major conclusions to be drawn from these studies were as follows:

1. Autoradiography of whole shoots after feeding of ^{14}C-labelled asparagine, glutamine, valine, serine or threonine, indicated that these

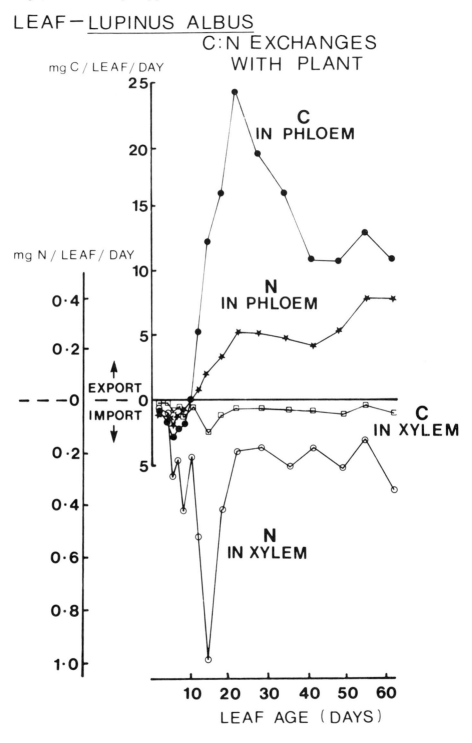

LEAF — LUPINUS ALBUS

C:N EXCHANGES WITH PLANT

mg C / LEAF / DAY

C IN PHLOEM

mg N / LEAF / DAY

N IN PHLOEM

EXPORT

IMPORT

C IN XYLEM

N IN XYLEM

LEAF AGE (DAYS)

C:N ECONOMY – LEAF OF LUPINUS ALBUS

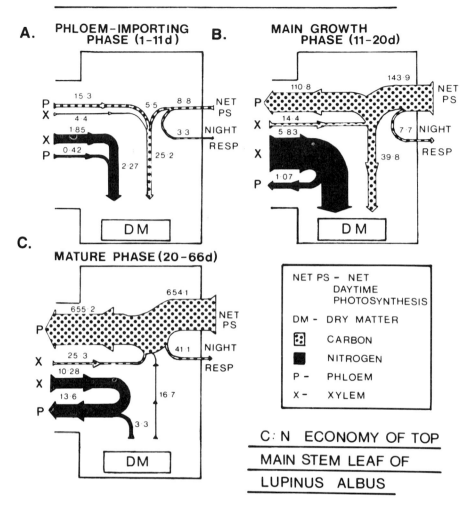

A. PHLOEM-IMPORTING PHASE (1-11d)

B. MAIN GROWTH PHASE (11-20d)

C. MATURE PHASE (20-66d)

NET PS - NET DAYTIME PHOTOSYNTHESIS
DM - DRY MATTER
CARBON
NITROGEN
P - PHLOEM
X - XYLEM

C:N ECONOMY OF TOP MAIN STEM LEAF OF LUPINUS ALBUS

Fig. 4.6 Uptake, utilization and export of C and N during three successive stages in the life of the uppermost main stem leaf of *Lupinus albus*. The economy of the leaf is depicted in terms of gaseous exchanges of C as CO_2, increments or losses of C and N in leaf dry matter, and net import or export of C and N through xylem and phloem connections with the parent plant. All values are given as mg C or mg N per leaf per study interval. Thicknesses of lines depicting xylem and phloem transport are drawn proportional to relative amounts of C or N transported (data from Pate & Atkins, 1983).

Fig. 4.5 Changes during leaf age in exchange of C and N by xylem and phloem between the uppermost main stem leaf and parent shoot of *Lupinus albus*. Note the major role of xylem in importing N for leaf growth and the increasing cycling of N through the leaf following xylem to phloem transfer (data from Pate & Atkins, 1983).

153

common xylem constituents were normally readily abstracted by stem tissue, presumably by parenchyma bordering the xylem. In each case large proportions of the fed amino acids were sequestered by leaf vein tissue. Time courses of ^{14}C labelling of the phloem sap collected from petioles of these leaves suggested rapid and effective transfer from xylem to phloem, with little evidence of breakdown associated with such transfer.

2. The basic amino acid arginine was almost entirely monopolized by stems, petioles and major leaf veins, with little autoradiographic evidence of it reaching leaf mesophyll. Virtually no transfer to phloem sap occurred still in the form of [^{14}C]arginine, but metabolic products of the compound were present in significant amounts. (Arginine is unusual in being present in significant amounts in the xylem, but not in the phloem, of *L. albus*.)

3. Certain other common xylem amino compounds, including aspartic acid, glutamic acid, γ-amino butyric acid and glycine were found to be absorbed only weakly by vascular tissue and thus reached leaf mesophyll in large amounts. There they appeared to be mostly metabolized, and metabolic products of the fed compounds, but not to any significant extent the compounds themselves, were recovered in ^{14}C-labelled form in the leaf phloem stream.

4. Experiments involving the application of double (^{14}C + ^{15}N) labelled amino compounds as substrates for xylem feeding enabled construction of a quantitatively accurate picture of the principal transfers of C and N operating in the xylem to phloem exchange processes within the mature white lupin leaf. These are illustrated in Fig. 4.7 in terms of relative importance for each donor xylem compound of: (a) transfer to xylem in unmetabolized state (C + N of compound transferred); (b) transfer of N of the xylem solute in question to N of other phloem amino acids; and (c) transfer of C of the xylem solute to other phloem amino acids. Solute traffic was found clearly to be dominated by massive direct cycling of unmetabolized asparagine and to a lesser extent of glutamine and valine, but a number of complicated cross-transfers of C and N were also in evidence relating to these and other solutes. For example, the phloem sap pool of asparagine contained not only molecules of the amide passing directly from the xylem but also those arising from metabolism within the leaf of glutamine, glutamic acid and aspartic acid, and amino butyric acid. Similarly, carbon from a wide range of amino acids and the nitrogen of aspartic acid, asparagine, glutamine and/or glutamic acid were shown to be involved in the synthesis of the glutamic acid, threonine and serine molecules appearing in phloem sap, but in these cases *de novo* synthesis in this manner in the leaf proved to be a more important source than direct transfer in an unmetabolized state from the xylem.

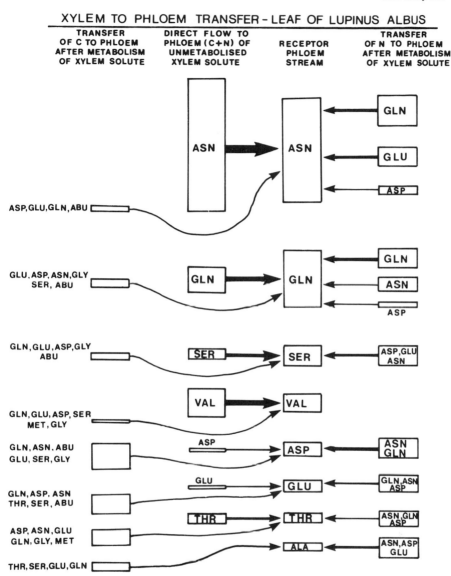

Fig. 4.7 Principal transfers of amino acid C and N from xylem to phloem in the mature *Lupinus albus* leaf, as constructed from studies involving xylem feeding of [^{14}C], [^{15}N] amino acids. Rectangles representing amino compounds are drawn to areas proportional to relative abundance in the phloem stream or to relative importance in the transfer of the C, N, or C + N of a particular amino compound from xylem to phloem (redrawn from Pate, 1986).

In summary, the origin of phloem solutes within the white lupin leaf involves a continuously changing balance in availability of sugar relative to amino compounds, as evidenced by a rise and then progressive fall in the ratio of carbohydrate to nitrogenous solutes (C:N ratio) during the course of the leaf's life. Compounded with this, xylem intake of amino compounds is at first geared largely to protein synthesis and the establishment of a soluble pool of N in the leaf, but then is increasingly geared to xylem to phloem transfer and thereby to the cycling of N to other younger organs in the plant. Spatial and metabolic compartmentation within the system is complex, leading to marked differences between the amino acid components of phloem with regard to their mode of origin within or outside the leaf.

4.5 Case study IV: The origin of solutes delivered to fruits in the phloem

The cowpea (*Vigna unguiculata*) will be selected for this case, partly because of the ease with which phloem sap can be obtained from its fruits, and also because it utilizes ureides (allantoin and allantoic acid) as principal xylem transport products following N_2 fixation. All of the experiments to be described involved analysis of phloem sap collected from cryopunctured first-formed attached fruits on plants subjected to specific nutrient regimes or isotopic labelling treatments (see Pate, 1984). Collectively the investigations were aimed at highlighting the major solutes translocated to the developing fruit and, where possible, identifying the principal source agencies for these solutes.

The sugars of cowpea phloem sap were shown to be present at levels ranging from 14–28% w/v and consisted almost entirely (95%) of sucrose. Sugars accounted for over 80% of the total C of the phloem sap. $^{14}CO_2$ feeding of leaves previously identified as primary sources to the fruits (see Pate *et al.*, 1983) resulted in intense labelling of phloem sucrose, and to a much lesser degree the organic acids and amino acids of the phloem sap (Pate *et al.*, 1984a). It also proved possible to label the sucrose of fruit phloem sap obtained following injection of $^{14}CO_2$ into the gas space or application of [^{14}C]urea to the lateral walls of an illuminated fruit (Pate *et al.*, 1984a). These studies indicated that the phloem stream to the seeds included some carbon fixed by the photosynthetic tissues of the pod wall (see Peoples *et al.*, 1985b).

Turning to the nitrogenous solutes transported to fruits in the phloem, the role of the vegetative organs of the shoot in processing N coming from the root was highlighted by comparing the amounts and proportions of different N solutes in xylem root bleeding sap and fruit cryopuncture sap of plants assimilating either N_2 or NO_3 as their sole source of nitrogen (Peoples *et al.*, 1985a). The data obtained are shown in Fig. 4.8. As had

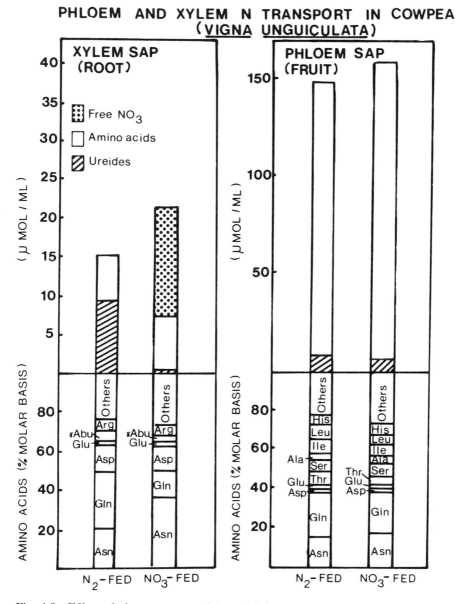

Fig. 4.8 Effect of nitrogen source (N_2 or NO_3) on composition of the nitrogenous fractions of root xylem bleeding sap and fruit cryopuncture phloem sap of cowpea (cv. Vita 3). The analyses are of pooled samples of sap from sets of 15 plants sampled at six times during fruit development (data redrawn with modification from Peoples *et al.*, 1985a).

Table 4.2 ^{15}N labelling of nitrogenous solutes of fruit cryopuncture phloem sap of cowpea (*Vigna unguiculata*) after feeding of ^{15}NO$_3$ to roots*

Solute	Total N (μg ml^{-1})	^{15}N excess	% distribution of ^{15}N
Ureide	220	0.43	0.8
Asparagine	515	10.23	42.2
Glutamine	536	7.76	33.3
Alanine	48	8.65	3.4
Threonine + serine	144	2.45	2.8
Valine	74	1.46	0.9
Isoleucine + leucine	111	1.93	1.7
Histidine	148	2.53	3.0

* Nodulated plants grown in sand culture on 10 mol m^{-3} NO$_3$ (unlabelled) and then fed ^{15}NO$_3$ (10 mol m^{-3}, atom % excess 98%) for 24 h, during which phloem sap was collected continuously from their 12–15 d fruits (data from Pate *et. al.*, 1984a).

been previously reported, the bulk of the xylem N of symbiotically dependent plants consisted of ureide, whereas the NO$_3$-fed plants transported to their shoots large amounts of NO$_3$, some amide and virtually no ureide. Similar absolute levels and relative proportions of amino compounds were present in the xylem of the two treatments, but with a noticeably higher ratio of asparagine: glutamine in the xylem of the NO$_3$-fed plants than in that of the N$_2$-dependent plants. Phloem sap levels of N solutes on a molar basis (Fig. 4.8) were 8–10 times higher than in corresponding xylem sap, reflecting the general ability of plants to concentrate N to a noticeable extent during transfer between the two transport channels (see Pate, 1976, 1980). Unlike xylem sap, however, fruit phloem sap was virtually identical in composition in the N$_2$- and NO$_3$-fed plants, with over 95% of the solutes (molar basis) consisting of amino compounds and only 5% as ureide. Nitrate was not detected in the phloem of either treatment.

It was concluded from the data that the vegetative parts of the fruiting shoots of cowpea are intensely active in metabolizing root-derived nitrate-N or ureide-N, and in mobilizing the resulting amino acids and amides to fruits via the phloem. Two isotope labelling experiments were then conducted to amplify the findings. In one, nodulated plants grown on 10 mol m^{-3} NO$_3$ (unlabelled) were fed through the rooting medium for 24 h with ^{15}NO$_3$ (10 mol m^{-3}, 98 atom % excess), during which time cryopuncture phloem sap was collected continuously from the fruits (Pate *et al.*, 1984a). Phloem sap nitrogenous solutes were separated and the ^{15}N excess of each determined. Knowing the total amounts of N carried by each solute, % distribution of ^{15}N between the sap solutes was determined (Table 4.2). Reduction of ^{15}NO$_3$ in vegetative parts of the shoot led to heavy labelling of the two commonest sap constituents, asparagine and

Table 4.3 ^{14}C labelling of major organic solutes of cryopuncture phloem sap of first-formed fruits of cowpea (*Vigna unguiculata*) after application of ^{14}C] asparagine or allantoin to their subtending leaves through the transpiration stream*

	Radiosubstrate applied	
	[14]C] asparagine	[14]C] allantoin
	% [14]C distribution amongst labelled solutes	
Aspartic acid	3.1	—
Threonine + serine	5.3	—
Asparagine	54.3	7.3
Glutamine	2.1	9.1
Alanine	6.3	5.5
Valine	—	6.8
Ureides (allantoin + allantoic acid)	—	21.3
Sugars ı organic acids	28.9	50.0

* Applications of radiosubstrates for 6 h via a midvein leaf flap with phloem sap collected continuously from cyropunctured fruits (data from Pate *et al.*, 1984a).

glutamine; ureide, the third next most abundant solute of phloem sap, contained only 0.8% of the total ^{15}N recovered, although comprising over 10% of the total sap (labelled and unlabelled) N. Of a range of minor sap constituents only alanine achieved a ^{15}N enrichment approaching that of the amides.

In the second set of experiments the relative abilities of leaf tissue to metabolize and translocate amide (^{14}C]asparagine) and ureide (^{14}C]allantoin) were examined. The radiosubstrates were dissolved in dilute unlabelled xylem sap and applied to an illuminated leaflet through a flap cut in its midvein (see Pate, 1984; Pate *et al.*, 1984a). After being absorbed by the transpiring leaf, the labelled solutes and their metabolic products were partly transferred to the exiting phloem stream and could therefore be intercepted simply by collecting cryopuncture phloem sap from the fruit subtended by the fed leaf. The data obtained for ^{14}C distribution among nitrogenous solutes of this phloem sap (Table 4.3) suggested lesser proportional breakdown of amide than ureide prior to phloem loading. Thus 54% of the ^{14}C in phloem was found to be still attached to asparagine following feeding of this amide, compared with only 21% in the case when feeding the ureide allantoin. High recovery (50%) of the ^{14}C of allantoin in the sugar and organic acid fractions of the phloem indicated complete reworking of the labelled molecule, but since ^{15}N]ureide was not available for use in the study, it was not possible to check whether the N atoms of the molecule had indeed been involved in amide synthesis.

Clearly, the cowpea plant, with large fruits easily bled from phloem, comprises a particularly useful system for monitoring phloem mobilization to fruits. However, while labelling studies have again been shown to help in elucidating the mode of origin of certain key solutes, they do little to show how sap flow and composition are regulated in response to different nutritional circumstances during fruiting. Herein may lie important clues to understanding how reproductive yield becomes fine-tuned to the environmental conditions which a plant may experience before or after its reproduction.

4.6 Case study V: Fate of phloem-borne solutes in developing seeds

The developing seeds of large-seeded grain legumes such as white lupin and cowpea collectively acquire from the parent plant substantial fractions of the plant's post-anthesis photosyntate (see Pate & Minchin, 1980; Pate *et al.*, 1979, 1983), variable proportions of its resources of various mineral elements (Hocking & Pate, 1977), and 70% or more of the N fixed over its growth cycle (see Pate & Minchin, 1980; Pate & Farquhar 1988).

Two principal sources of evidence point to phloem being instrumental in the final mobilization of the above materials from vegetative parts of the plant and pods to seeds. Firstly, seed final composition resembles closely that of phloem sap (Hocking & Pate, 1977; Peoples *et al.*, 1985c). Secondly, it has been demonstrated for cowpea (Pate *et al.*, 1985) that when cut shoots are fed via the transpiration stream with the apoplast markers acid fuchsin or [^{3}H]inulin, the tracers never enter seeds during the cotyledon-filling stage. This is consistent with them receiving excess water continuously by mass import in phloem and using their xylem connection with the placental suture as a means of voiding this excess to the fruit, and, in certain situations, even back to the parent plant. This conclusion is of course not new, having been suggested generally for fruits by Münch (1930) and demonstrated experimentally by several workers, especially in the classic ^{45}Ca feeding studies on fruits described by Ziegler (1963) and Mix and Marschner (1976).

Developing seeds thus provide an ideal experimental system for following the intake and utilization of phloem-borne solutes. We have examined the processes involved in both white lupin (Atkins *et al.*, 1975; Pate *et al.*, 1977, 1978) and cowpea (Peoples *et al.*, 1985b; Pate *et al.*, 1985). The basic approach has first been to obtain data on the total C, N and H$_2$O economy of both fruit and seed and on the C:N weight ratios of phloem and xylem streams serving the fruit throughout development. One can then construct budgets for proportional intake of C, N and water through xylem and phloem for different stages of fruit development (see Pate, 1984). It was thus demonstrated that seeds of both species were in

Table 4.4 Percentage distribution of ^{15}N in fruit phloem sap and seed protein residues of white lupin (*Lupinus albus*) after 6 h feeding of [^{15}N] (amide label) asparagine to cut fruiting shoots through the transpiration stream (data from Atkins *et al.*, 1975).

	% of total ^{15}N recovered	
Compound	Phloem sap	Seed protein
Asparagine (sap)	92.1	—
Aspartyl (protein)	—	8.9
Glutamine (sap)	4.4	—
Glutamyl (protein)	—	7.0
NH$_3$ (amide-N protein)	—	39.0
Valine	1.2	5.6
Glutamate (sap)	0.7	—
Alanine	0.7	3.7
Serine	0.7	2.5
Lysine	0.1	4.0
Arginine	—	14.2
Leucine	—	7.4
Glycine	—	4.6
Isoleucine	—	1.4
Tyrosine	—	0.9
Phenylalanine	—	0.5
Histidine	—	0.4

net positive balance for water through phloem import over the second half of fruit development. In the case of cowpea, the whole fruit was shown to have imported an amount of phloem-derived water greater than its daytime transpiration loss and was therefore presumed to have returned this excess of water to the parent plant via the xylem of the fruit stalk. The models thus corroborated the earlier mentioned direct tracer studies on the inability of seeds to import through xylem (see Pate *et al.*, 1985).

The second phase of the study involved comparisons of the spectra of nitrogenous solutes coming into the fruit via the phloem, and being incorporated into solute pools and protein of the seed coat and embryo. Additionally, analyses of embryo sac fluid or solute collections using an *in vivo* seed cup leakage technique (see Wolswinkel & Ammerlaan, 1983; Rainbird *et al.*, 1984; Peoples *et al.*, 1985c) were used to identify the nitrogenous solutes traversing the apoplastic junction between seed coat and embryo. Then, knowing the net amounts of each solute incorporated into the seed coat and embryo, it became possible to determine which solutes were being supplied to the seeds in excess of current requirement and which were being delivered in significantly deficient amounts.

The net balance sheet for the developing seed of white lupin (Atkins *et al.*, 1975) showed that rates of supply of the amide asparagine – a solute which accounted for 50–70% of the N supplied to the seed – were always greatly in excess of current direct incorporation of this compound into

FATE OF PHLOEM – BORNE NITROGENOUS SOLUTES IN SEEDS OF COWPEA (VIGNA UNGUICULATA)

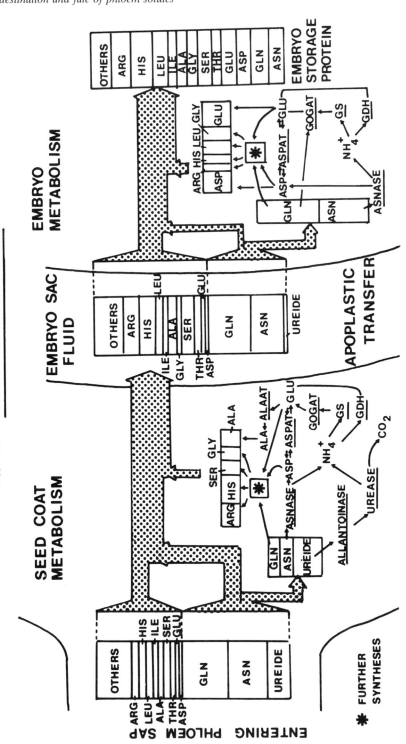

soluble pools or protein of the seed. In complete contrast, arginine, which comprised 15% of the N of seed storage protein, was present in only trace amounts in the phloem, and clearly must have been synthesized entirely by the seed itself. Labelling studies in which [^{15}N](amide label) asparagine was fed to fruiting shoots through the xylem showed patterns of labelling in fruit phloem sap and seed protein as shown in Table 4.4. Of the ^{15}N collected in the phloem 92% was still attached to asparagine, indicating direct transfer from xylem to phloem and little accompanying breakdown of the amide along the pathway to the fruit. In seed protein, however, the ^{15}N was distributed widely across a number of amino acid residues, albeit still mostly present in the amide-N residues of glutamine and asparagine. Significantly, arginine was heavily labelled, suggesting that its synthesis within the seed had utilized the amide-N of asparagine (see Table 4.4).

Embryo sac fluid of *Lupinus albus* proved to be somewhat unusual in having 20–30% of its N in the form of free ammonium (Atkins *et al.*, 1975), and in the above-mentioned labelling study this ammonium and glutamine showed higher levels of ^{15}N labelling in embryo sac fluid than the fed asparagine. It was therefore concluded that in early seed development the seed coat had processed the incoming asparagine, secreting ammonium and other products derived from its metabolism (e.g. alanine, glutamine, aspartic acid, serine, histidine) into the embryo sac, there to act as potential sources of N for protein synthesis in the young embryo. Later, when embryo sac fluid had been absorbed, the embryo developed its own capacity to utilize asparagine, as evidenced by high rates of asparaginase activity. From then until seed maturity it was assumed that asparagine entering through the phloem crossed into the embryo largely in an unmetabolized form and that pre-processing by the seed coat had waned to insignificant proportions (see Atkins *et al.*, 1983).

The parallel situation in cowpea (Fig. 4.9) differed principally in relation to developing seeds having access to ureide and glutamine as well as asparagine. These three compounds were all delivered in the phloem to the seed coat in excess of current requirements, and upon metabolism their N was utilized for synthesis of other compounds, notably arginine, histidine, serine, glycine and alanine. This group of compounds arrived in deficient amounts through phloem import. Consistent with such conclusions, the seed coat showed high *in vitro* activities of allantoinase,

Fig. 4.9 Fate of phloem-delivered nitrogenous solutes in 12–18 d old filling seeds of symbiotically dependent cowpea (*Vigna unguiculata*). The scheme assumes that the seed is fed exclusively through the phloem. It incorporates data on composition of phloem cryopuncture sap and *in vitro* enzyme activities of seed coat and seed, and depicts the embryo as being fed by solutes in proportions indicated by analyses of leakage products of seed coats. Standard abbreviations for amino acids: ASNASE – asparaginase; ALAAT – alanine amino transferase; ASPAT – aspartate aminotransferase; GS – glutamine synthetase; GOGAT – glutamate synthase glutamate oxidoreductase; GDH – glutamate dehydrogenase. (Data from Peoples, *et al.*, 1985b.)

urease, asparaginase, and other N-transforming enzymes likely to be involved in the reworking of the N arriving in the phloem (see Fig. 4.9).

Solutes intercepted at the apoplastic interface between seed coat and embryo (Fig. 4.9) showed lower proportional amounts of ureide, but higher relative amounts of histidine and arginine than did the entering phloem stream, due, one would suppose, to differing extents of catabolism of these compounds by the seed coat. The embryo was predicted to engage in further metabolic rearrangements prior to incorporation of amino compounds into protein, notably further metabolism of amide and net synthesis of aspartic acid, arginine, histidine, leucine, glycine, and glutamic acid.

In further studies of this kind (e.g. Peoples *et al.*, 1985a) it has proved possible to alter N in seed dry weight simply by physiological manipulation of the inputs of C and N during fruiting. Phloem sap composition then reflects changing circumstances at sites of loading and unloading. By further study of these interrelated processes, it is to be hoped that we will understand the signalling and control systems which modulate rates of flow and quality of assimilates and which specific phenomena dictate how certain privileged fruits and seeds are able to monopolize the limited translocate resources of a senescing parent plant.

References

Atkins, C. A., Pate, J. S. & McNeil, D. L. (1980). Phloem loading and metabolism of xylem-borne amino compounds in fruiting shoots of a legume. *Journal of Experimental Botany* **31**, 1509–20.

Atkins, C. A., Pate, J. S., Peoples, M. B. & Joy, K. W. (1983). Amino acid transport and metabolism in relation to the nitrogen economy of a legume leaf. *Plant Physiology* **71**, 841–8.

Atkins, C. A., Pate, J. S., & Sharkey, P. J. (1975). Asparagine metabolism – key to the nitrogen nutrition of developing legume seeds. *Plant Physiology* **56**, 807–11.

Hocking, P. J. & Pate, J. S. (1977). Mobilization of minerals to developing seeds of legumes. *Annals of Botany* **41**, 1259–78.

Jeschke, W. D., Atkins, C. A. & Pate, J. S. (1985). Ion circulation via phloem and xylem between root and shoot of nodulated white lupin. *Journal of Plant Physiology* **117**, 319–30.

Jeschke, W. D., Pate, J. S. & Atkins, C. A. (1986). Effects of NaCl salinity on growth, development, ion transport and ion storage in white lupin (*Lupinus albus* L. cv. Ultra). *Journal of Plant Physiology* **124**, 257–74.

Jeschke, W. D., Pate, J. S. & Atkins, C. A. (1987). Partitioning of K^+, Na^+, Mg^{++}, and Ca^{++} through xylem and phloem to component organs of nodulated white lupin under mild salinity. *Journal of Plant Physiology* **128**, 77–93.

Larson, P. R., Isebrands, J. G. & Dickson, R. E. (1980). Sink to source transition of poplar leaves. *Berichte der deutschen botanischen Gessellschaft* **93**, 79–90.

Layzell, D. B., Pate, J. S., Atkins, C. A. & Canvin, D. T. (1981). Partitioning of carbon and nitrogen and the nutrition of root and shoot apex in a nodulated legume. *Plant Physiology* **67**, 30–6.

McNeil, D. L., Atkins, C. A. & Pate, J. S. (1979). Uptake and utilization of xylem-borne amino compounds by shoot organs of a legume. *Plant Physiology* **63**, 1076–81.

Mix, G. P., Marschner, H. (1976). Calcium umlgerun in Bohnen Fruchten während des Samen wachstums (Redistribution of calcium in bean fruits during seed development). *Zeitchrift für Pflanzenphysiologie* **80**, 354–66.

Münch, E. (1930). *Die Stoffbewegungen in der Pflanze.* Jena, *Gustav Fischer.*

Pate, J. S. (1976). Nutrient mobilization and cycling: Case studies for carbon and nitrogen in organs of a legume. In Wardlaw, I. F. & Passioura, J. B. (eds.). *Transport and Transfer Processes in Plants*, pp. 447–61. New York, Academic Press.

Pate, J. S. (1980). Transport and partitioning of nitrogenous solutes. *Annual Review of Plant Physiology* **31**, 313–40.

Pate, J. S. (1984). The carbon and nitrogen nutrition of fruit and seed – Case studies of selected legumes. In Murray, D. R. (ed.). *Seed Physiology. Vol. 1. Development*, pp. 41–81. New York, Academic Press.

Pate, J. S. (1986). Xylem-to-phloem transfer – Vital component of the nitrogen-partitioning system of a nodulated legume. In Cronshaw, J., Lucas, W. J. & Giaquinta, R. T. (eds.). *Phloem Transport*, pp. 445–62. New York, Alan R. Liss.

Pate, J. S. & Farquhar, G. D. (1988). Role of the crop plant in cycling of nitrogen. In Wilson, J. R. (ed.). *Advances in nitrogen cycling in agricultural ecosystems*, pp. 23–45. Oxford, *CAB International.*

Pate, J. S. & Atkins, C. A. (1983). Xylem and phloem transport and the functional economy of carbon and nitrogen of a legume leaf. *Plant Physiology* **71**, 835–40.

Pate, J. S., Atkins, C. A., Hamel, K., McNeil, D. L. & Layzell, D. B. (1979). Transport of organic solutes in phloem and xylem of a nodulated legume. *Plant Physiology* **63**, 1082–8.

Pate, J. S., Atkins, C. A., Layzell, D. B. & Shelp, B. J. (1984a). Effects of N₂ deficiency on transport and partitioning of C and N in a nodulated legume. *Plant Physiology* **76**, 59–64.

Pate, J. S., Kuo, J. & Hocking, P. J. (1978). Functioning of conducting elements of phloem and xylem in the stalk of the developing fruit of *Lupinus albus* L. *Plant Physiology* **5**, 321–6.

Pate, J. S. & Layzell, D. B. (1981). The basic necessity for differential partitioning of C and N. In Bewley, J. D. (ed.). *Nitrogen and Carbon Metabolism*, Ch. 4, *Carbon and Nitrogen Partitioning in the Whole Plant – A Thesis based on Empirical Modelling*, pp. 94–134. The Hague, M. Nijhoff/Dr W. Junk.

Pate, J. S. & Minchin, F. R. (1980). Comparative studies of carbon and nitrogen nutrition of selected grain legumes. In Summerfield, R. J. & Bunting, A. H. (eds.). *Advances in Legume Science*, pp. 105–13. Kew, Royal Botanic Gardens.

Pate, J. S., Peoples, M. B. & Atkins, C. A. (1983). Post-anthesis economy of carbon in a cultivar of cowpea. *Journal of Experimental Botany* **34**, 544–62.

Pate, J. S., Peoples, M. B. & Atkins, C. A. (1984b). Spontaneous phloem bleeding from cryopunctured fruits of a ureide-producing legume. *Plant Physiology* **74**, 499–505.

Pate, J. S., Peoples, M. B., Van Bel, A. J. E., Kuo, J. & Atkins, C. A. (1985). Diurnal water balance of the cowpea fruit. *Plant Physiology* **77**, 148–56.

Pate, J. S., Sharkey, P. J. & Atkins, C.A. (1977). Nutrition of a developing legume fruit. *Plant Physiology.* **59**, 506–10.

Pate, J. S., Sharkey, P. J. & Lewis, O. A. M. (1974). Phloem bleeding from legume fruits. A technique for study of fruit nutrition. *Planta* **120**, 229–43.

Peoples, M. B., Atkins, C. A., Pate, J. S. & Murray, D. R. (1985c). Nitrogen nutrition and metabolic interconversions of nitrogenous solutes in developing cowpea fruits. *Plant Physiology* **77**, 382–8.

Peoples, M. B., Pate, J. S. & Atkins, C. A. (1985a). The effect of nitrogen source on transport and metabolism of nitrogen in fruiting plants of cowpea (*Vigna unguiculata* L. Walp.). *Journal of Experimental Botany* **165**, 567–82.

Peoples, M. B., Pate, J. S., Atkins, C. A. & Murray, D. R. (1985b). Economy of water, carbon, and nitrogen in the developing cowpea fruit. *Plant Physiology* **77**, 142–7.

Rainbird, R. M., Thorne, J. H. & Hardy, R. W. F. (1984). Role of amides, amino acids and ureides in the nutrition of developing soybean seeds. *Plant Physiology* **67**, 973–6.

Sharkey, P. J. & Pate, J. S. (1975). Selectivity in xylem to phloem transfer of amino acids in fruiting shoots of white lupin (*Lupinus albus* L.). *Planta* **127**, 251–62.

Turgeon, R. & Webb, J. A. (1976). Leaf development and phloem transport. *Cucurbita pepo*: maturation of the minor veins. *Planta* **129**, 265–9.

Wolswinkel, P. & Ammerlaan, A. (1983). Phloem unloading in developing seeds of *Vicia faba* L. *Planta* **158**, 205–15.

Ziegler, H. (1963). Verwendüng von [45]Calcium zur Analyse der Stoffversorgung wachsender Fruchte. *Planta* **60**, 41–5.

5 Loading of photoassimilates

S. Delrot

5.1 General background

5.1.1 Importance and definition of phloem loading

It is generally accepted that the source controls the intensity of long-distance transport, while the sinks control its direction. The control of the source is exerted through the rate of photosynthesis, the metabolic compartmentation of assimilates between mobile (sucrose, amino acids) and non-mobile compounds (starch), and through the spatial compartmentation of mobile compounds in different organelles and cells. The assimilates undergo a lateral transport from the mesophyll to the conducting bundle, and are then loaded into the phloem. Depending on the structural features of the species considered, lateral transport may be symplastic or apoplastic. Phloem loading, which will be reviewed here, is 'the process by which the major translocated substances are selectively and actively delivered to the sieve tubes in the source region prior to translocation' (Geiger, 1975). This process is essential not only for the transport of sugars but also for the translocation of all other compounds present in the phloem, including amino acids and xenobiotics. Indeed, the active accumulation of sugars in the conducting bundle generates an osmotic gradient which results in the influx of water into the phloem (Minchin & Thorpe, 1982), and thus provides the 'push' for longitudinal transport by mass-flow. Phloem loading may be considered both as the terminal step of lateral transport and as the initial step of longitudinal transport, and it thus plays an important role in sink/source interactions.

The topic has been the subject of reviews (Ho & Baker, 1982; Giaquinta, 1983; Delrot & Bonnemain, 1984), but currently research in this area is active, sometimes yielding contradictory data. Moreover, recent approaches have disputed the validity of the apoplast concept of loading.

167

This chapter will attempt to summarize the present status of what is known and, unfortunately, still unknown. Indeed, in spite of some progress, phloem loading is still far from being completely understood, because there is no experimental model allowing an easy and unequivocal study of the process *sensu stricto*. It is worth recalling briefly the limits of the models used, before summarizing the main data yielded with these models.

5.1.2 Limits of the experimental approaches

5.1.2.1 Whole plants

Studies of loading of [^{14}C]assimilates on whole plants, which are the closest to normal physiology, have been scarce (Geiger *et al.*, 1974; Sovonick *et al.*, 1974; Housley *et al.*, 1977; Malek & Baker, 1977, 1978; Servaites *et al.*, 1979; Atkins *et al.*, 1980). These types of experiment cannot be repeated many times because they are time-consuming. The data are sometimes difficult to interpret because of the variability from plant to plant, and because they result from a combination of lateral transport, loading, and export.

5.1.2.2 Isolated organs

Geiger *et al.* (1974) showed that sucrose was the main mobile sugar in the apoplast of sugar-beet leaf and that exogenous [^{14}C]sucrose fed to the leaf was concentrated in the veins. The same group (Sovonick *et al.*, 1974) also showed that the kinetics of uptake of exogenous sucrose by isolated leaf discs were similar to the kinetics of export of [^{14}C]assimilates by a leaf still attached to the plant. These data provided the background for a study of phloem loading by measuring sucrose uptake in isolated leaf discs of sugar-beet (Giaquinta 1976, 1977a, b, c), and of other species (Delrot, 1981; Van Bel & Ammerlaan, 1981). This approach may be criticized because the leaf discs contain not only veins, but also numerous parenchyma cells able to take up sucrose, and because the label observed in the veins might come from a by-pass via the mesophyll, if loading is symplastic.

The possibility of uptake by non-phloem cells seems particularly important in studies with *Ricinus* cotyledons (Komor, 1977; Hutchings, 1978). With this material, the electrical (H$^+$ and K$^+$) changes induced by sucrose appear almost immediate (Cho & Komor, 1980), but autoradiographs (Martin & Komor, 1980) show that within five minutes, at least, only the parenchyma cells are labelled. Studies on sucrose uptake by suspension cells cultures of *Ricinus* cotyledons suggest that the parenchyma cells are unable to take up sucrose without prior hydrolysis (Cho & Komor, 1985), but it has not been established whether this invertase activity is present *in vivo*, or whether it developed after processing of the cells for *in vitro* culture.

5.1.2.3 Isolated veins, isolated cells

Several attempts have been made to study the uptake of exogenous sugars by veins isolated either enzymatically (Cataldo, 1974; Van Bel & Koops, 1985; Wilson *et al.*, 1985) or mechanically (Daie, 1985).

Basically, this approach confirms that the contribution of parenchyma to the uptake of solutes by whole tissues or leaf discs cannot be neglected, although some of the data yielded by this method must be treated with caution. It is not known how the maceration of the tissues in enzymes for several hours at high osmoticum concentration may affect the functional status of the very fragile conducting elements. In some cases the physiological status of the isolated material has been checked by ultrastructural studies (Wilson *et al.*, 1985) or by tests using vital dyes (Van Bel & Koops, 1985), but these controls give no information at the membrane level. For example, high osmoticum concentrations and enzymatic treatments may result in the depolarization of the transmembrane potential difference (Cornel *et al.*, 1983). The qualitative contents and the concentrations of solutes in protoplasts are very different from those of the cells from which they are derived (Renaudin *et al.*, 1986). Cataldo (1974) reported that sucrose was a competitive inhibitor of hexose uptake by isolated minor veins of tobacco. These results suggest that some invertase activity may have developed in the digested samples. In intact tissues there are separate carriers for sucrose and for hexoses, and, in addition, sucrose is not recognized by the hexose carrier (see Sect. 5.4.1). The fact that isolated veins take up hexoses (Cataldo, 1974; Wilson *et al.*, 1985) also shows that this material cannot be simply equated to the conducting elements of the phloem, where hexoses are loaded at a very slow rate (Fondy & Geiger, 1977; Delrot, 1981).

Uptake by veins is sometimes calculated as the difference between intact tissues and isolated mesophyll (Van Bel *et al.*, 1986b) or between intact tissues and tissues deprived of veins by enzymatic or mechanical treatments (Wilson *et al.*, 1985). This type of calculation is potentially misleading because it relies on a number of unverified assumptions. In addition to the possible physiological stresses mentioned above, it is unlikely that the access of the label to the various materials which are compared is the same, thus making direct comparisons hazardous.

5.2 The pathway of phloem loading

Many of the data available suggest that phloem loading occurs directly from the apoplast. However, some authors have recently criticized the experimental approaches that led to this conclusion and suggest that phloem loading might be symplastic. While the criticisms presented are mostly valid, the alternative conceptual experimental approaches suggested are also equivocal, and no clear mechanism has been proposed to explain a symplastic pathway consistent with the main characteristics of phloem loading.

5.2.1 Evidence for loading from the apoplast

5.2.1.1 Structural evidence

Several anatomical and cytological features provide strong support for the concept that phloem loading occurs from the apoplast. Firstly, in many species (*Beta vulgaris, Vicia faba, Commelina benghalensis, Zea mays*), there is a decrease in plasmodesmatal frequency towards the vein region, and the companion cell/sieve tube complex appears as an isolated unit. Secondly, the development in many species of transfer cells at the time of sugar export by the leaf appears as an adaptative mechanism favouring membrane exchanges. These cells greatly increase the surface area of membrane available for exchanges with the apoplast, whilst their symplastic connections with the outer parenchyma are poor or absent. Thirdly, the apoplastic pathway is also supported by the steep osmotic gradient found at the boundary of the conducting complex of all species investigated so far (Geiger *et al.*, 1973; Fisher & Evert, 1982a,b; Russell & Evert, 1985; Russin & Evert, 1985). In the case of symplastic loading, one would rather expect a progressive build-up of concentration from the periphery of the phloem tissues to the sieve tube.

5.2.1.2 Presence of sugars in the leaf apoplast

If phloem loading occurs from the apoplast, then the mesophyll cells should be able to leak assimilates into this compartment, and these assimilates should be demonstrable there. There is little doubt that assimilates are present in the leaf apoplast. By washing (Kursanov & Brovchenko, 1969) or by isotope trapping (Geiger *et al.*, 1974), sugars have been recovered from cut or abraded fragments of sugar-beet leaf. Increased translocation rates induced by 4 mol m^{-3} ATP or increased light were accompanied by an increased recovery of sucrose (Geiger *et al.*, 1974). Sugars have also been collected in the centrifugation fluid from sugar-beet leaves (Fondy & Geiger, 1977) and squash leaf fragments (Madore & Webb, 1981), in the washing medium of broad bean leaf fragments (Delrot *et al.*, 1983) and in the xylem exudate of maize leaf strips (Heyser *et al.*, 1978).

In broad bean, the sucrose concentration of the apoplast was shown to vary fivefold during a day-night cycle (Delrot *et al.*, 1983). Results obtained with whole leaves (Geiger *et al.*, 1974) or large leaf fragments (Madore & Webb, 1981; Delrot *et al.*, 1983; Ntsika & Delrot, 1986) show that these sugars recovered are not simply contaminants coming from broken cells or cut veins. Estimates of the overall concentration of sucrose in the apoplast gave the following values: 0.07 mol m^{-3} (sugar beet, Fondy & Geiger, 1977); 0.15–1.5 mol m^{-3} (maize, Heyser et al., 1978); 1–5 mol m^{-3} (broad bean, Delrot et al., 1983); 0.02 mol m^{-3} (squash, Madore & Webb, 1981). Sucrose is accompanied by appreciable amounts of glucose (0.2 mol m^{-3} in sugar-beet; 3–5 mol m^{-3} in broad bean; 0.03 mol m^{-3} in squash), but fructose is almost absent in the apoplast. Amino acids are also present in the wash solutions of sugar-beet (Brovchenko, 1963) and broad bean (Despeghel, 1981) leaf fragments. For example, the overall apoplastic concentrations of asparagine and glutamine in broad bean leaf were 9 and 1.5 mol m^{-3} respectively.

The site of assimilate release into the free space is not known exactly. Geiger (1975) suggested that this release occurs in the vicinity of the conducting bundle, most probably from phloem parenchyma cells. Several considerations support this idea. Firstly, assimilates released throughout the whole mesophyll might be dragged away from the phloem by the transpiration stream. Secondly, in plants possessing a water-impermeable suberin layer in the bundle-sheath, release from all the mesophyll cells would not be efficient. Thirdly, the restriction of plasmodesmatal number in the vein area also suggests a preferential release into this region of the apoplast. Nevertheless, there is still no direct experimental evidence to support this hypothesis, which has been made on logical considerations. Some data suggest that the adverse effect of the transpiration stream may not be as strong as expected (Gunning et al., 1974; Boyer, 1985). Huber & Moreland (1980, 1981) reported that assimilates are released from wheat and tobacco protoplasts, and concluded that this release may occur throughout the whole mesophyll. However, the efflux of assimilates from isolated mesophyll cells (with their cell wall) is much less. The mesophyll cells possess strong uptake systems for sugars and amino acids (Cheruel et al., 1979; Van Bel & Ammerlaan, 1981). However, this does not preclude the possibility that assimilates may flow out of the cell if the internal sugar concentration reaches a high level during photosynthesis. The uptake properties of both the mesophyll cells and phloem tissue may vary widely depending on the external concentration of the substrate (see Sect. 5.3.3). Therefore, it seems important to develop techniques which are able to map the sites of assimilate release (and uptake) in the leaf tissues, because these processes may largely affect the local concentrations of sugars. At present, we have little indication that measurements of 'overall' apoplastic sugar concentrations reflect any physiological situation.

The mechanism of assimilate release from the mesophyll is still unknown. A K$^+$-sucrose symport (co-transport) mechanism has been postulated for assimilate efflux from wheat and tobacco protoplasts, but the evidence presented is not unequivocal (Huber & Moreland, 1981). Sugar efflux from the mesophyll is stimulated by KCl additions smaller than 30 mol m^{-3} (Doman & Geiger, 1979), and is also sensitive to light, although the light effects reported are contradictory. Brovchenko & Riabushkina (1972) have suggested that sugar efflux from sugar-beet mesophyll was promoted by light, and more particularly by non-cyclic photophosphorylation. Van Bel *et al.* (1986a) found that, in *Commelina*, light exerted a dual effect in that it inhibited the efflux of sugars and amino acids from the mesophyll but promoted a greater efflux from the veins, with a net stimulation resulting. However, because of the high concentration of substrate added in the efflux medium, the latter authors were dealing with sucrose and amino acid exchange rather than with efflux *per se*. In broad bean, light promoted the uptake of threonine and of α-aminoisobutyric acid by both the mesophyll and the veins (Despeghel *et al.*, 1980).

5.2.1.3 Effects of non-permeant inhibitors on sucrose uptake and phloem loading

It is well known that the uptake of exogenous sucrose by leaf tissues is inhibited by PCMBS (Giaquinta, 1976; Delrot *et al.*, 1980). This fact does not show that loading is apoplastic, since the inhibition may also affect the uptake of sucrose by mesophyll cells. However, the inhibition of phloem loading of [^{14}C]assimilates by PCMBS, without concurrent inhibition of photosynthesis, provides strong, less equivocal support for apoplastic loading in sugar-beet (Giaquinta, 1976).

5.2.2 *Possibility of symplastic loading*

5.2.2.1 Structural evidence

Relatively large numbers of plasmodesmata are visible between the companion cells and the surrounding cells in several species (*Populus deltoides, Fraxinus excelsior*; see Chs. 2,3). Proponents of symplastic loading suggest that these plasmodesmata may offer a symplastic route towards the conducting bundles. Symplastic loading would require three major conditions: (a) that the plasmodesmata seen at the boundary of the conducting complex are open; (b) that they are able to accumulate sucrose and some other organic compounds on one side of the membrane; and (c) that this accumulation is selective (hexoses, for example, are excluded from the phloem).

5.2.2.2 How do plasmodesmata work?

Micro-injection of fluorescent dyes of different molecular weights has been used to trace symplastic connections in plants and to assess the molecular size cut-off of plasmodesmata (Goodwin, 1983; Erwee & Goodwin, 1985; Erwee *et al.*, 1985). The data suggest that there are different symplastic domains within a plant, limited by plasmodesmata differing in their molecular size cut-off. The permeability of plasmodesmata is strongly reduced by injection of calcium into the cells. Hyperosmotic treatments have the same effect, although it is transient (Erwee & Goodwin, 1984). The lowest MW recorded for the size cut-off of leaf plasmodesmata was 376, the most frequent values being around 700–900.

Madore *et al.* (1986) have recently addressed the question of symplastic connections towards vascular bundles by injecting Lucifer Yellow CH (MW 457) into *Ipomoea tricolor* and into sugar-beet leaves (see Ch. 2). In *Ipomoea* the dye reached the conducting bundle, and this movement was inhibited by alkaline pH but not by PCMBS. In sugar-beet the authors failed to observe movement of the dye in the veins and ascribed this failure to the presence of dense mesophyll layers hampering the visualization of the veins. In *Ipomoea* it is not known whether the dye actually reached the sieve tube/companion cell complex because the specific phloem cells containing the dye could not be discerned (Madore *et al.*, 1986). Assuming it actually does, the movement of a charged dye along its concentration gradient may not be representative of the movement of sucrose (uncharged) against its concentration gradient. Possible contributions of physico-chemical phenomena (e.g. interfacial flow), which may be pH-sensitive, in the movement of the dye are not known. Last, but not least, assuming that these data demonstrate symplastic continuity from the mesophyll to the sieve tube, given the size of the dye used, one may wonder how sucrose (MW 352) or Lucifer Yellow CH (MW 457) would pass, while hexoses would not. In summary, if these data suggest possible symplastic continuity in some species, there is still no evidence that plasmodesmata may function as 'one-way' valves in plant tissues. Cytochemical detection of ATPase activity associated with plasmodesmata has been reported in several instances, but this does not necessarily indicate a 'one-way' valve function.

Another way of achieving symplastic loading is to hypothesize that the accumulation step is performed not by the plasmodesmata, but by the ER which may cross the plasmodesmata. There is little, and only indirect, evidence for this in the literature (Altus & Canny, 1985). A loading mechanism involving ER would not be sensitive to the non-permeant reagent PCMBS, and loading of [^{14}C]assimilates is inhibited by this reagent in sugar-beet (Giaquinta, 1976). Also, it is believed that the movement of solutes through plasmodesmata occurs through the cytoplasmic annulus rather than the central core (Gunning & Overall, 1983).

Finally, one would predict that with symplastic loading, transport to the veins would be inhibited when symplastic connections were broken. Osmotic treatments, which effectively decrease the permeability of plasmodesmata (Erwee & Goodwin, 1984) either have no effect, or more generally promote loading of [^{14}C]assimilates and of exogenous [^{14}C]sucrose (Geiger et al., 1974; Servaites *et al.*, 1979; Geiger & Fondy,

1980; Hagège *et al.*, 1984; Daie & Wyse, 1985; Lemoine & Delrot, 1987), suggesting that the symplastic pathway is not of physiological importance for loading, and also that in experiments with leaf discs, by-pass of the label via the mesophyll to the veins does not occur to any significant extent.

5.2.3 Conclusions

Several pathways for phloem loading are summarized in Fig. 5.1 (lateral transport from the mesophyll to the vein region is not included). In plants where symplastic connections at the boundary of the conducting complex are few (e.g. sugar-beet, maize), loading is likely to occur from the apoplast. In this case (Fig. 5.1A), there is no major objection to this pathway and it accounts for all the characteristics of loading. In plants where plasmodesmata are found in large numbers (e.g. *Fraxinus, Cucurbita*), several possibilities remain, depending on the way that these plasmodesmata function.

If these plasmodesmata are closed (Fig. 5.1B), one is brought back to the first model. If they are open, they may be active and selective and symplastic loading of the phloem may occur, with uptake through the plasmalemma involved 'only' as a retrieval mechanism which limits leakage (Fig. 5.1C). Open plasmodesmata may also behave passively as 'holes' limited only by a molecular size cut-off but unable to create a concentration gradient. In this case, loading must again be envisaged as apoplastic, and the plasmodesmata would behave as leak points where unloading occurs (Fig. 5.1D). There is evidence for symplastic unloading in young leaves (Gougler-Schmalstig & Geiger, 1985, 1987) and in roots (Giaquinta *et al.*, 1983) (see Ch. 6). This model may appear as wasteful in energy terms, but this depends on the relative capacities of the carrier-mediated input and of the output through the plasmodesmata. The advantage of the model is that it requires neither selectivity nor active functioning of the plasmodesmata. The fluorescent dye data (see above) suggest no other selectivity than that due to molecular size.

Fig. 5.1 Different possibilities for the pathway of phloem loading. *A*: Apoplastic loading, absence of symplastic connections and apoplastic unloading. *B*: Plasmodesmata exist between the companion cell and the surrounding cells, but they are closed; loading is apoplastic. *C*: Plasmodesmata between the companion cell and the surrounding cells are open and able to create a concentration gradient. Symplastic loading is prevalent and apoplastic uptake only occurs as a retrieval mechanism. *D*: Plasmodesmata are open, but unable to accumulate solutes from the apoplast. Loading is apoplastic and plasmodesmata behave as leakage, or unloading, sites. In all cases, transfer from the companion cell to the sieve tube is thought to be symplastic, and active uptake may also occur through the membrane of the sieve tube, although it may be less efficient than in the companion (or transfer) cell. CC = companion cell; P = plasmodesma; ST = sieve tube.

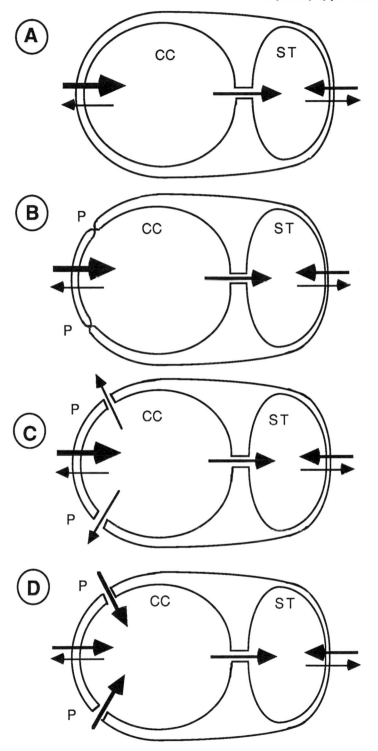

In the present context, given our knowledge of carrier-mediated transport on the one hand and of plasmodesmata on the other, the active and selective nature of loading is more readily interpreted in terms of membrane transport from the apoplast. However, this may only reflect our lack of understanding of plasmodesmatal structure and function. Structural and uptake data for some species, particularly *Cucurbita* (Hendrix, 1973, 1977; Turgeon *et al.*, 1975) may suggest symplastic loading, but membrane uptake is certainly of physiological importance, either as the primary accumulating process or as a retrieval mechanism controlling the efficiency of loading and unloading in all cases.

5.3 The mechanism of loading

5.3.1 Loading is active

There is overwhelming evidence that phloem loading and uptake of exogenous sucrose are inhibited by appropriate physical and chemical treatments. For example, in sugar-beet and in soybean, phloem loading is sensitive to anoxia and to uncouplers (Sovonick *et al.*, 1974; Servaites *et al.*, 1979). The effects of anoxia vary with the plant and with the experimental conditions used. Thorpe *et al.* (1979) found that, in wheat and in *Panicum milioides*, anoxia resulted in a 70% decrease of phloem loading within 1–6 min after the beginning of treatment, while it did not affect loading in maize and in *Panicum maximum*. They concluded that C_3 and C_4 plants may differ in their short-term oxygen requirements for loading. This difference may only be due to the speed with which anoxia is transmitted at the cellular level, depending on the structure of the leaf and on its metabolism, and not reflecting differences in the mechanisms of loading *per se*. Loading and sucrose uptake are mediated by a saturable system and an ill-defined linear system (see Sect. 5.3.3). In sugar-beet, anaerobiosis inhibits only the linear system (Maynard & Lucas, 1982b), while in onion, both systems are sensitive to anoxia (Wilson *et al.*, 1985).

In sugar-beet, uptake of exogenous sucrose by leaf tissues is inhibited by low temperature, uncouplers (Giaquinta, 1977b) and permeant thiol reagents such as NEM and PCMB (Giaquinta, 1976). Similar results have been observed with castor bean cotyledons (Komor, 1977) and petioles (Malek & Baker, 1977), leaf discs of broad bean (Delrot & Bonnemain, 1981) and of *Phaseolus* (Daie, 1985), and isolated vascular bundles of celery (Daie, 1986). In broad bean, uptake of exogenous sucrose (Delrot, 1981) and threonine (Despeghel & Delrot, 1983) is also sensitive to the ATPase inhibitors, diethylstilbestrol and dicyclohexylcarbodiimide (DCCD).

The active nature of loading is also apparent from the high concentration of solutes, particularly sucrose, in the phloem, compared with that of the leaf apoplast (see Ch. 4). The transmembrane gradient of amino acids is less important, but still evident. For example, McNeil (1979) calculated that phloem loading of valine in the leaf of *Lupinus albus* occurred against a concentration gradient of about 10.

5.3.2 Sucrose is not hydrolysed during phloem loading

Autoradiographs (Fondy & Geiger, 1977; Delrot & Bonnemain, 1981) and the use of asymmetrically labelled [^{14}C] sucrose (Giaquinta, 1977c; Daie, 1985) have shown that sucrose is not hydrolysed into hexoses during phloem loading in *Beta*, *Vicia* and *Phaseolus*.

5.3.3 Loading is carrier-mediated

Several lines of evidence show that phloem loading of sugars and amino acids is carrier-mediated.

5.3.3.1 Loading is selective

The loading of sugars is selective, since hexoses are absent in all species and sucrose is the major phloem sugar in many species (see Ch. 4). This composition is not due to metabolism of sugars in the phloem, since autoradiographic studies have shown that exogenous L-glucose or 3-O-MeG are very poorly phloem loaded in leaf discs, compared with sucrose (Fondy & Geiger, 1977; Delrot, 1981).

Although the amino acid composition of the phloem sap is somewhat different from that of the leaf apoplast, this does not provide strong evidence for carrier-mediated uptake of amino acids by the conducting complex, because of possible metabolism of amino acids in the phloem.

5.3.3.2 Saturation kinetics

Studies on the concentration dependence of sucrose uptake and phloem loading have been numerous and heterogeneous in their experimental designs, in their results and in their interpretations, making comparisons and generalizations almost impossible. The basic interpretation of uptake in terms of Michaelis–Menten kinetics is criticizable in itself, since these kinetics were designed for soluble enzymes, not for membrane-embedded carriers. Although it is not necessarily relevant in physiological terms, this interpretation will be adopted in the following discussion, for operational convenience. It must also be stressed that the diffusion barrier due to the presence of the epidermis and/or of the cell wall is not taken into account in these studies. For example, autoradiographs have shown that, in leaf discs neither peeled nor abraded, which have sometimes been used to study 'phloem loading' (Maynard & Lucas, 1982b; Sturgis & Rubery, 1982), uptake of sucrose occurs only at the periphery of the discs (Delrot & Bonnemain, 1978).

5.3.3.2.1 Sugars Concentration dependence studies on phloem loading of sugars *in situ* are very few. Sovonick *et al.* (1974) studied the rates of sugar export after supplying different concentrations of sucrose through the abraded epidermis of a sugar-beet leaf still attached to the plant. They found dual isotherms involving two saturable components. Similar experiments led to similar results with soybean (Servaites *et al.*, 1979).

Table 5.1 shows the variety of interpretations given to multi-phasic isotherms of sugar uptake. A saturable system with a relatively high affinity is found in all cases, but the kinetic parameters associated with this system vary widely. There is good evidence that this phase is mediated by a proton-substrate symport (see Sect. 5.3.4). Beyond this system, the interpretations differ largely: saturable system, linear non-saturable system, or saturable + linear system working together.

The high-affinity system is always sensitive to PCMBS, whilst, depending on the material, the low-affinity system is not (sugar-beet leaf discs, Maynard & Lucas, 1982b; discs of *Phaseolus coccinius*, Daie & Wyse, 1985) or only partially sensitive to this inhibitor (leaf discs of broad bean, M'Batchi & Delrot, 1984; protoplasts of soybean cotyledon, Schmitt *et al.*, 1984). In the latter cases, sensitivity to PCMBS suggests that the low-affinity system is not purely diffusional, although this criterion is not unequivocal because of possible side-effects of the inhibitor. Competition by maltose in sugar-beet (Maynard & Lucas, 1982a) and by phloridzin in broad bean (Lemoine & Delrot, 1987) also show that in these plants the low-affinity system is not purely diffusional.

The physiological basis of multi-phasic isotherms cannot be assessed easily, due to the scattered evidence found in the literature. Multiple isotherms may be ascribed to the superimposition of non-membrane events (cell wall invertase, intracellular metabolism) on a carrier-mediated system, or to the co-operation of several carriers. Intracellular or extracellular metabolism of sucrose as a function of its concentration has not been systematically investigated. However, it is not likely as a general explanation of dual isotherms, since such isotherms are also found with metabolizable or non-metabolizable sugars and amino acids (Caussin *et al.*, 1979; Despeghel, 1981; Van Bel *et al.*, 1982, 1986b; Daie, 1986).

Interpretation of multi-phasic isotherms only in terms of membrane events leads to several possibilities: (a) different carriers located in different cells; (b) different carriers located in different membranes of one cellular type; (c) different carriers located in the same membrane; or (d) a single carrier undergoing allosteric changes leading to different kinetics. There is little doubt that the high-affinity system is present in the phloem cells. This system is found in isolated vascular bundles of *Heracleum*

(Sokholova *et al.*, 1979) and *Commelina* (Van Bel & Koops, 1985) and is the only one detected in isolated phloem tissue of celery (Daie, 1986). In the latter case, hypothesis (a) may therefore be valid. However, a variant of this hypothesis assuming that different carriers are located, not exclusively, but preferentially in various kinds of cells seems generally closer to the experimental results. Indeed, many data have shown that the presence of two uptake components is a general characteristic of many plant cells, but there is evidence that the high-affinity system is more concentrated in the conducting complex than in the other cells. The presence of the high-affinity system in phloem means neither that this system is totally absent in other cells nor that phloem cells do not contain the low-affinity component. In sugar-beet autoradiographs suggest that the low-affinity component of uptake is preferentially associated with mesophyll cells (Sovonick *et al.*, 1974). In the same species, the relative contribution of the saturable component to total uptake is much larger in isolated vascular bundles than in petiole slices (Maynard & Lucas, 1982b). Few data are available to consider hypotheses (c) and (d). The possibility of a protonated and a non-protonated form of the carrier has been suggested to explain hexose uptake in *Chlorella* (Komor & Tanner, 1974). However, this model could not account for all the data concerning sucrose uptake in broad bean leaf discs (Delrot & Bonnemain, 1981). Recently, Borstlap (1986) obtained transport mutants for amino acids which were affected on the low-affinity system only, suggesting the presence of distinct proteins.

As with its nature and localization, the energetics of the low-affinity system are also poorly understood. It is not directly powered by the proton motive force (Giaquinta, 1977b; Delrot, 1981; Delrot & Bonnemain, 1981). In sugar-beet, the low-affinity system is not sensitive to anoxia, leading Maynard & Lucas (1982b) to consider whether it is not linked to the functioning of a terminal oxidase. However, in onion this system is inhibited by anoxia (Wilson *et al.*, 1985).

The relative part played by each component in the uptake of sucrose by the conducting complex *in situ* is dependent on the actual concentration of sucrose in the apoplast in this region. In sugar-beet, Sovonick *et al.* (1974) found that a free space sucrose concentration in the region of minor vein phloem of approximately 20 mol m^{-3} can support translocation at the rates commonly observed for photosynthetically produced sugars. This value must be compared with the overall sucrose concentration of 0.07 mol m^{-3} estimated by Fondy and Geiger (1977) for the leaf apoplast of the same species. Although they differ by a range of about 30, both values concern the high-affinity system. In broad bean, measurements of the overall apoplastic concentration of sucrose (1 to 5 mol m^{-3}) suggest that both the high- and the low-affinity systems may be of physiological significance (Delrot *et al.*, 1983). In celery, phloem loading must occur through the high-affinity system, which is the only one present in this tissue (Daie, 1985). However, the inclusion (generally admitted) of the high-

Table 5.1 Kinetic properties of the uptake (or export) of sucrose and amino acids in various materials

Species	Parameter investigated	Material	Phases	K_{m1} (mol m^{-3})	High affinity
	A. Sugars				
Beta vulgaris	Export of sucrose	abraded leaf attached to the plant	2S	16	4.9
	uptake of sucrose	abraded leaf discs	2S	88	4.6
Glycine max	export of sucrose	abraded leaf attached to the plant	2S	35	2.9
Beta vulgaris	uptake of sucrose	abraded leaf discs	2S	25	9.1
			2S	8	2.8
Ricinus communis		cotyledon	S + L	25	1883
			S + L	34	2299
Vicia faba		peeled leaf discs	2S + L	2.6	0.6
Phaseolus vulgaris		non-abraded discs	S	11	0.2
Phaseolus coccinius			S + L	15	1.25
Beta vulgaris			S + L	12	ND
Nicotiana tabaccum		peeled discs	2S	32	18.2
		mesophyll cells	?	31	1.5
		isolated veins	?	16	2.0
Heracleum sosnowskyi		isolated phloem	S	7.6	98.3
		isolated xylem	?	14.3	313.3
Beta vulgaris		isolated bundles from petiole	?	0.01	155
Plantago major			?	0.01	322
Heracleum sp.		isolated phloem	?	0.006	73
		isolated xylem	?	0.005	110

	Kinetic Parameters			
V_{max1}	K_{m2} (mol m^{-3})	Low affinity	V_{max2}	
nmol cm^{-2} min^{-1}	620	23.1	nmol cm^{-2} min^{-1}	Sovonick *et al.*, 1974
—	270	9.8	—	—
—	169	14.2	—	Servaites *et al.*, 1979
—	n.d.	n.d.		Giaquinta, 1977
—				Maynard and Lucas, 1982a
nmol gFW^{-1} min^{-1}				Komor, 1977
—				Hutchings, 1978
nmol cm^{-2} min^{-1}	35	3.1	—	Delrot and Bonnemain, 1981
—				Sturgis and Rubery, 1982
—				Daie and Wyse, 1985
				Turkina and Sokholova, 1972
—	n.d.	n.d.		Cataldo, 1974
—				—
—				—
nmol gFW^{-1} min^{-1}				Sokholova *et al.*, 1979
—				—
—				Grimm *et al.*, 1983
—				—
—				—
—				—

Table 5.1 (continued)

Species	Parameter investigated	Material	Phases	K_{m1} (mol m^{-3})	High affinity
	A. Sugars				
Cyclamen persicum		isolated bundles from petiole	S + L	0.003	55
Commelina benghalensis		isolated veins	S + L	0.5	53
		peeled discs	2S	1.2	7.5
		mesophyll cells	2S	0.16	5.0
Apium graveoleus		isolated veins	S + L	14.3	33.0
		isolated phloem	S	4.0	20.0
Glycine max		protoplasts of cotyledon cells	S + 2L	1.5	5
	B. Amino acids				
Glycine max	Export of leucine	abraded leaf attached to the plant	3S	1.0	0.015
Ricinus communis	uptake of glutamine	cotyledons	S	11	1000
Vicia faba	uptake of glycine	peeled leaf discs	2S	4.2	3.53
	uptake of threonine		2S	4.7	2.52
	uptake of α-AIB		2S	6.0	1.33
Commelina benghalensis	uptake of valine	peeled discs	2S + L	0.04	2
		mesophyll cells	2S + L	0.04	10

Values were determined at pH 5.0 or 6.0. S: 1 saturable phase; 2S: 2 saturable phases; S + L: 1 saturable + 1 linear component; S + 2L: 1 saturable + 2 linear components.

Kinetic Parameters				
V_{max1}	K_{m2} (mol m^{-3})	Low affinity	V_{max2}	
—				—
—				Van Bel and Koops, 1985
—	150	220	nmol gFW^{-1} min^{-1}	Van Bel et al., 1986
—	340	250	—	—
—				Daie, 1985
—				—
nmol 10^6 protoplasts^{-1} min^{-1}				Schmitt et al., 1984
nmol cm^{-2} min^{-1}	15	0.67	nmol cm^{-2} min^{-1}	Servaites et al., 1979
nmol gFW^{-1} min^{-1}	+ 45	2	—	Robinson and Beevers, 1981
nmol cm^{-2} min^{-1}	49.5	13.6	—	Despeghel, 1981
—	42.7	12	—	—
—	121.0	15.4	—	—
nmol gFW^{-1} min^{-1}	2.4	53	nmol gFW^{-1} min^{-1}	Van Bel et al., 1986
—	3.3	30	—	—

affinity system in phloem loading of sucrose should await extensive measurements of the local concentration of sucrose at the phloem boundary. Estimations of overall apoplastic sucrose concentrations obtained to date are of limited use, because numerous parameters may locally increase or decrease this concentration.

5.3.3.2.2 Amino acids McNeil (1979) fed valine to detached shoots of *Lupinus albus* and found evidence for a carrier-mediated phloem loading of this amino acid in the leaf, although, unfortunately, the kinetic parameters of loading cannot be easily derived from the data presented. Studies on the export of exogenous leucine from abraded leaves still attached to the plant showed the existence of three saturable components (Servaites *et al.*, 1979). Multi-phasic isotherms are also commonly found in studies with isolated materials (Table 5.1). There is evidence that the high-affinity system works with proton symport (Robinson & Beevers, 1981; Despeghel & Delrot, 1983). In leaf disc experiments autoradiographs show that amino acids are concentrated by the phloem, but are also absorbed by the parenchyma much more than sucrose (Despeghel *et al.*, 1980), making such experiments of limited use to study phloem loading of amino acids. With a few exceptions, the amino acid content of the phloem sap in rice (Fukumorita & Chino, 1982) and broad bean (Despeghel, 1981) parallels that of the xylem sap or of the leaf apoplast, suggesting that loading of amino acids is not very selective (Servaites *et al.*, 1979). This is not surprising since studies on the specificities of carriers have shown that three different carriers are sufficient to transport the range of the 20 most common amino acids (Kinraide & Etherton, 1980; Wyse & Komor, 1984). One system is specific for all neutral amino acids including glutamine, asparagine and histidine. The basic amino acids are transported by another system, and the acidic amino acids by a third one.

The location of the high-affinity and low-affinity systems of amino acid uptake is still conjectural. Van Bel (1986) proposed that, in *Commelina*, the amino acids are maintained in the mesophyll by the high-affinity component. Then they would be concentrated and released near the conducting complex, where the low-affinity system would take them up efficiently. This model, based on indirect estimations of the contributions of the mesophyll and of the veins in total uptake, needs further investigation. It does not fit well with the effect of hyperosmotic treatments on the uptake of amino acids and sucrose. Autoradiographs show that uptake and phloem loading of amino acids are stimulated by plasmolysis (Hagège *et al.*, 1984). Likewise, phloem loading of sucrose is either not sensitive (Sovonick *et al.*, 1974) or strongly stimulated by high osmoticum concentrations which plasmolyse the parenchyma cells (Geiger & Fondy, 1980; Delrot, 1981; Hagège *et al.*, 1984; Lemoine & Delrot, 1987), and only the high-affinity component of sucrose uptake, present in the veins (see above), is turgor-sensitive (Daie & Wyse, 1985; Delrot, unpublished).

5.3.4 Evidence for proton-sucrose co-transport

In several species, there is overwhelming evidence that, at low concentrations, sucrose is taken up by symport with protons (Giaquinta, 1977a). For example, a summary of the data obtained in the author's laboratory illustrating loading in broad bean leaf is presented in Fig. 5.2. The evidence can be summarized as follows. In leaf discs of sugar-beet (Giaquinta, 1977b) and of broad bean (Delrot & Bonnemain, 1981), acidic pH decreases the K_m of the high-affinity component with little effect on V_{max}. A detailed study of the effect of pH showed that the kinetics of uptake were best explained by two-substrate kinetics, with the proton and the sucrose molecule as substrates (Delrot & Bonnemain, 1981). The K_m for proton uptake suggests that the carrier is half-maximally protonated at pH 9.0. The low-affinity system is not dependent on the pH (Giaquinta, 1977b; Delrot & Bonnemain, 1981), although this may depend on the experimental conditions (Maynard & Lucas, 1982a; Lemoine *et al.*, 1984). In protoplasts of soybean cotyledon, the sucrose carrier has a different sensitivity, since acidic pH decreased the K_m but also increased the V_{max} (Schmitt *et al.*, 1984). Whether this different pH sensitivity is due to the species, to the organ, or to the fact that isolated protoplasts may react differently from intact cells, is not known.

The evidence summarized above shows that protons interact with the sucrose carrier. That protons are actually moved through the membrane during sucrose uptake is also well documented. Addition of sucrose to the bathing medium of broad bean leaf fragments is followed by a pH transient dependent on the concentration of sucrose added (Delrot & Bonnemain, 1979; Delrot, 1981). This pH transient is due to the entry of protons into the tissues, with a constant stoichiometry of about 1.9 sucrose taken up/proton absorbed, for sucrose concentrations below 5 mol m^{-3}. Above 5 mol m^{-3} sucrose, the stoichiometry increases steadily, due to the operation of the low-affinity system, independent of protons. The K_m for sucrose uptake by the high-affinity component (2.1 mol m^{-3}) is in good agreement with the K_m for proton uptake (2.6 mol m^{-3}). The actual stoichiometry of the symport is difficult to assess, because of the biasing effect of the cell wall and of the medium, and because of the recycling of protons by the proton pump. Sucrose-induced alkalizations have also been reported after perfusion of castor bean petiole (Malek & Baker, 1977) and during sucrose uptake by cotyledons of the same species (Komor, 1977; Hutchings, 1978). The latter authors found an apparent stoichiometry of about 1 between sucrose and protons. There is also good evidence for proton-sucrose symport during phloem loading in monocotyledons. Heyser (1980) observed sucrose-induced alkalizations on perfusion of sucrose in the vascular bundles of corn. Since the suberized lamella prevents the movement of the perfusate in the mesophyll and since the conducting complex is symplastically isolated from the vascular parenchyma, the proton fluxes observed can be ascribed to phloem cells.

185

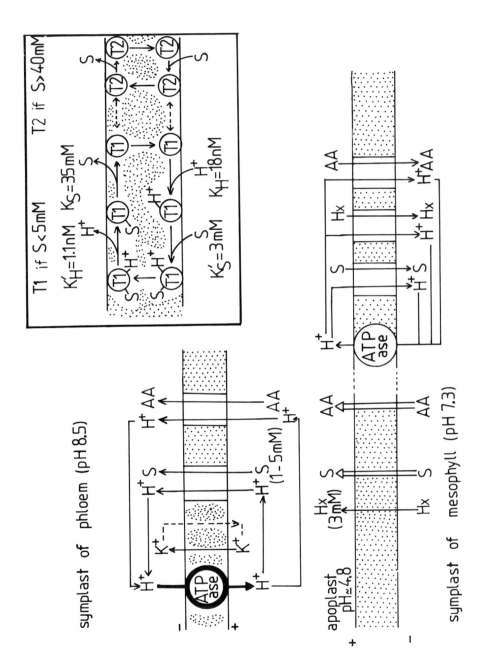

Several authors have shown that sucrose uptake is accompanied by K^+ efflux (or by a reduction of K^+ influx) for charge compensation of the protons entering into the cells (Cho & Komor, 1980; Delrot, 1981). Smith & Milburn (1980a) found an inverse relationship between the sucrose and the K^+ content of the phloem exudate of *Ricinus* plants in the dark.

Wright & Fisher (1981) used the stylet of an aphid as a micro-electrode to measure the transmembrane potential difference of the sieve tube and they were able to show sucrose-induced depolarizations. However, hexoses were almost as efficient as sucrose in eliciting these depolarizations. Evidence for co-transport mechanism has also been found with non-phloem transfer cells (Jones *et al.*, 1975; Caussin *et al.*, 1979; Pichelin *et al.*, 1984).

High KCl concentrations (100 mol m^{-3}) which depolarize the membrane also inhibit phloem loading (Servaites *et al.*, 1979; Vyskrebentseva & Semenov, 1975), showing the importance of the proton motive force for this phenomenon. The last piece of evidence favouring a mechanism of proton-sucrose symport is given by the promoting effects of fusicoccin, a well-known activator of the plasmalemma proton pump (Delrot & Bonnemain, 1979). This effect can be easily explained by an increase in the proton motive force energizing the proton-sucrose symport.

The effects of pH on the kinetics of amino acid uptake by leaf discs (Despeghel, 1981) and amino acid-induced alkalizations show that amino acid uptake also occurs with proton symport (Baker *et al.*, 1980; Despeghel & Delrot, 1983), but unlike sucrose, both the high- and the low-affinity systems are pH-dependent. Because a large part of amino acid uptake in leaf discs occurs in the parenchyma, it is more difficult to ascribe these phenomena to phloem loading than it was for sucrose, which is mainly taken up by the veins. However, there is also very strong evidence for an amino acid-proton symport in transfer cells of mosses, which are a convenient material for pH and electrophysiological measurements (Caussin *et al.*, 1979; Pichelin *et al.*, 1984). Altogether, these data suggest that phloem loading of amino acids is also a proton co-transport process.

Fig. 5.2 Proton-sucrose co-transport during phloem loading in *Vicia faba* leaf. A mesophyll cell (bottom) and a cell of the conducting complex (top) are represented. A plasmalemma proton pump, more concentrated or more active in the phloem than in the mesophyll, generates a proton motive force (PMF). The PMF energizes the uptake of various solutes from the apoplast with proton symport (AA = amino acid; Hx = hexose; S = sucrose). The proton pump also drives uptake of K^+, and less K^+ is taken up when the symports are active. The mechanism of solute efflux from the mesophyll is unknown. Inset (top right) gives computed values of kinetic parameters. Below 5 mol m^{-3} sucrose, a proton-sucrose symport (T1) is active, and above this concentration another apparent system (T2), much less dependent on protons, may work. The exact nature of T2, as well as its relationship with T1, are unknown. K_S and K'_S are the affinities of the unprotonated and protonated forms of T1 for sucrose, and K_H and K'_H are the affinities of the unloaded carrier and of the carrier loaded with sucrose for the protons. These numbers and the representation of the carrier have only operational values.

5.4 Molecular aspects of sucrose uptake and phloem loading

5.4.1 Stereochemical requirements of the sucrose carrier

Relatively few detailed experiments have been devoted to the study of the substrate specificity of the sucrose carrier. Recent work based on counter-exchange experiments (M'Batchi & Delrot, submitted) has given some insight into the specificity of the sucrose carrier in broad bean leaf. Comparisons were made with the hexose carrier of the same tissue (Table 5.2). The kinetic properties of the sucrose carriers are such that, in the presence of a transported sugar in the medium, the efflux of pre-loaded [^{14}C]sucrose is markedly stimulated, compared with the efflux in a medium without sugar or with a sugar not transported by the carrier. The same holds for the efflux of [^{14}C]3-O-MeG in the presence of sugars transported by the hexose carrier. The data (Table 5.2) show that nitrophenyl glucosides or nitrophenyl galactosides are readily transported by the sucrose carrier, with βnitrophenyl galactose being a substrate almost as good as sucrose itself, and with α-D-phenyl glucoside much better recognized than β-D-phenyl glucoside. The hydrophobic nature of the phenyl ring seems important for recognition, since methylglucoside is not transported. These data confirm recent results obtained by another approach on the sucrose carrier of soybean cotyledon protoplasts (see below). Amongst the other molecules efficient in the test are maltose and hexose-phosphates (G1P, G6P, F6P). Maltose recognition is in agreement with results from other approaches (M'Batchi *et al.*, 1985) or from other materials (Maynard & Lucas, 1982a), although the affinity of the carrier for maltose is weaker than for sucrose (Komor, 1977; Schmitt *et al.*, 1984). The sucrose analogues palatinose and turanose are inefficient in the test, also confirming the previous suggestion that they are not transported (M'Batchi & Delrot, 1984; Schmitt *et al.*, 1984). Interestingly, raffinose, a trisaccharide derived from sucrose, is not transported by the sucrose carrier, although some data suggest that it is recognized (M'Batchi *et al.*, 1985; Pichelin-Poitevin *et al.*, 1987). Stachyose is not efficient in driving the counter-exchange of sucrose, and likewise it is not phloem-loaded in sugar-beet (Fondy & Geiger, 1977). The same remark applies to phloridzin, which is recognized (Lemoine & Delrot, 1987), but not transported by the sucrose carrier (Table 5.2.).

The hexose carrier transports to a significant extent the following sugars: 2-deoxyglucose, fructose, glucose, galactose, mannose, fucose, xylose, G6P, G1P, and α and β nitrophenyl galactose. Sucrose and other disaccharides were not transported, except maltose to some extent. The possibility that the maltose effect may be mediated through invertase hydrolysis cannot be excluded at the present time. Raffinose and stachyose were inefficient in driving the counter-exchange of 3-O-MeG. In summary, the sucrose carrier of broad bean leaf exhibits a specificity markedly different

188

Table 5.2 Specificity of the sucrose carrier and of the hexose carrier of broad bean leaf as estimated by counter-exchange. Leaf discs pre-loaded with [^{14}C] sucrose or [^{14}C] 3-0-MeG were rinsed, and effluxed in a medium containing no sugar (control) or the sugar tested. The numbers reported (means of 2 experiments ± s.e.) are the ratio 'rate of efflux in the presence of sugar/rate of efflux in control'. For other explanations see the text. Sugars were tested at 50 mol m^{-3}, except * (30 mol m^{-3}) and ** (5 mol m^{-3}). Sugars in italic are recognized by the sucrose carrier

Substrate tested	Efflux (test/control)	
	sucrose	3-0-MeG
D-arabinose	1.25 ± 0.21	1.70 ± 0.21
D-ribose	0.85 ± 0.20	0.79 ± 0.14
D-xylose	1.27 ± 0.24	11.75 ± 0.94
D-xylitol	1.20 ± 0.07	1.37 ± 0.14
Ascorbic acid	2.48 ± 0.22	2.79 ± 0.07
D-fructose	1.10 ± 0.26	3.23 ± 0.23
F6P	2.87 ± 0.18	3.74 ± 0.16
D-fucose	1.14 ± 0.31	12.56 ± 0.53
L-fucose	1.42 ± 0.18	3.08 ± 1.23
D-galactose	1.34 ± 0.05	10.22 ± 0.61
G-glucose	1.21 ± 0.07	8.42 ± 1.82
D-glucose 25 mol m^{-3} + D-fructose 25 mol m^{-3}	1.13 ± 0.13	7.99 ± 0.89
G1P	2.92 ± 0.13	5.54 ± 2.63
G6P	2.40 ± 0.06	15.90 ± 3.41
2-deoxyglucose	1.41 ± 0.06	14.68 ± 1.64
L-glucose	0.92 ± 0.06	4.45 ± 0.08
D-glucose amine	1.61 ± 0.01	6.72 ± 0.22
N-methylglucamine		2.59 ± 0.15
Mannose	0.95 ± 0.04	10.37 ± 0.94
Myo-inositol	0.89 ± 0.23	1.05 ± 0.21
3-0-MeG	1.08 ± 0.02	9.75 ± 1.53
L-rhamnose	0.92 ± 0.20	0.51 ± 0.11
D-sorbitol	0.94 ± 0.01	1.39 ± 0.33
L-sorbose	1.32 ± 0.20	1.55 ± 0.43
αmethyl glucose	1.27 ± 0.13	3.07 ± 0.06
βmethyl glucose	1.04 ± 0.01	3.30 ± 0.19
*αnitrophenyl galactose**	2.43 ± 0.51	4.53 ± 0.48
*βnitrophenyl galactose**	3.07 ± 0.40	11.51 ± 0.55
*αnitrophenyl glucose**	4.16 ± 0.57	2.23 ± 0.67
αD-phenyl glucoside	3.79 ± 0.31	1.16 ± 0.10
βD-phenyl glucoside	1.69 ± 0.68	1.46 ± 0.10
Phloridzin (5 mol m^{-3})**	1.21 ± 0.37	
Lactitol	0.67 ± 0.15	
αD-lactose	0.85 ± 0.13	1.91 ± 0.16
βD-lactose	0.79 ± 0.03	2.71 ± 0.40
Lactulose	0.78 ± 0.03	1.45 ± 0.02
Maltose	3.36 ± 0.02	3.16 ± 0.42
Melibiose	0.81 ± 0.23	0.81 ± 0.02
Palatinit	0.63 ± 0.04	1.22 ± 0.06
Palatinose	0.99 ± 0.03	1.23 ± 0.27
Saccharose	4.81 ± 1.59	0.96 ± 0.22
Turanose	1.32 ± 0.14	1.26 ± 0.18
Raffinose	1.69 ± 0.11	1.62 ± 0.27
Stachyose	1.10 ± 0.08	1.44 ± 0.01

from that of the hexose carrier, although some molecules might be transported by both carriers (G1P, G6P, F6P, and β nitrophenyl galactose). Non-specific effects of the latter molecules cannot be ruled out at the present time.

Although it is not known if the sucrose carrier of protoplasts of soybean cotyledon cells is the same as that involved in phloem loading in the leaves, it is worth mentioning the elegant studies of Hitz and co-workers (Card *et al.*, 1986; Hitz *et al.*, 1986). These authors combined chemistry and enzymology to synthesize a number of sucrose derivatives (O-glycosides and S-glycosides) varying by the degree of substitution and the position of the substituents (fluor, benzyl, phenyl). These derivatives were assayed as alternate substrates in a transport assay with the protoplasts. The data suggest that a large portion of substrate recognition by the sucrose carrier of this material arises from the interaction of a relatively hydrophobic portion of the sucrose molecule and a hydrophobic region of the carrier protein binding site. The fructose moiety of sucrose represents a large portion of the hydrophobic surface interacting with the binding site. With the possible exception of the OH-3′, none of the OH groups of fructose appear to participate in intermolecular bonding with the protein. The OH-1′ is intramolecularly hydrogen-bonded and thus presents a non-polar surface for interaction with the carrier protein. The glucosyl hydroxyls at positions 3, 4 and 6 are also involved in substrate recognition, apparently by donating their hydrogen to the protein in hydrogen bonding. The S-glucosides were better substrates than the O-glucosides. Studies made with broad bean (Table 5.2) and with soybean (Hitz *et al.*, 1986) agree that the interactions between sucrose and its carrier are mainly hydrophobic.

5.4.2 Evidence for the presence of a thiol in (or close to) the active site of the sucrose carrier

In 1976, Giaquinta showed that phloem loading of assimilates and uptake of exogenous sucrose were sensitive to the non-permeant thiol reagent PCMBS, while non-permeant amino reagents had no effect. In the context of proton- sucrose symport, PCMBS inhibition can be envisaged as an effect on the proton pump, on the sucrose carrier, or both. Based on measurements of the rate of acidification of the medium by sugar-beet leaf tissues, Giaquinta (1979) concluded that the PCMBS effect on loading was mediated via the proton pump. Yet, similar experiments conducted on broad bean led to the conclusion that, at relatively low concentrations and with relatively short times of treatment, PCMBS directly inhibits the sucrose carrier without affecting the proton pump (Delrot *et al.*, 1980). This direct effect of PCMBS was confirmed by electrophysiological measurements performed on various materials, including sugar-beet and broad bean leaves, and soybean and broad bean cotyledons (Lichtner & Spanswick, 1981; M'Batchi *et al.*, 1986). Nevertheless the data suggest that PCMBS may enter the cells, particularly cotyledon cells, if treatments longer than 20 min with 0.5 mol m^{-3} inhibitor are used. Kinetic studies showed that the inhibition of sucrose uptake by PCMBS under mild conditions is of a competitive type, suggesting that the inhibitor binds to a thiol group present in the active site of the carrier (Delrot & M'Batchi, 1984). Additions of sucrose, maltose, raffinose (M'Batchi *et al.*, 1985) or αD-phenyl glucoside (M'Batchi & Delrot, unpublished) during pre-treatment of the tissue with PCMBS are able partially to protect the sucrose carrier against PCMBS inhibition. In contrast, hexoses or non-transported sucrose analogues (palatinose, turanose) were not protective. It was assumed that saturation of the active site by high concentrations of transported sugars during the pre-treatment to some extent prevented the access of PCMBS to its target, and thus resulted in a partial protection (M'Batchi *et al.*, 1985). This interpretation was strengthened by the fact that only the protective sugars were able to decrease the binding of [^{203}Hg]PCMBS to the leaf discs (M'Batchi *et al.*, 1985). The proteins differentially labelled by PCMBS in the presence of sucrose are preferentially localized in the 120 000 g microsomal pellet after cell fractionation. As expected for a membrane carrier, they are solubilized by detergents solubilizing intrinsic proteins, but not by treatments solubilizing only peripheral proteins (M'Batchi *et al.*, 1987). This indirect but concurring evidence for the existence of a sucrose- protected PCMBS-sensitive group in the active site of the sucrose carrier of broad bean leaf gives rise to the possibility of differential labelling of this protein by thiol reagents, as has been done successfully with several sugar carriers in bacteria.

5.4.3 Recent advances towards the isolation of sucrose carrier proteins

Compared with bacterial systems, very little is known about the identity of the sugar carriers in plants. Preparation of defective or overproducing mutants for carriers, which is very helpful in these kinds of studies, is much easier in bacteria than in plants. Only recently have efficient methods appeared for the routine preparation of plasma membranes with sufficient yield and purity, even from green tissues (Larsson, 1985). The characterization of solute carriers devoid of enzymatic (chemical) activity must therefore rely on the use of indirect probes such as substrate derivatives or competitive inhibitors. Nevertheless, the characterization and isolation of the sucrose carrier in a functional form would open up a wide range of new studies, including the molecular functioning of the sucrose carrier in reconstituted proteoliposomes and its distribution amongst different membranes, cells, organs and plants by immunocytochemistry. This protein may be considered as an essential part of the machinery allowing long-distance transport not only of sucrose, but also of compounds transported by mass-flow in the phloem sap. In this regard, it may be both a potential and a selective target for herbicides (animal cells, which transport glucose and not sucrose, are devoid of a sucrose carrier), or a 'door' into the phloem for fungicides or insecticides which could possess sites recognized by the carrier by their spatial structure.

The first steps towards a better knowledge of this protein have been made recently, using photo-affinity labelling and differential labelling. In a preliminary abstract, Liu *et al.* (1986) reported the synthesis of a photolabile derivative of sucrose (6′-deoxy-6′-(2-hydroxy-4-azido)-benzamido sucrose) and its use to label a polypeptide induced by sucrose in an inducible strain of *Pseudomonas saccharophilae*. Overproducing strains of this 84 kD polypeptide took up sucrose at higher rates. The same photo-affinity probe labelled a 63 kD polypeptide in protoplasts of soybean cotyledons (Ripp *et al.*, 1986).

Based on the presence of a protectable thiol in the active site of the sucrose carrier, Pichelin-Poitevin & Delrot (1987), and Pichelin-Poitevin *et al.* (1987) identified, in microsomal fractions and in purified plasma membranes of broad bean leaf, a 42 kD polypeptide, differentially labelled by the irreversible thiol reagent *N*-ethylmaleimide (NEM). The differential labelling was observed only when transported sugars (sucrose, αD-phenyl glucoside) were present in the incubation medium, but not in the presence of the non-transported analogues palatinose and turanose. In the author's laboratory the 42 kD polypeptide has been differentially labelled in sugar beet and the antiserum to this polypeptide selectively inhibited sucrose uptake.

5.5 Loading of xenobiotics

The transport of xenobiotics, and particularly their phloem mobility, is a critical step for their activity *in situ*. Numerous compounds active *in vitro* may be inefficient *in vivo* due to their poor transport. For this reason, phloem loading of xenobiotics is receiving more and more attention. Phloem transport is particularly important for those compounds which must exert their effect in areas of new growth (fungicides, insecticides), for herbicides inhibiting cell division in meristems, or for plant growth regulators active in sink regions. Therefore, it seems worth briefly comparing the present information available on phloem loading of pesticides and on phloem loading of photoassimilates. At the present time, both phenomena have little, if any, relationship, because loading of assimilates is carrier-mediated whilst loading of xenobiotics is thought to occur by diffusion through the lipid bilayer of the membrane. Until now, all attempts (saturation kinetics, competition experiments, use of inhibitors) to demonstrate that all or part of the uptake of systemic pesticides is carrier-mediated have failed. Therefore, although this kind of mechanism cannot be excluded, there is still no evidence for it. It must be stressed that, in spite of numerous papers on membrane transport of pesticides, only a small number are relevant to phloem loading. Many studies have dealt with cells in culture or with fragments of parenchyma in which the conducting elements of phloem were almost, or completely, absent. When studying membrane transport, parenchyma cells cannot be equated with the cells of the conducting complex. Many peculiarities of the conducting cells have been mentioned throughout this chapter. A striking example with xenobiotics is given by the comparison of [^{14}C]glyphosate (phloem-mobile herbicide) and [^{14}C]iprodione (non-phloem-mobile fungicide). Iprodione is taken up in broad bean leaf discs at a much higher rate than glyphosate. Yet, autoradiographs show that glyphosate is phloem-loaded while iprodione remains excluded from the phloem (El Ibaoui *et al.*, 1984).

Two popular theories explain, at least in part, phloem systemy of pesticides.

5.5.1 *The acid trap mechanism*

For compounds possessing a weak acid function (carboxyl) or a structure possessing a weak acid character, it is well known that the undissociated form of the acid moves easily through the lipophilic membrane, because of its neutrality, while the corresponding anion is much less permeant. At diffusion equilibrium, the concentration of the undissociated form is the same on both sides of the membrane. Given the respective pH of the cell wall (5.0) and of the cytoplasm (7.0–7.5), this implies that the compound will accumulate strongly in the intracellular compartment. This model applies to the natural hormones auxin and abscisic acid (Everat-Bourbouloux *et al.*, 1984) as well as to 2,4-D analogues or derivatives of benzoic acid (Grimm *et al.*, 1983, 1985; Neumann *et al.*, 1985). Nevertheless many phloem-mobile agrochemicals are neutral or zwitterionic, and their systemy cannot be explained in terms of the acid trap mechanism.

5.5.2 *The 'intermediate permeability' hypothesis*

According to this model, developed by Tyree *et al.* (1979), phloem mobility of neutral xenobiotics depends on their permeability coefficient (*P*) through the membrane. The authors designed a linearized model of the conducting bundle of the plant and derived equations predicting the concentration of the xenobiotic at a given distance from the source, as a function of *P*, the length of the plant, the radius of the sieve tube, and the velocity of the phloem sap. One main feature stressed by the model is that the water flow displaced by the transpiration stream in the apoplast is much stronger than, and opposite to, the phloem flow. The permeability coefficient is generally estimated by compartmental studies of pesticide fluxes in potato tuber tissue (Tyree *et al.*, 1979; Lichtner, 1986). Chemicals with

193

a low P are assumed to be washed away into the xylem stream without any possibility of entry into the phloem. Pesticides highly permeable through the membranes of the conducting complex will rapidly equilibrate across these membranes in the loading zone. Since the phloem sap moves the xenobiotics out of the loading area, continuous re-equilibration occurs between the phloem and the apoplast. The large water flux of xylem, assumed to be in direct continuity with the leaf apoplast, moves the xenobiotic back into the leaf, up to the leaf margins, and export of the chemical does not occur efficiently. Pesticides with an intermediate permeability will accumulate slowly in the phloem, but will also exit out of the phloem slowly enough to allow export out of the leaf.

Several sets of data support this theory (Gougler & Geiger, 1981; Jachetta *et al.*, 1986; Lichtner, 1986). For example, the systemy of glyphosate is generally interpreted by the model of intermediate permeability. The permeability coefficient found for this compound was $9.6 \; 10^{-10}$ m s^{-1} in sugar-beet (Gougler & Geiger, 1981) and $2.5 \; 10^{-10}$ m s^{-1} (not $1.5 \; 10^{-8}$ m s^{-1}, as incorrectly stated in the original paper; El Ibaoui *et al.*, 1984) in broad bean. Amitrole has a permeability coefficient of $2.1 \; 10^{-9}$ m s^{-1} in isolated cells in roots of *Phaseolus vulgaris* (Lichtner, 1983) and of $5.9 \; 10^{-10}$ m s^{-1} in leaf tissues of broad bean (El Ibaoui, unpublished). These data fit rather well with the theory, but P is not always an unequivocal parameter. For example, in broad bean, the non-systemic fungicide iprodione has a P value of $2.5 \; 10^{-9}$ m s^{-1} (El Ibaoui *et al.*, 1986) which seems to be still within the range of P considered to be favourable to phloem systemy (Tyree *et al.*, 1979). Although it proves a good tool for the study of phloem systemy of agrochemicals, the model of intermediate permeability still suffers from potential pitfalls. The model does not take into account the asymmetry often found between the entry and the exit of pesticides into or out of, the cells (Martin & Edgington, 1981). The efflux is often slower than the influx (Singer & McDaniel, 1982; El Ibaoui *et al.*, 1984; Jachetta *et al.*, 1986). The permeability coefficient is estimated with potato tuber tissue. Although these tissues derive from phloem, they have lost most of their conducting properties and thus may not be an ideal material for such studies.

A diffusion model unifying the acid trap and the intermediate permeability mechanism has been recently developed and tested (Kleier, 1988; Hsu *et al.*, 1988). The model predicts that for a given acid dissociation constant there is an optimum permeability (and vice versa), and that membrane permeability is the overriding factor in determining phloem mobility.

5.6 Control of loading

Regulation of phloem loading can be envisaged at various levels, which all come to affect basically the same parameters, i.e. the transmembrane sucrose gradient, and the energization status of the membrane. This status

194

can be estimated by the magnitude of the proton motive force (PMF), with PMF = $\triangle\psi - z\,\triangle$pH. Changes in the transmembrane distribution of ions, particularly H^+ and K^+, may potentially affect phloem loading. The efficiency of phloem loading may also depend on the number and on the permeability of the plasmodesmata at the boundary of the conducting complex. The actual regulation *in vivo* results from all the factors which can affect these parameters, some of which are briefly reviewed below.

5.6.1 Points of control

5.6.1.1 The transmembrane sucrose gradient

In *Ricinus*, cutting of the stem and exudation of the phloem sap promotes phloem loading of sucrose (Smith & Milburn, 1980a). This effect can be interpreted as the result of a decrease in turgor (see Sect. 5.4.4) or of a decrease in the sucrose concentration of the conducting elements. Conversely, in broad bean, blocking of sugar export results in the inhibition of phloem loading within five minutes (Ntsika & Delrot, 1986). At least part of this inhibition may be ascribed to an increase in the internal sucrose concentration of the sieve tube, sucrose uptake being inhibited in tissues pre-loaded with high sucrose concentrations (Komor, 1977; Giaquinta, 1980; M'Batchi & Delrot, 1984).

Little is known about the possible variations of apoplastic sucrose concentrations except that they exhibit nycthemeral changes, with a maximum (5 times higher) 6 h after the beginning of photosynthesis in broad bean, and a minimum at the end of the night (Delrot *et al.*, 1983). Since the sucrose concentrations measured in this work were overall apoplastic concentrations, it is not known to what extent these changes reflect what occurs in the apoplast of the conducting complex.

5.6.1.2 The proton motive force

All factors affecting the proton pumping activity of the membrane may be part of the complex regulation of phloem loading. Among these factors are all those able to affect the ATP supply of the conducting cells, which will not be detailed here.

5.6.1.3 Passive component

The composition of the cell wall may affect its Donnan potential and modify the external ionic composition perceived by the membrane. For example, in broad bean stem, the Donnan potential of the cell walls is such that the membrane 'sees' a pH of 4.7 when the pH of the bathing medium is 5.4 (Everat-Bourbouloux *et al.*, 1984). This effect is absent in the leaf of the same species (Delrot *et al.*, 1981). In this regard, the particular composition of the cell wall of the sieve tube (see Ch. 3) and the large volume of cell wall in the transfer cell may create unique conditions for uptake.

5.6.1.4 Active component

Rapid changes in the internal pH are unlikely to be significant in the control of loading, because the cytoplasmic pH is strongly regulated. However, the alkaline pH of the phloem, compared to the neutral pH of the cytoplasm of parenchyma cells, may be one of the reasons why sucrose accumulates preferentially in the sieve tube, although the proton-sucrose co-transport mechanism is also present in the parenchyma cells. Autoradiographs show that markers of alkaline pH (dimethyloxazolidine dione Li & Delrot, unpublished; ABA, Delrot *et al.*, 1981) accumulated by the acid trap mechanism are more efficiently absorbed in the phloem than in the parenchyma.

5.6.2 Factors of control

5.6.2.1 Hormones

Applications of IAA and kinetin to isolated bark strips of willow increase the concentration of sucrose in exudates of aphid, suggesting a hormonal regulation of loading (Lepp & Peel, 1970). Fusicoccin (FC) or IAA (10 mmol m^{-3}) rapidly promote the uptake of sucrose from a solution perfusing the hollow petioles of *Ricinus*, while ABA inhibits it (Malek & Baker, 1978). Loading of [^{14}C]sucrose in peeled broad bean leaf discs is strongly promoted by 10 mmol m^{-3} FC within 1 h (Delrot & Bonnemain, 1978), but is not sensitive to 10 mmol m^{-3} IAA or ABA (Delrot, unpublished). Sturgis and Rubery (1982) have reported that, in leaf discs of *Phaseolus vulgaris*, sucrose uptake was promoted by FC or IAA in the presence of 25 mol m^{-3} KCl. The stimulation by 10 mmol m^{-3} FC appeared at all sucrose concentrations tested (5 mol m^{-3} to 100 mol m^{-3}), but stimulation by 1 mmol m^{-3} IAA was only apparent for sucrose concentrations greater than 5–10 mol m^{-3}. Unfortunately, lack of autoradiographs and use of unabraded and unpeeled leaf discs make the relevance of the latter data to phloem loading uncertain. The same remark holds for the inhibition of sucrose uptake by ABA, reported for cotyledons of castor bean and for various leaf tissues (Vreugdenhil, 1983).

Aloni *et al.* (1986) recently reported that applications of gibberellic acid (GA$_3$) to mature leaves of trimmed plants of broad bean promoted export of assimilates. The effect could not be explained by changes in photosynthetic rate or in the chemical partitioning of carbon. It appeared with GA$_3$ treatment as short as 10 min duration, and was interpreted as a promotion of phloem loading by a mechanism still unknown. In the long term (24 h), indirect effects of GA$_3$ on export, mediated by changes in sink activitiy, were also involved. In spite of the lack of any statistical analysis, the paper may suggest a direct effect of GA$_3$ on phloem loading under '*in vivo*' conditions. Strong hormonal effects were also found with isolated vascular bundles of celery (Daie *et al.*, 1986). In this material low concentrations of GA_3 or IAA (1 mmol m^{-3}) added to unbuffered media doubled the uptake of sucrose and mannitol (translocated sugars) within 90 min, but uptake of 3-O-MeG (not translocated) was unaffected. The hormonal effect could not be explained by invertase activity, and was accompanied by increased export *in situ*. No hormonal increase of uptake was observed in the uptake of sucrose by sink tissues. The authors attributed these results to a direct effect of GA_3 and IAA on phloem loading.

5.6.2.2 Turgor control

Possible involvement of turgor in the regulation of loading is supported by data from different materials. In castor bean, reduction of sieve tube turgor by stem incision stimulated phloem loading (Smith & Milburn, 1980a,b). These data suggest that an increase in sink demand provokes a decrease in sieve tube turgor rapidly propagated to the source and induces a promotion of loading. Experiments with isolated leaf discs of *Ricinus* (Smith & Milburn, 1980a), of *Beta* (Geiger & Fondy, 1980), of *Vicia* (Hagège *et al.*, 1984; Lemoine & Delrot, 1987), and of *Phaseolus* (Daie & Wyse, 1985) also showed that a decrease in turgor, due to incubation of the discs in high osmoticum concentrations, promoted phloem loading. In sugar-beet, it was proposed that this effect may be due to plasmolysis of the mesophyll and easier access of the label to the vein apoplast (Geiger & Fondy, 1980). This explanation does not seem to hold for the lacunous tissues of broad bean (Lemoine & Delrot, 1987). Use of permeant and non-permeant osmotic buffers shows that sucrose uptake is sensitive to turgor rather than to osmotic potential (Geiger & Fondy, 1980; Smith & Milburn, 1980a; Daie & Wyse, 1985). Kinetic studies with *Phaseolus* leaf discs demonstrated that turgor affects only the saturable component of sucrose uptake, whereas the linear component was unchanged. The proton pump seems a good candidate to be the sensor and the transducer of osmotic conditions. Proton pumping, as measured by the rate of acidification of the medium, is sensitive to the osmotic conditions of the medium (Delrot & Bonnemain, 1979; Reinhold *et al.*, 1984), and this sensitivity is confirmed by electrophysiological measurements (Li & Delrot, 1987). Hagège *et al.* (1984) found that, in broad bean leaf discs, optimum uptake of sucrose occurred at a mannitol concentration (750 mol m^{-3}) different from the optimum for 3-O-MeG uptake (500 mol m^{-3}), the latter sugar being taken up almost exclusively by the parenchyma. These data suggest that the turgor of parenchyma cells does not respond in the same way as that of the sieve tube, as predicted from their respective osmotic potentials (Geiger *et al.*, 1973).

5.7 Effects of pollutants

Various pollutants may affect directly or indirectly phloem loading, a vital function for the plant. For example, excess Co^{2+} (400 mmol m^{-3}), Ni^{2+} (200 mmol m^{-3}) or Zn^{2+} (200 mmol m^{-3}) increases sucrose uptake by *Phaseolus* leaf discs, but autoradiographs show that phloem loading is inhibited (Rauser & Samarakoon, 1980). The effect was evident within 24 h of exposure to excess metal. Heavy metals such as Hg^{2+} and Pb^{2+} inhibit phloem loading in *Vicia* leaf discs. Kinetic studies indicate that the inhibition by Hg^{2+} is immediate and of a competitive type. The inhibition by Pb^{2+} only appears with treatments longer than 24 h, and at metal concentrations higher than 5 mmol m^{-3} (Maurousset, unpublished).

In *Phaseolus vulgaris*, exposure of the leaf to SO_2 for 2 h results in an inhibition of net photosynthesis (75%) and of loading (45%). However, the effect on photosynthesis is reversible, while the effect on loading is not restored upon removal of SO_2 (Teh & Swanson, 1982). The authors suggest that SO_2 or its derivatives (sulphites) may directly affect the sucrose carrier. SO_2 inhibition of phloem loading is immediate in C_3 plants, but does not appear in the short term with C_4 plants (Minchin & Gould, 1986). This difference is similar to that seen with anoxia, and it is proposed that the Kranz anatomy of the C_4 plants in the vicinity of the vascular bundles provides a barrier for SO_2.

References

Aloni, B., Daie, J. & Wyse, R. E. (1986). Enhancement of [^{14}C]sucrose export from source leaves of *Vicia faba* by gibberellic acid. *Plant Physiology* **82**, 962–7.

Altus, D. P. & Canny, M. J. (1985). Loading of assimilates in wheat leaves. II. The path from chloroplast to vein. *Plant, Cell and Environment* **8**, 275–85.

Atkins, C. A., Pate, J. S. & McNeil, D. L. (1980). Phloem loading and metabolism of xylem-borne amino compounds in fruiting shoots of a legume. *Journal of Experimental Botany* **31**, 1505–20.

Baker, D. A., Malek, F. & Dehvar, F. D. (1980). Phloem loading of amino acids from the petioles of *Ricinus* leaves. *Berichte der Deutschen botanischen Gesellschaft* **93**, 203–9.

Boyer, J. S. (1985). Water transport. *Annual Review of Plant Physiology* **36**, 473–516.

Brovchenko, M. I. (1963). The uptake of amino acids by the conducting tissue of leaves. *Soviet Plant Physiology* **10**, 416–25.

Brovchenko, M. I. & Ryabushkina, N. A. (1972). Possible coupling between sugar transfer from mesophyll cells into free space with energy of photophosphorylation. *Soviet Plant Physiology* **25**, 1144–50.

Card, P. J., Hitz, W. D. & Ripp, K. G. (1986). Chemoenzymatic syntheses of fructose-modified sucroses via multienzyme systems. Some topographical aspects of the binding of sucrose to a sucrose carrier protein. *Journal of the American Chemical Society* **108**, 158–61.

Cataldo, D. A. (1974). Vein loading: the role of the symplast in intercellular transport of carbohydrate between the mesophyll and minor veins of tobacco leaves. *Plant Physiology* **53**, 912–17.

Caussin, C., Despeghel, J. P., Faucher, M., Léger, A. & Bonnemain, J. L. (1979). Etude du mécanisme des échanges entre le gamétophyte et le sporophyte chez les Bryophytes. *Compte rendu hebdomadaire des séances de l'Académie des Sciences* **289**, 1329–34.

Cheruel, J., Jullien, M. & Surdin-Kerjan, Y. (1979). Amino acid uptake into cultivated mesophyll cells from *Asparagus officinalis* L. *Plant Physiology* **63**, 621–6.

Cho, B. H. & Komor, E. (1980). The role of potassium in charge compensation for sucrose-proton symport by cotyledons of *Ricinus communis*. *Plant Science Letters* **17**, 425–35.

Cho, B. H. & Komor, E. (1985). Comparison of suspension cells and cotyledons of *Ricinus* with respect to sugar uptake. *Plant Physiology* **118**, 381–90.

Cornel, D., Grignon, C., Rona, J. P. & Heller, R. (1983). Measurement of intracellular potassium activity in protoplasts of *Acer pseudoplatanus*: origin of their electropositivity. *Physiologia Plantarum* **57**, 203–9.

Daie, J. (1985). Sugar transport in leaf discs of *Phaseolus coccinius*. *Physiologia Plantarum* **64**, 553–8.

Daie, J. (1986). Kinetics of sugar transport in isolated vascular bundles and phloem tissue of celery. *Journal of the American Society of Horticultural Science* **111**, 216–20.

Daie, J., Watts, M., Aloni, B. & Wyse, R. E. (1986). *In vitro* and *in vivo* modification of sugar transport and translocation in celery by phytohormones. *Plant Sciences* **46**, 35–41.

Daie, J. & Wyse, R. E. (1985). Evidence on the mechanism of enhanced sucrose uptake at low cell turgor in leaf discs of *Phaseolus coccinius*. *Physiologia Plantarum* **64**, 547–52.

Delrot, S. (1981). Proton fluxes associated with sugar uptake in *Vicia faba* leaf tissues. *Plant Physiology* **68**, 706–11.

Delrot, S. & Bonnemain, J. L. (1978). Etude du mécanisme de l'accumulation des produits de la photosynthèse dans les nervures. *Compte rendu hebdomadaire des séances de l'Académie des sciences* **287**, 125–30.

Delrot, S. & Bonnemain, J. L. (1979). Echanges H⁺-Rb⁺ et co-transport H⁺-glucide dans les tissues foliaires de *Vicia faba* L. *Compte rendu hebdomadaire des séances de l'Académie des sciences* **288**, 71–6.

Delrot, S. & Bonnemain, J. L. (1981). Involvement of protons as a substrate for the sucrose carrier during phloem loading in *Vicia faba* leaves. *Plant Physiology* **67**, 560–4.

Delrot, S. & Bonnemain, J. L. (1984). Mechanism and control of phloem transport. *Physiologie Végétale* **23**, 199–200.

Delrot, S., Despeghel, J. P. & Bonneman, J. L. (1980). Phloem loading in *Vicia faba* leaves: effects of N-ethylmaleimide and parachloromercuribenzenesulphonic acid on H⁺ extrusion, K⁺ and sucrose uptake. *Planta* **149**, 144–8.

Delrot, S., Everat-Bourbouloux, A. & Bonnemain, J. L. (1981). Effet du pH sur l'absorption et l'exportation de l'acide abscissique par les tissus foliaires du *Vicia faba* L. *Compte rendu hebdomadaire des séances de l'Académie des sciences* **293D**, 659–63.

Delrot, S., Faucher, M., Bonnemain, J. L. & Bonmort, J. (1983). Nycthermeral changes in intracellular and apoplastic sugars in *Vicia faba* leaves. *Physiologie Végétale* **21**, 459–67.

Delrot, S. & M'Batchi, B. (1984). Existence possible d'un groupement thiol dans le site actif du transporteur de saccharose dans le feuille du *Vicia faba* L.

Compte rendu hebdomadaire des séances de l'Académie des sciences **298D**, 103–6.

Despeghel, J. P. (1981). Etude des mécanismes de l'absorption des acides aminés neutres par les tissus foliaires et de leur accumulation dans les nervures. *Thèse Doct. 3ème cycle*, University of Poitiers.

Despeghel, J. P. & Delrot, S. (1983). Energetics of amino acid uptake by *Vicia faba* leaf tissue. *Plant Physiology* **71**, 1–6.

Despeghel, J. P., Delrot, S., Bonnemain, J. L. & Besson, J. (1980). Etude du mécanisme de l'absorption des acides aminés par les tissus foliaires du *Vicia faba* L. *Compte rendu hebdomadaire des séances de l'Académie des sciences* **290D**, 609–14.

Doman, D. C. & Geiger, D. R. (1979). Effect of exogenously supplied foliar potassium on phloem loading in *Beta vulgaris* L. *Plant Physiology* **64**, 528–33.

El Ibaoui, H., Delrot, S., Besson, J. & Bonnemain, J. L. (1984). Uptake and release of a phloem-mobile (glyphosate) and of a non-phloem mobile (iprodione) xenobiotic by broad bean leaf tissues. *Physiologie Végétale* **24**, 431–42.

Erwee, M. G. & Goodwin, P. B. (1984). Characterization of the *Egeria densa* leaf symplast: response to plasmolysis, deplasmolysis and to aromatic amino acids. *Protoplasma* **122**, 162–8.

Erwee, M. G. & Goodwin, P. B. (1985). Symplast domains in extrastelar tissues of *Egeria densa* Planch. *Planta* **163**, 9–19.

Erwee, M. G., Goodwin, P. B. & Van Bel, A. J. E. (1985). Cell-cell communication in the leaves of *Commelina cyanea* and other plants. *Plant, Cell and Environment* **8**, 173–8.

Everat-Bourbouloux, A., Delrot, S.& Bonnemain, J. L. (1984). Propriétés de l'absorption et distribution de l'acide abscissique dans les tissus caulinaires du *Vicia faba* L. *Physiologie Végétale* **22**, 37–46.

Fisher, D. G. & Evert, R. F. (1982a). Studies on the leaf of *Amaranthus retroflexus* (Amaranthaceae): ultrastructure, plasmodesmatal frequency, and solute concentration in relation to phloem loading. *Planta* **155**, 377–87.

Fisher, D. G. & Evert, R. F. (1982b). Studies on the leaf of *Amaranthus retroflexus* (Amaranthaceae): quantitative aspects, and solute concentration in the phloem. *American Journal of Botany* **69**, 1375–88.

Fondy, B. R. & Geiger, D. R. (1977). Sugar selectivity and other characteristics of phloem loading in *Beta vulgaris* L. *Plant Physiology* **59**, 953–60.

Fukumorita, T. & Chino, M. (1982). Sugar, amino acid and inorganic contents in rice phloem sap. *Plant and Cell Physiology* **23**, 273–83.

Geiger, D. R. (1975). Phloem loading. In Zimmermann, M. H. & Milburn, J. A. (eds.). *Transport in Plants* 1. Phloem Transport, Encyclopedia of Plant Physiology, pp. 395–431. Berlin, Heidelberg, New York, Springer-Verlag.

Geiger, D. R. & Fondy, B. R. (1980). Response of phloem loading and export to rapid changes in sink demand. *Berichte der Deutschen botanischen Gesellschaft* **93**, 177–86.

Geiger, D. R., Giaquinta, R. T., Sovonick, S. A. & Fellows, R. J. (1973). Solute distribution in sugar beet leaves in relation to phloem loading and translocation. *Plant Physiology* **52**, 585–9.

Geiger, D. R., Sovonick, S. A., Shock, T. L. & Fellows, R. J. (1974). Role of free space in translocation in sugar beet. *Plant Physiology* **54**, 892–8.

Giaquinta, R. T. (1976). Evidence for phloem loading from the apoplast. Chemical modification of membrane sulfhydryl groups. *Plant Physiology* **57**, 872–5.

Giaquinta, R. T. (1977a). Possible role of pH gradient and membrane ATPase in the loading of sucrose into the sieve tubes. *Nature* **267**, 369–70.

Giaquinta, R. T. (1977b). Phloem loading of sucrose. pH dependence and selectivity. *Plant Physiology* **59**, 750–5.

Giaquinta, R. T. (1977c). Sucrose hydrolysis in relation to phloem translocation in *Beta vulgaris*. *Plant Physiology* **60**, 339–43.

Giaquinta, R. T. (1979). Phloem loading of sucrose. Involvement of membrane ATPase and proton cotransport. *Plant Physiology* **63**, 743–8.

Giaquinta, R. T. (1980). Mechanism and control of phloem loading of sucrose. *Berichte der Deutschen botanischen Gesellschaft* **93**, 187–201.

Giaquinta, R. T. (1983). Phloem loading of sucrose. *Annual Review of Plant Physiology* **34**, 347–87.

Giaquinta, R. T., Lin, W., Sadler, N. L. & Franceschi, V. R. (1983). Pathway of phloem unloading of sucrose in corn roots. *Plant Physiology* **72**, 362–7.

Goodwin, P. B. (1983). Molecular size limit for movement in the symplast of the *Elodea* leaf. *Planta* **157**, 124–30.

Gougler, J. & Geiger, D. R. (1981). Uptake and distribution of N-phosphonomethylglycine in sugar beet plants. *Plant Physiology* **68**, 668–72.

Gougler-Schmalstig, J. & Geiger, D. R. (1985). Phloem unloading in developing leaves of sugar beet. I. Evidence for pathway through the symplast. *Plant Physiology* **79**, 237–41.

Gougler-Schmalstig, J. & Geiger, D. R. (1987). Phloem unloading in developing leaves of sugar beet. II. Termination of phloem unloading. *Plant Physiology* **83**, 49–52.

Grimm, E., Neumann, S. & Jacob, F. (1983). Uptake of sucrose and xenobiotics into conducting tissue of *Cyclamen*. *Biochemie und Physiologie der Pflanzen* **178**, 29–42.

Grimm, E., Neumann, S. & Jacob, F. (1985). Transport of xenobiotics in higher plants. II. Absorption of defenuron, carboxyphenylmethylurea, and maleic hydrazide by isolated conducting tissue of *Cyclamen*. *Biochemie und Physiologie der Pflanzen* **180**, 383–92.

Gunning, B. E. S. & Overall, R. L. (1983). Plasmodesmata and cell-to-cell transport in plants. *Bioscience* **33**, 260–5.

Gunning, B. E. S., Pate, J. S., Minchin, F. R. & Marks, I. (1974). Quantitative aspects of transfer cell structure in relation to vein loading in leaves and solute transport in legume nodules. *Symposium of Society for Experimental Biology* **28**, 87–126.

Hagège, I., Hagège, D., Delrot, S., Despeghel, J. P. & Bonnemain, J. L. (1984). Effet de la pression osmotique sur l'excrétion des protons et l'absorption de K⁺(Rb⁺), de glucides et d'acides aminés par les tissus foliaires de *Vicia faba* L. *Compte rendu hebdomadaire des séances de l'Académie des sciences* **299D**, 435–40.

Hendrix, J. E. (1973). Translocation of sucrose by squash plants. *Plant Physiology* **52**, 688–9.

Hendrix, J. E. (1977). Phloem loading in squash. *Plant Physiology* **60**, 567–9.

Heyser, W. (1980). Phloem loading in the maize leaf. *Berichte der Deutschen botanischen Gesellschaft* **93**, 221–8.

Heyser, W., Evert, R. F., Fritz, E. & Eschrich, W. (1978). Sucrose in the free space of translocating maize leaf bundles. *Plant Physiology* **62**, 491–4.

Hitz, W. D., Card, P. J. & Ripp, K. G. (1986). Substrate recognition by a sucrose transporting protein. *Journal of Biological Chemistry* **261**, 11986–91.

Ho, L. C. & Baker, D. A. (1982). Regulation of loading and unloading in long distance transport systems. *Physiologia Plantarum* **56**, 225–30.

Housley, T. L., Peterson, D. M. & Schrader, L. E. (1977). Long distance translocation of sucrose, serine, leucine, lysine, and CO_2 assimilates. *Plant Physiology* **59**, 217–20.

Hsu, F. C., Kleier, D. A. & Melander, W. R. (1988). Phloem mobility of xeno-biotics. II. Bioassay testing of the unified mathematical model. *Plant Physiology* **86**, 811–16.

Huber, S. C. & Moreland, D. E. (1980). Efflux of sugars across the plasmalemma of mesophyll protoplasts. *Plant Physiology* **65**, 560–2.

Huber, S. C. & Moreland, D. E. (1981). Cotransport of potassium and sugars across the plasmalemma of mesophyll protoplasts. *Plant Physiology* **67**, 163–9.

Hutchings, V. M. (1978). Sucrose and proton cotransport in *Ricinus* cotyledons. I. H$^+$ influx associated with sucrose uptake. *Planta* **138**, 229–35.

Jachetta, J. J., Appleby, A. P. & Boersma, L. (1986). Apoplastic and symplastic pathways of atrazine and glyphosate transport in shoots of seedling sunflower. *Plant Physiology* **82**, 1000–7.

Jones, M. G. K., Novacky, A. & Dropkin, V. H. (1975). Transmembrane potentials of parenchyma cells and nematode-induced transfer cells. *Protoplasma* **85**, 15–37.

Kinraide, T. B. & Etherton, B. (1980). Electrical evidence for different mechanisms of uptake for basic, neutral and acidic amino acids in oat coleoptiles. *Plant Physiology* **65**, 1085–9.

Kleier, D. A. (1988). Phloem mobility of xenobiotics. I. Mathematical model unifying the weak acid and intermediate permeability theories. *Plant Physiology* **86**, 803–10.

Komor, E. (1977). Sucrose uptake by cotyledons of *Ricinus communis* L.: characteristics, mechanism and regulation. *Planta* **187**, 119–31.

Komor, E. & Tanner, W. (1974). The hexose-proton symport system of *Chlorella vulgaris*. pH dependent changes in K_m values and translocation constants of the uptake system. *Journal of General Physiology* **64**, 568–81.

Kursanov, A. L. & Brovchenko, M. I. (1969). Free space as an intermediate zone between photosynthesizing and conducting cells in leaves. *Soviet Plant Physiology* **16**, 801–7.

Larsson, C. (1985). Plasma membranes. In Linskens, H. F. & Jackson, J. F. (eds.). *Modern Methods of Plant Analysis*, pp. 85–104. Berlin, Springer-Verlag.

Lemoine, R. & Delrot, S. (1987). Recognition of phlorizin by the carriers of sucrose and hexose in broad bean leaves. *Physiologia Plantarum* **69**, 639–44.

Lemoine, R., Delrot, S. & Auger, E. (1984). Development of pH sensitivity of sucrose uptake during ageing of *Vicia faba* leaf discs. *Physiologia Plantarum* **61**, 571–6.

Lepp, N. W. & Peel, A. J. (1970). Some effects of IAA and kinetin on the movement of sugars in the phloem of willow. *Planta* **90**, 230–5.

Li, Z. S. & Delrot, S. (1987). Osmotic dependence of the transmembrane potential difference of broad bean mesocarp cells. *Plant Physiology* **84**, 895–9.

Lichtner, F. T. (1983). Amitrole absorption by bean (*Phaseolus vulgaris* L. cv. Red Kidney) roots. *Plant Physiology* **71**, 307–12.

Lichtner, F. T. (1986). Phloem transport of agricultural chemicals. In Cronshaw, J., Lucas, W. J. & Giaquinta, R. T. (eds.). *Phloem Transport*, pp. 601–8. New York, Alan R. Liss.

Lichtner, F. T. & Spanswick, R. M. (1981). Electrogenic sucrose transport in developing soybean cotyledons. *Plant Physiology* **67**, 869–74.

Liu, D. F., Hitz, W. D., Giaquinta, R. T. & Viitanen, P. V. (1986). Sucrose transport in *Pseudomonas saccharophila*. *Plant Physiology* **80**, S-464.

Madore, M. A., Oross, J. W. & Lucas, W. J. (1986). Symplastic connections in *Ipomoea tricolor* source leaves. Demonstration of functional symplastic connections from mesophyll to minor veins by a novel dye-tracer method. *Plant Physiology* **82**, 432–42.

Madore, M. A. & Webb, J. A. (1981). Leaf free space analysis and vein loading in *Cucurbita pepo*. *Canadian Journal of Botany* **59**, 2250–7.

Malek, F. & Baker, D. A. (1977). Proton co-transport of sugars in phloem loading. *Planta* **135**, 297–9.

Malek, F. & Baker, D. A. (1978). Effect of fusicoccin on proton co-transport of sugars in the phloem loading of *Ricinus communis* L. *Plant Science Letters* **11**, 233–9.

Martin, R. A, & Edgington, L. V. (1981). Comparative systemic translocation of several xenobiotics and sucrose. *Pesticide Biochemistry and Physiology* **16**, 87–96.

Martin, E. & Komor, E. (1980). Role of phloem in sucrose transport by *Ricinus* cotyledons. *Planta* **148**, 367–73.

Maynard, J. W. & Lucas, W. J. (1982a). A reanalysis of the two component phloem loading systems in *Beta vulgaris*. *Plant Physiology* **69**, 734–9.

Maynard, J. W. & Lucas, W. J. (1982b). Sucrose and glucose uptake into *Beta vulgaris* leaf tissue: a case for general (apoplastic) retrieval systems. *Plant Physiology* **70**, 1436–43.

M'Batchi, B. & Delrot, S. (1984). Parachloromercuribenzene sulfonic acid. A potential tool for differential affinity labelling of the sucrose transporter. *Plant Physiology* **68**, 391–5.

M'Batchi, B., El Ayadi, R., Delrot, S. & Bonnemain, J. L. (1986). Direct versus indirect effects of p-chloromercuri benzene-sulfonic acid on sucrose uptake by plant tissues: the electrophysiological evidence. *Physiologia Plantarum* **68**, 391–5.

M'Batchi, B., Pichelin, D. & Delrot, S. (1985). The effects of sugars on the binding of [^{203}Hg]-p-chloromercuri benzenesulfonic acid to leaf tissues. *Plant Physiology* **79**, 537–42.

M'Batchi, B., Pichelin, D. & Delrot, S. (1987). Selective solubilization of membrane proteins differentially labelled by p-chloromercuribenzenesulfonic acid in the presence of sucrose. *Plant Physiology* **83**

McNeil, D. L. (1979). The kinetics of phloem loading of valine in the shoot of a nodulated legume (*Lupinus albus* L. cv. Ultra). *Journal of Experimental Botany* **118**, 1003–12.

Minchin, P. E. H. & Gould, R. (1986). Effect of SO$_2$ on phloem loading. *Plant Science Letters* **43**, 179–83.

Minchin, P. E. H. & Thorpe, M. R. (1982). Evidence for a flow of water into sieve tubes associated with phloem loading. *Journal of Experimental Botany* **33**, 233–40.

Neumann, S., Grimm, E. & Jacob, F. (1985). Transport of xenobiotics in higher plants. I. Structural prerequisites for translocation in the phloem. *Biochemie und Physiologie der Pflanzen* **180**, 257–68.

Ntsika, G. & Delrot, S. (1986). Changes in apoplastic and intracellular leaf sugars induced by the blocking of export in *Vicia faba*. *Physiologia Plantarum* **68**, 145–53.

Pichelin, D., Mounoury, G., Delrot, S. & Bonnemain, J. L. (1984). Cotransport proton-glycine dans la cellule de transfert de l'haustorium des Bryophytes: les données de l'électrophysiologie. *Compte rendu hebdomadaire des séances de l'Académie des sciences* **298D**, 439–44.

Pichelin-Poitevin, D. & Delrot, S. (1987). Differential labelling of a 42 kD membrane polypeptide in the presence of sucrose. *Compte rendu hebdomadaire des séances de l'Académie des sciences* **304**, 371–4.

Pichelin-Poitevin, D., Delrot, S., M'Batchi, B. & Everat-Bourbouloux, A. (1987). Differential labelling of membrane proteins by N-ethylmaleimide in the presence of sucrose. *Plant Physiology and Biochemistry* **25**, 597–607.

Rauser, W. E.& Samarakoon, A. B. (1980). Vein loading in seedlings of *Phaseolus vulgaris* exposed to excess cobalt, nickel, and zinc. *Plant Physiology* **65**, 578–83.

Reinhold, L., Seiden, A. & Volokita, M. (1984). Is modulation of the rate of proton-pumping a key event in osmoregulation? *Plant Physiology* **75**, 846–9.

Renaudin, J. P., Brown, S. C., Barbier-Brygoo, H, & Guern, J. (1986). Quantitative characterization of protoplasts and vacuoles from suspension-cultured cells of *Catharanthus roseus*. *Physiologia Plantarum* **68**, 695–703.

Ripp, K. G., Liu, D. F., Viitanen, P. & Hitz, W. D. (1986). Identification of sucrose binding, membrane proteins using a photolyzable sucrose analog. *Plant Physiology* **80**, S-465.

Robinson, S. P.& Beevers, H. (1981). Amino acid transport in germinating castor bean seedlings. *Plant Physiology* **68**, 560–6.

Russell, S. H. & Evert, R. F. (1985). Leaf vasculature in *Zea mays* L. *Planta* **164**, 448–58.

Russin, W. A. & Evert, R. F. (1985). Studies on the leaf of *Populus deltoides* (Salicaceae): quantitative aspects, and solute concentrations of the sieve tube members. *American Journal of Botany* **72**, 487–500.

Schmitt, M. R., Hitz, W. D., Lin, W. & Giaquinta, R. T. (1984). Sugar transport into protoplasts isolated from developing soybean cotyledons. II. Sucrose transport kinetics, selectivity and modelling studies. *Plant Physiology* **75**, 941–6.

Servaites, J. C. Schrader, L. E. & Jung, D. M. (1979). Energy-dependent loading of amino acids and sucrose into the phloem of soybean. *Plant Physiology* **64**, 546–50.

Singer, S. R. & McDaniel, C. N. (1982). Transport of the herbicide 3-amino-1,2,3,4-triazole by cultured tobacco cells and leaf protoplasts. *Plant Physiology* **69**, 1382–6.

Smith, J. A. C. & Milburn, J. A. (1980a). Phloem transport, solute flux and the kinetics of sap exudation in *Ricinus communis* L. *Planta* **148**, 35–41.

Smith, J. A. C. & Milburn, J. A. (1980b). Phloem turgor and the regulation of sucrose loading in *Ricinus communis* L. *Planta* **148**, 42–8.

Sokolova, S. V., Krasavina, M. S. & Lushchikov, S. B. (1979). Sucrose absorption by various conducting bundle tissues. *Soviet Plant Physiology* **26**, 721–7.

Sovonick, S. A., Geiger, D. R. & Fellows, R. J. (1974). Evidence for active phloem loading in the minor veins of sugar beet. *Plant Physiology* **54**, 886–91.

Sturgis, J. N. & Rubery, P. H. (1982). The effects of indol-3-yl-acetic acid and fusicoccin on the kinetic parameters of sucrose uptake by discs from expanded primary leaves of *Phaseolus vulgaris*. *Plant Science Letters* **24**, 319–26.

Teh, K. H. & Swanson, C. A. (1982). Sulfur dioxide inhibition of translocation in bean plants. *Plant Physiology* **69**, 88–92.

Thorpe, M. R., Minchin, P. E. H. & Dye, E. A. (1979). Oxygen effects on phloem loading. *Plant Science Letters* **15**, 345–50.

Turgeon, R., Webb, J. A. & Evert, R. F. (1975). Ultrastructure of minor veins in *Cucurbita pepo* leaves. *Protoplasma* **83**, 217–32.

Tyree, M. T., Peterson, C. A. & Edgington, L. V. (1979). A simple theory regarding ambimobility of xenobiotics with special reference to the nematicide, oxamyl. *Plant Physiology* **63**, 367–74.

Van Bel, A. J. E. (1986). Amino acid loading by minor veins of *Commelina benghalensis*: an integration of structural and physiological aspects. In Lambers, H., Necteson, J. J. & Stulen, I. (eds.). *Fundamental, Ecological and Agri-*

cultural Aspects of Nitrogen Metabolism in Higher Plants, pp. 111–14. The Hague, M. Nijhof.

Van Bel, A. J. E. & Ammerlaan, A. (1981). Light-promoted diffusional amino-acid efflux from *Commelina* leaf disks. Indirect control by proton pump activities. *Planta* **152**, 115–23.

Van Bel, A. J. E., Ammerlaan, A. & Blaauw-Jansen, G. (1986b). Preferential accumulation by mesophyll cells at low and by veins at high exogenous amino acid and sugar concentrations in *Commelina benghalensis* leaves. *Journal of Experimental Botany* **37**, 1899–910.

Van Bel, A. J. E., Borstlap, A. C., Van Pinxteren-Bazuine, A. & Ammerlaan, A. (1982). Analysis of valine uptake by *Commelina* mesophyll cells in a biphasic active and a diffusional component. *Planta* **155**, 355–61.

Van Bel, A. J. E. & Koops, A. J. (1985). Uptake of [^{14}C]sucrose in isolated minor vein networks of *Commelina benghalensis* L. *Planta* **164**, 362–9.

Van Bel, A. J. E., Koops, A. J. & Dueck, T. (1986a). Does light-promoted export from *Commelina benghalensis* leaves result from differential light sensitivity of the cells in the mesophyll-to-sieve tube path? *Physiologia Plantarum* **71**, 227–34.

Vreugdenhil, D. (1983). Abscisic acid inhibits phloem loading of sucrose. *Physiologia Plantarum* **57**, 463–7.

Vyskrebentseva, E. I. & Semenov, I. L. (1975). Effect of univalent cations on sugar transport in sugar beet tissues. *Soviet Plant Physiology* **6**, 120: 6–12.

Wilson, C., Oross, J. W. & Lucas, W. J. (1985). Sugar uptake in *Allium cepa* leaf tissue: an integrated approach. *Planta* **164**, 227–40.

Wright, J. P. & Fisher, D. B. (1981). Measurement of the sieve tube membrane potential. *Plant Physiology* **67**, 845–8.

Wyse, R. E. & Komor, E. (1984). Mechanism of amino acid uptake by sugarcane suspension cells. *Plant Physiology* **76**, 865–70.

6 Phloem unloading of photoassimilates

W. Eschrich

6.1 Definition of the subject

6.1.1 The pathways

Phloem unloading is the reverse process of phloem loading (Eschrich, 1970). This process is primarily related to sucrose, the first non-phosphorylated product of photosynthesis, which provides the carbon skeleton for syntheses of all organic compounds in vascular plants. Sucrose is electroneutral, chemically inert and extremely soluble in water, which means it is especially suitable for being passively distributed in the system of sieve elements in plants. Since sucrose also occurs in parenchymatous tissue (vascular parenchyma, mesophyll, storage parenchyma), cells or cell compartments in which sucrose is not attacked by hydrolytic or phosphorolytic enzymes, can also belong to the sucrose transport system. In many plants, sucrose is stored in the vacuoles of parenchyma cells (sugar-cane, sugar-beet, many fruits) showing that the tonoplast must be regarded as a barrier between the transport path and the store. In some plants (mosses) long-distance transport of sucrose is directed entirely through plasmodesmata. Consequently, these compartments should be included in the transport system for sucrose.

This would mean that unloading comprises the escape of sucrose through the plasmalemma into the apoplast as well as its release from the endoplasmic reticulum (ER) into the cytoplasm. However, it is not known which pathway sucrose follows through the plasmodesmata, in the cytoplasmic annulus or in the desmotubule, and it therefore remains open whether the ER carries sucrose or not.

6.1.2 The materials

The structure of the pathway can be recognized, but the moving matter remains obscure. Since sieve tube sap contains amino compounds and ions besides sucrose (see Ch. 4), phloem unloading also refers to such substances. It is not known, but can be postulated that they are unloaded by the same mechanisms which are applicable for sucrose unloading.

It is generally agreed that sucrose transport inside sieve tubes is a mass movement, depending on sink activities (see Ch. 7). Amphoteric amino acids certainly will move together with the sucrose. Differential delivery, therefore, can be managed only during or after phloem unloading. This means that to satisfy the demand for a particular amino acid, sucrose unloaded together with this amino acid may be reloaded while the amino acid is metabolized. It is not known whether charged compounds, such as inorganic ions, are transported in sieve tubes in a different way from sucrose.

Photosynthesis occurs in the chloroplasts of green leaves and shoots. However, sucrose is synthesized not in the chloroplasts but in the cytoplasm (cytosol, see Ch. 1). The transport path of sucrose thus begins in the cytoplasm of the photosynthetic cells. This path may be very short for some sucrose molecules, when the vacuole of the photosynthetic cell stores sucrose. Sucrose is also a precursor for starch synthesis, which strictly occurs in plastids. If a photosynthetic cell in the cortex of a shoot ceases photosynthesis and starts to accumulate starch, the chloroplasts are converted into amyloplasts. In this case, photosynthetically synthesized and deposited sucrose would be used to build starch in the same cell. The reverse process occurs in potato tubers on exposure to light. Here the starch of the amyloplasts is hydrolysed, the plastids green, and eventually produce starch by photosynthesis. In such a case, neither leaves nor sieve tubes are involved in the delivery of sucrose. It has not been shown, but seems possible, that sucrose in the cytoplasm may be both a breakdown product and a precursor of starch at the same time.

When the photosynthetic production (source) and use of sucrose (sink) are separated by other cells, phloem serves as the interconnecting tissue in most cases. When the activity of a sink shows no diurnal fluctuations, sucrose will be removed from the phloem continuously, irrespective of its production, which is restricted to the light hours. For continuous growth to occur other sources of sucrose must be made available during the night (hydrolysis of photosynthetic starch accumulated during the day; vacuolar sucrose pools; see Ch. 1).

6.1.3 The mechanisms

Sieve tubes are always turgid; they show a high affinity for sucrose. As phloem loading affords a supply of energy, unloading should be able to proceed without further energy supply. However, unloading is regulated because sucrose efflux can differ in intensity at any given place along the

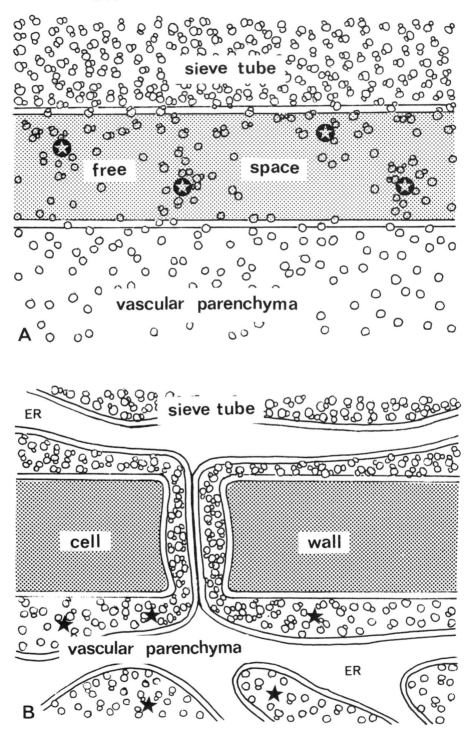

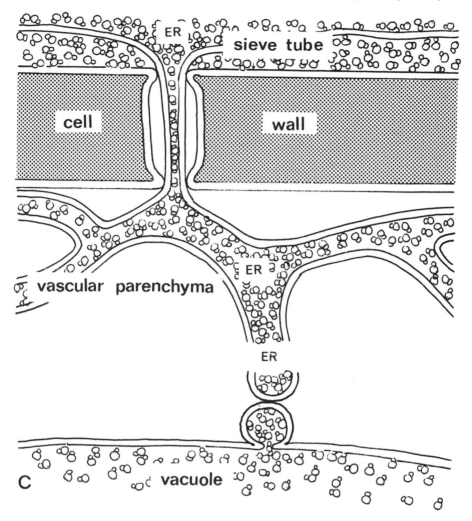

Fig. 6.1 Methods of phloem unloading, schematically. *A*: Unloading of sucrose (8-shaped symbols) into the free space, in which particle-bound acid invertase (open stars) is active. The resulting hexoses (open circles) cannot be reloaded into the sieve tube; instead they will be taken up into adjacent vascular parenchyma cells. Active invertase thus causes a lot of continuous unloading of sucrose. *B*: Symplastic unloading of sucrose (8-shaped symbols) through the cytoplasmic annulus of plasmodesmata. When the cytoplasm of an adjacent vascular parenchyma cell carries active soluble invertase (black stars), sucrose will be hydrolysed and the hexoses (open circles) will be metabolised. *C*: Compartmented symplastic unloading of sucrose, which is taken up into the ER. Sucrose (8-shaped symbols) is translocated through the desmotubule of plasmodesmata, and inside the ER of an adjacent vascular parenchyma cell to the tonoplast. Storage of sucrose in the vacuole evokes continuous unloading until the vacuolar sucrose pool is full.

system of sieve tubes. Since turgor is high in the sieve tubes, based mainly on the high sucrose concentration, unloading pressure will be high everywhere in the sieve tube system. The unloading pressure is counteracted by the sieve tube's affinity for sucrose, and thus sucrose molecules released into the apoplast will be reloaded if they are not taken out of the pathway by hydrolysis. This eddy current of sucrose occurs throughout the length of the phloem path. The sucrose concentration in the apoplast of mature maize leaves depends on illumination; it can rise to 7.3 mol m^{-3} (Heyser *et al.*, 1978). Much higher apoplastic sucrose concentrations, of 90–200 mol m^{-3}, have been reported for the apoplastic interface between the seed coat and embryo in growing seeds of legumes (Gifford & Thorne, 1985).

Sink activity develops when the sucrose, released into the apoplastic free space, is drawn away into adjacent parenchyma cells. However, the hydrolysis of the unloaded sucrose into glucose and fructose is just sufficient to create sink activity, because these hexoses are not loaded into the sieve tubes. It follows that unloading of sucrose is forced by steepening the concentration gradient between sucrose$_i$:sucrose$_0$, irrespective of the steadily increasing hexose concentration in the free space. If these hexose molecules are not continuously taken up into parenchyma cells, they will increasingly lower the water potential of the free space. Under these circumstances, sieve tubes will react by lowering their osmotic potential to counteract the water potential outside the plasmalemma. Sucrose is thus drawn to those sinks with increasing concentrations of sugar, such as growing fruits (Lang *et al.*, 1986), or branches in the crown of tall trees which stay under water stress.

Another possible means of increasing sucrose unloading is to deposit symplastically transported sucrose in the vacuoles of adjacent vascular parenchyma cells. Since experiments with asymmetrically labelled sucrose have shown that the molecule is not randomized during transport, it can be accepted that sucrose moves along compartmented pathways from the sieve tube to the tonoplast of the parenchyma cell, probably via the ER. This seems to be the way to protect the sucrose against attack by hydrolysing or phosphorolytically splitting enzymes of the cytoplasm.

If this protection is not available, a third kind of unloading would be the symplastic movement of sucrose into the cytoplasm of vascular parenchyma cells. The metabolization of the sucrose will then create the sink effect, and more sucrose will move into the cytoplasm. Widely distributed in storage sinks is the conversion of sucrose into precursors of starch or fructans. By deposition of osmotically ineffective starch, the concentration gradient of soluble sugars between sieve tubes and parenchyma cells will be steepened. These sink mechanisms mentioned above are shown schematically in Fig. 6.1.

There may be combinations of the proposed mechanisms of unloading. For example, growing maize kernels are provided with sucrose by symplastic transport from the sieve tubes of the pedicel into the vascular parenchyma and from there into the apoplast of the placento-chalazal region which rapidly disintegrates after pollination. The transfer cells of the endosperm, which develop later, will take up the sucrose from the apoplast (Felker & Shannon, 1980). This symplast–apoplast–symplast transfer is possible because of the lack of invertase in the apoplastic part of this pathway. Combinations of symplastic and apoplastic pathways of sucrose seem to occur frequently during unloading in growing seeds (Patrick & McDonald, 1980).

6.2 Developmental aspects

For phloem unloading, tubes of living but enucleate sieve elements are required. To understand unloading from such highly specialized living cells requires some knowledge of the developmental and anatomical features of the transport system (see Ch. 3). When mature, its elements are unable to divide or to fuse with other elements. Sieve tubes cannot regenerate. When wounded, phloem bundles have to differentiate complete new strands of sieve tubes by redifferentiation of longitudinally oriented vascular parenchyma cells (Eschrich, 1953). Although newly formed wound sieve tubes transport labelled photoassimilates derived from a previously existing source leaf (Jacobsen, 1986), it is still unknown where and how the wound sieve tubes are attached to the pre-existing obviously mature sieve tubes.

In this section the most probable method of such interconnections will be described, despite the fact that our knowledge of functional anatomy of the transport system is fragmentary. During development of a plant the growth processes are the most important triggers for phloem unloading. When a seed germinates sucrose is already available in the protophloem, synthesized from mobilization of reserve substances of the cotyledons or the endosperm. The first sinks are probably the apical meristems of the epicotyl and of the radicle. The apical meristems show sink activity throughout the later stages of plant development. Macro-autoradiographs of ^{14}C-labelled plants show labelled shoot tips as well as labelled root tips (Fig. 6.2). Since sieve elements differentiate from procambial cells, phloem unloading must occur in this subapical primary meristem. Micro-autoradiography of root tips of a monocotyledon has shown that the main unloading occurs at the level of the first differentiating metaxylem elements (Fig. 6.3). From there, movement of unloaded sucrose to the apical meristem can occur in both symplastic and apoplastic compartments. It can be assumed that sucrose is being continuously metabolized in zones of nuclear division and extension growth, because of the synthesis of cytoplasmic and cell wall constituents. By conversion of sucrose into insoluble

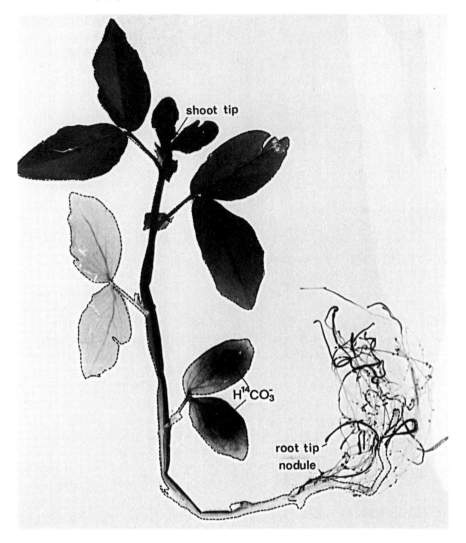

Fig. 6.2 Macro-autoradiography of a *Vicia faba* plant, which was labelled with ^{14}C on one leaf, and allowed to photosynthesize for 12 h. The freeze-dried plant was exposed to X-ray film; ^{14}C label is indicated by grey to black tinted photoemulsion.

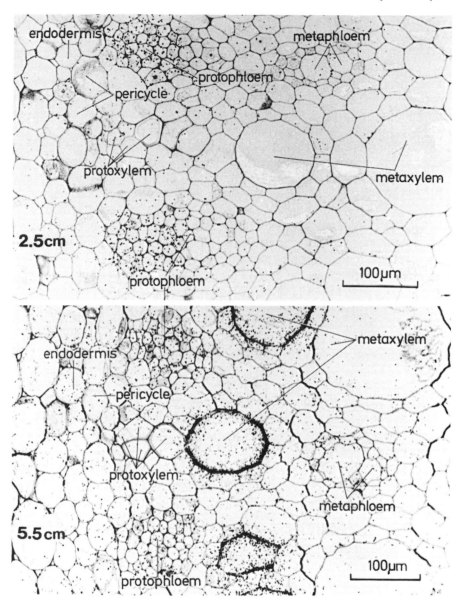

Fig. 6.3 Micro-autoradiographs of cross-sections of the root tip of a *Monstera deliciosa* aerial root. Upper section 2.5 cm, lower section 5.5 cm distant from the tip. ^{14}C-labelled photoassimilates from a source leaf are unloaded at the level of metaxylem differentiation. Small black spots are silver grains of the photoemulsion, which indicate the localization of soluble and incorporated ^{14}C activity (Eschrich, 1983).

polymers a concentration gradient of sucrose, extending backwards to the functional sieve elements, will keep unloading going.

This agrees with the fact that the sieve tube system is composed of young and mature sieve elements, and is complemented continuously, either by adding new elements at the procambial level of the apical regions or by interspersing new elements at sites of intercalary growth. The functional continuity of the sieve tube system is ensured by the strategy of replacement of proto-elements by meta-elements after extension growth has ceased. Communication between both types of sieve elements is possible because these elements belong to the same strand of vascular tissue.

The same pattern of phloem differentiation occurs in leaves. They start with apical growth, and sooner or later switch over to intercalary growth. Leaves with a petiole or rachis have more than one intercalary meristem. The lamina emerges, depending on the activity of marginal meristems, and is complemented by the activity of interspersed laminar meristems. The marginal meristems determine the shape and thickness of the lamina, and the laminar meristems take care of the proportions of veins and areoles. The ratio of vein area to intercostal area (the area occupied by areoles) varies between mature leaves of a single plant. Computer-aided area integrations of vein areas versus areole areas in leaves of different beech trees (*Fagus sylvatica*) showed values between 18.5% vein area in a shade leaf, and 31.4% vein area in a sun leaf (Fig. 6.4). The maturation of the laminar phloem can be observed by the import/export pattern in macro-autoradiographs (Turgeon & Webb, 1973; Turgeon, 1980); the phloem normally matures from the tip of the blade to its base, and mature stretches of phloem can be connected by immature phloem (Turgeon & Webb, 1975).

Leaves are anchored in the stem with leaf traces. There are different patterns of leaf traces, which mainly correspond to the phyllotaxis, but can vanish at the level of a node. Much less information is available on the developmental stage of leaf traces at the different levels of the stem. Fig. 6.5 shows the skeleton of vascular bundles in the shoot tip of *Helianthus annuus*. This skeleton based on four consecutive cross-sections is taken from Esau (1945). The leaf traces and leaf trace complexes are either procambial (black), or contain only phloem (stippled), or consist of phloem and xylem (unshaded). Some leaf traces are mature collateral bundles from apex to base, for example all traces from leaf 6. In leaf 8 the median leaf trace consists of phloem and xylem, but basipetally it becomes pure phloem. The lateral leaf traces of leaf 8 are phloem bundles. The median leaf trace of leaf 9 divides basipetally into one phloem branch and one procambium branch. The latter stays in contact with a procambial leaf trace from leaf 12, but in the next node following basipetally, it joins a mature lateral leaf trace from leaf 6.

Altogether, the tendency appears to be that leaf traces are getting 'younger', or less differentiated, in the basipetal direction. The further down that leaf traces can be recognized, the less differentiated they may

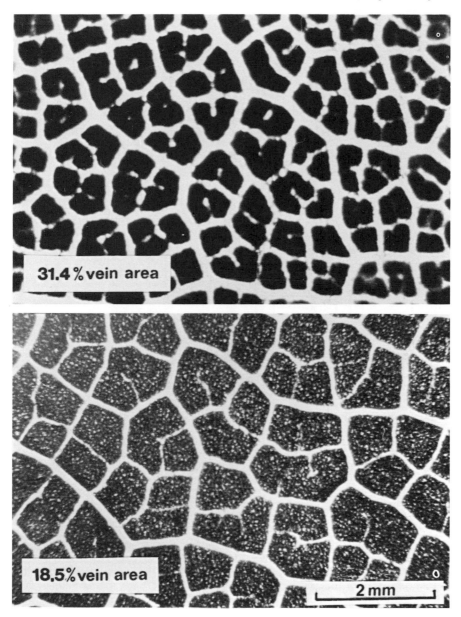

Fig. 6.4 The ratios of vein area to chlorenchyma area as computed with the aid of a video camera in mature beech leaves (*Fagus sylvatica*) in a sun leaf (upper) and a shade leaf (lower). Areoles differ in density of staining because the leaves differ in thickness. The (lower) shade leaf has only one layer of palisade cells in which the intercellular spaces appear translucent. Photographs taken at the same magnification.

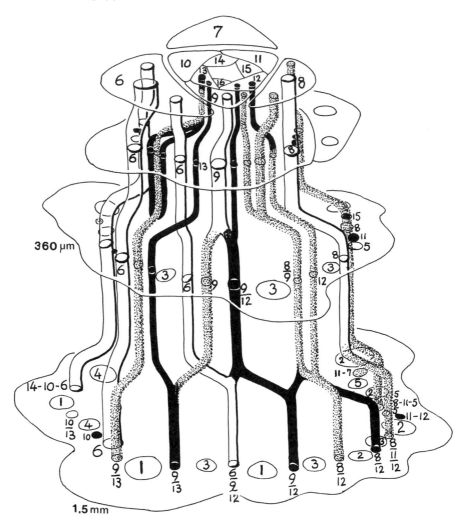

Fig. 6.5 Leaf traces of the shoot tip of *Helianthus annuus*, reconstructed diagrammatically from consecutive cross-sections (after Esau, 1945). The leaves are numbered 1 to 16. More than one number indicates leaf trace complexes. The latter can be composed of procambium (black), only phloem (stippled) and phloem with xylem (unshaded). Maximal distance from the shoot tip is 1.5 mm.

Fig. 6.6 Schematic pattern of leaf insertions of one orthostichy with their leaf traces. Black leaf traces have mature vascular tissue. Open-stippled leaf traces are meristematic; they can fuse with meristematic leaf traces from other leaves, as indicated by bridges of open squares (after Eschrich, 1975a).

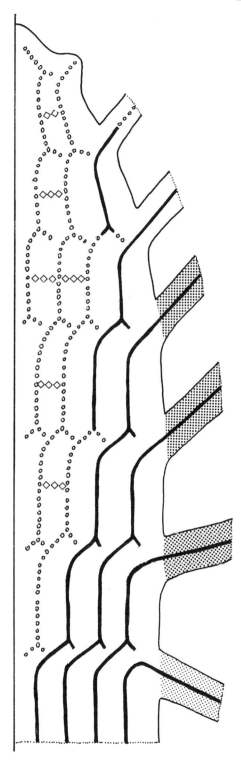

be. Thus one can postulate that leaf traces are not eliminated basipetally, but instead they are initiated in this direction. Such leaf trace initials are present prior to the emerging leaf to which the trace will be attached (Fig. 6.6).

The strategy of the plant could be interpreted as follows. Sieve tubes remain living as long as they are functional, and since the mature sieve elements are lacking their nuclei, they can neither divide nor regenerate. When leaf traces are meristematic at their basal end, they can merge with leaf traces of the same developmental stage. This makes the exchange of photoassimilates possible between leaves of different ages. Only between meristematic sieve elements is the opening of sieve pores possible, because the formation of a sieve plate between two cells is comparable to a fusion of protoplasts.

In gymnosperms and dicotyledons, which have secondary growth, leaf traces merge with the cylinder of secondary vascular tissues. Even here, the junction between phloem elements of the leaf trace and those of the secondary phloem occurs at a level where both are still in a meristematic stage of development.

In trees leaf traces are often incorporated into the cylinder of secondary phloem just a few millimetres below the apex (Larson, 1975). The cambial phloem increments are strictly separated from each other in the radial direction. However, in the tangential direction, sieve tubes are usually symplastically corresponding, because they are of the same developmental age. If photoassimilates have to move in the radial direction, either towards the cortex or towards the pith, they have to transfer into the rays. Thus it follows that sieve tubes of leaf traces (primary phloem), of cambial increment (secondary phloem), and ray cells of the radial pathway must necessarily be of the same developmental age when remaining in contact via cytoplasmic bridges.

Although the atactostele of monocotyledons lacking secondary growth (maize, palms) has a completely different structure, the fusion of the sieve tubes of leaf traces seems to be similar to that found in perennials with secondary growth. Zimmermann (1973), combining serial cross-sections of the palm *Rhapis excelsa* and micro-autoradiography, found that sieve tubes of leaf trace complexes can remain functional over periods of many years. With regard to our problem, one piece of information is still missing, namely whether the basal ends of the leaf traces, which may be buried in vascular complexes, still keep their meristematic character. This has so far been investigated in palms, where only mature sieve elements lacking a nucleus were described (Parthasarathy, 1980).

During the transition from vegetative growth to the reproductive phase, phloem differentiation in flower parts is similar to that in foliage leaves. It can be imagined that ripening growth of a fruit and development of the ovules are processes of intensive phloem unloading. In the pericarp the network of sieve tubes is often extended by anastomoses or extrafascicular sieve tubes (*Cucurbita*). In ovules, the vascular bundles can extend into the seed coat and form a fine-meshed network (Esau, 1965, 1969).

With regard to unloading, one conclusion which can be drawn from these studies of comparative anatomy of the phloem, is that all plants use a developmental strategy which leads to symplastic continuity between all functioning elements of that tissue.

6.3 Physiological aspects

During evolution land plants created an effective living transport system for foodstuffs between the leaves and all other parts of the plant. The integration of the xylem into the transport system for food confirms the assumption that water is not only necessary for sugar transport in the phloem, but also for the transport of compounds which primarily occur in the water vessels and sooner or later are transferred into the phloem.

Although the uptake of anions like NO_3^-, PO_4^{3-}, and SO_4^{2-} is small compared with the uptake of photoassimilates, it cannot be ignored that for the incorporation of N, P, and S into organic compounds, a carbon skeleton is necessary, which is provided by sucrose unloaded in the roots (Dick & ap Rees, 1975).

6.3.1 Water pumping

Probably the most important physiological consequence of unloading sucrose can be seen in the local decrease of water potential in the free space by the action of acid invertase. A high hexose content in the free space of the phloem must lead to an increased supply of sucrose in the sieve tubes for reasons of equilibration. Otherwise, sieve tubes would lose turgor. Plant organs which are under water stress, especially the crown twigs of tall trees, can obtain the concentrations of sucrose necessary for growth in this way. In turn, an increased sucrose concentration in the sieve tubes of crown twigs will induce more water to move up. Deciduous trees must decrease their water potential in the free space, especially when spring reactivation takes place. The same principle is applied in sweet, fleshy fruits for accumulating sugars (Lang *et al.*, 1986).

A model for this mechanism has been described by Heyser *et al.* (1976): 25 cm long strips of mature maize leaves were put with one end in 75 mol m^{-3} mannitol, the other end immersed in diluted Hoagland solution. When a small area of the middle of the strip is exposed to $^{14}CO_2$ in light, the labelled photoassimilates move in the direction of the mannitol solution (which seems to attract the label). If mannitol is replaced by sucrose, glucose, fructose, raffinose or trehalose, the labelled photoassimilates are 'pushed' to the end with the Hoagland solution.

Mannitol and some other compounds (2-deoxyglucose; Na$_2$-EDTA) are not taken up into the symplast of the vascular bundles. They decrease the water potential of the apoplastic free space, and trigger the movement of sucrose in the sieve tubes to compensate for the imbalance in osmotic and water potentials.

6.4 Economical aspects

6.4.1 Economy of storage

In biennial and hapaxanth (bloom only once, then die) plants, the economical significance of phloem unloading can be clearly seen. A carrot, for example, builds only a storage root and foliage leaves in the first season. In the second year the plant produces flowers, and the stored material of the root is used up (Fig. 6.7). After fruit ripening the plant dies. Sugar-beet has a similar life cycle. In cultivated sugar-cane, man has selected cultivars which have almost completely lost their ability to produce flowers and which have to be propagated by cuttings. Thus with this technique man's economic strategy has conflicted with the economy of the plant.

A typical hapaxanth plant is *Agave americana*, a species which has not only a storage axis in the ground but also succulent rosette leaves which store photoassimilates. *Agave* can live for tens of years in a vegetative state, until it starts to bloom with a luxurious flower stalk carrying up to 7000 flowers; but then the plant dies. The arm-thick flower stalk, several metres high, and the 5 cm long growing fruits use up all the storage material which has been accumulated over the years. Similar processes can be observed when bulbous plants start to bloom.

6.4.2 Storage dynamics in perennials

In trees the economy of phloem unloading is shown best by the deposition of starch in bark, xylem rays and pith, a pattern which changes depending on the season. The meaning of this fluctuation in starch content, which is different in each storage compartment, is not fully understood. Figure 6.8 shows the annual cycle of starch content in five-year-old crown branches of a beech tree, 120 years old. Note that between December and February starch in the bark is hydrolysed completely, but is deposited again long before leaves could contribute the necessary sugar. In addition to the data shown in Fig. 6.8, the starch contents of younger and older parts of the branches, back to the seventh year have been analysed. It was found that nearly every year's increment had a somewhat different starch-storage dynamic; the five phases, however, were easy to recognize (Essiamah &

Fig. 6.7 The carrot plant (*Daucus carota*) in the first (left) and the second (right) year of its life. The storage root of the first year is emptied in the second year during bloom and fruit ripening.

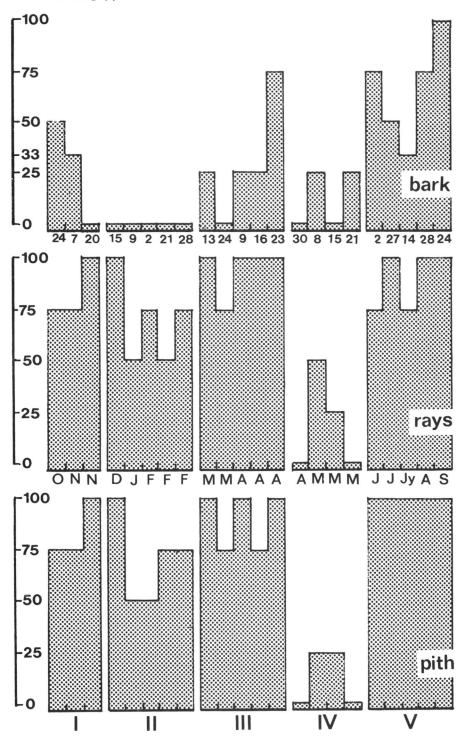

Eschrich, 1985). Obviously, starch deposition also depends on climatic conditions.

As can be deduced from Fig. 6.8, parenchyma cells in cortex and pith undergo a periodic hydrolysis and redeposition of starch. It appears that sugar transport to and from these five-year-old reservoirs goes symplastically via the ray system. Again, the above-mentioned principle is valid, that lateral (in tangential direction) symplastic connections occur only between sieve elements and ray cells of the same developmental stage. However, the ray in its longitudinal (radial) direction is composed of elements of different age. Thus, xylem parenchyma in any year's increment will have contact somewhere with a ray cell of its own age.

The functioning of this combined transport system has been shown in young spruce plants by Langenfeld-Heyser (1987). Fig. 6.9 shows a microautoradiograph extending over half the cross-section of the stem in its second year. Its radioactivity has been provided by the younger stem above, which was exposed to $^{14}CO_2$. It can be seen that the complete current year's secondary phloem is used for assimilate transport. From these sieve cells, labelled photoassimilates were distributed radially via adjacent ray cells, spreading in both directions towards the cortex and towards the pith. All distinctly labelled cells, except sieve elements and ray initials, contain plastids with either chlorophyll or (and) starch.

6.4.3 *Biomass turnover*

Economy of phloem unloading in trees is also shown by other examples. In spring the new leaves, twigs and the developing buds are potential sinks. After complete unfolding the leaves function as sources until they are shed in the autumn. Prior to the shedding of leaves, a phase of intense remobilization utilizes all the materials in the leaf which can be converted to sucrose. Proteins and nucleic acids are hydrolysed and chlorophyll is decomposed into transportable compounds. The transfer of all this material into the stem is obviously made possible by continuous production of sucrose and by photosynthetic energy for phloem loading. Salts such as Ca-oxalates and polyphenols remain in the leaves, as do the skeletons of the cell walls, and are lost to the plant at leaf fall.

Fig. 6.8 Dynamics of starch deposition and hydrolysis in bark, rays and pith of five-year-old crown branches from a 120-year-old beech tree. Filling of storage cells (shaded blocks) in per cent (scales on the left) as judged from permanent-stained cross-sections. Crown branches were collected weekly to monthly during a whole year at noon by shooting with a rifle. The annual starch cycle is divided into five phases: I, autumnal filling phase; II, winter hydrolysis (probably) for bark protection (drought, frost); III, refilling for biomass production; IV, bud break and biomass production; V, summer replenishment. (Data in part from Essiamah & Eschrich, 1985.)

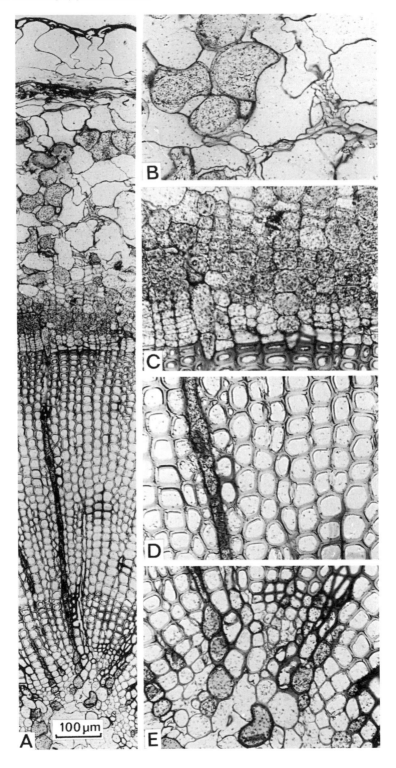

This biomass turnover is planned and regulated, as evidenced by the abscission layers which are preformed. Similar events can be predicted when flower parts are shed, and post-floral protein decomposition has been shown in ephemeral flowers (Schumacher, 1932).

6.5 Photoassimilates

The unique characteristic which plants have in common is sucrose as the principal photoassimilate. Between synthesis and metabolic conversion of sucrose, phloem unloading is the event which provides a steady and balanced supply of sucrose, always enough to satisfy extreme demands. In addition, sucrose is not only the basic foodstuff, it also is the principal osmoticum of turgid cells, especially the sieve tubes. This dual role of sucrose, being kept as an osmotic solute on the one hand, and being unloaded to satisfy the demands of sink sites on the other, is a 'concerted action', and this must be kept in mind.

6.5.1 Allocation

Unloading of sucrose in sinks requires allocation and loading of photoassimilates. Starved maize plants (which stayed in the dark for two days), require 30–45 min re-illumination to increase the sucrose level in mature leaves before resuming export in the phloem. The resumption of export starts when the sucrose level in the source tissue has increased to 25 mol m^{-3} (Eschrich & Burchardt, 1982).

Short-term illumination of darkened mature leaves of *Gomphrena globosa* (C$_4$ plant) in a $^{14}CO_2$-atmosphere results in the first labelled sucrose being present after 2 min (Fig. 6.10). Micro-autoradiographs of such leaves show that the chloroplasts of the bundle sheath of the minor and tertiary veins are heavily labelled, and that some label also occurs in the phloem (Fig. 6.11).

A micro-autoradiographic study on the movement of ^{14}C-labelled photoassimilates has shown that 5 min after labelling maize leaf strips with $^{14}CO_2$, in addition to sieve elements, the vascular parenchyma was also

Fig. 6.9 Micro-autoradiography of a two-year-old spruce stem (*Picea abies*), which was exposed to $^{14}CO_2$ on the younger part of the stem. ^{14}C-labelled photoassimilates were translocated basipetally and appear in sieve cells of the present year's secondary phloem. From there, ray cells were provided with labelled photoassimilates, which were distributed into chlorophyllous cells of the cortex and the pith. *A*: Overall view of the cross-section. *B*: Enlarged detail of the cortex. *C*: Enlarged detail of the cambial region with secondary phloem. *D*: Enlarged detail of secondary xylem with conducting rays. *E*: Enlarged detail of the primary xylem and pith. (Courtesy Dr Langenfeld-Heyser).

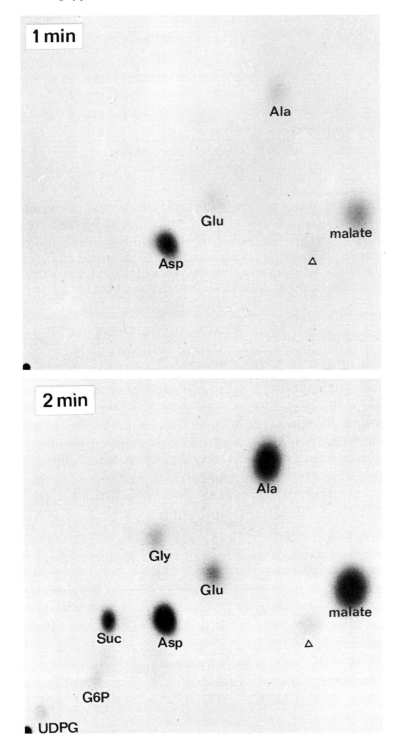

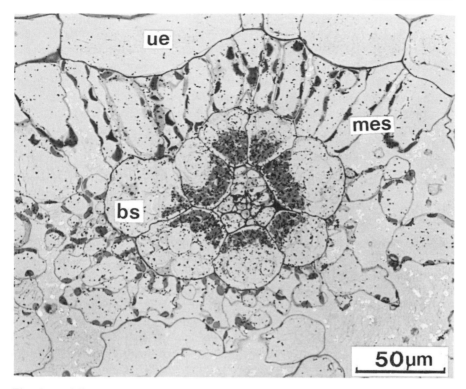

Fig. 6.11 Micro-autoradiography of the cross-section of a mature *Gomphrena* leaf with minor vein after 2 min of photosynthesis in $^{14}CO_2$ atmosphere. Label is spread most densely over the bundle-sheath (bs), and some has reached the vascular tissue. mes, mesophyll; ue, upper epidermis.

labelled. Also [^{14}C] sucrose, which was fed to the leaf bundles through the xylem, was accumulated both in sieve elements and in vascular parenchyma (Fritz *et al.*, 1983). Obviously, the vascular parenchyma can serve as a temporary sucrose storage pool. Probably the strength of the sink attached to the source will determine the magnitude of such pools.

Results obtained with $^{14}CO_2$ treatment of a sugar-beet leaf under steady state conditions of labelling indicated that a sucrose pool exists in the leaf which is exchangeable with the sucrose destined for export (Geiger *et al.*, 1983). During the night, allocation of photoassimilates must fall back upon reserves which have been deposited during the day. These reserves are the

Fig. 6.10 Thin-layer chromatograms of extracts of mature *Gomphrena globosa* leaves which were exposed to $^{14}CO_2$. Upper TLC: 1 min illumination; lower TCL: 2 min illumination. The chromatograms were developed according to Eschrich (1984). Open triangle, unidentified organic acid; Ala, alanine; Asp. asparagine; Gly, glycine; G6P, glucose-6-phosphate; Glu, glutamine; Suc, sucrose; UDPG, uridinediphosphate glucose.

sucrose pools in photosynthetic cells and the starch in the chloroplasts of the leaves.

Since the sieve tube system is interconnected, it allows sucrose deficits occurring at low light intensities (shaded leaves) to be corrected by allocation of additional sucrose from leaves with more optimal illumination. In this way a mature leaf will develop a sink capacity when kept dark, and can thus be provided with photoassimilates as if it were a typical sink leaf (see also Sect. 6.7.2, Sink limitation).

6.5.2 Partitioning

It is clear that the partitioning of photoassimilates depends on the amount and the velocity of their consumption at sink sites.

Macro-autoradiographs of whole plants obtained after pulse labelling of a single source leaf show that some parts of the plant are preferentially supplied with labelled photoassimilates. This is shown, for example, in Fig. 6.2, where some root tips and root nodules of the bean plant appear densely labelled. In similar experiments when such preferred sinks were importing labelled photoassimilates, the transport velocity was calculated as 7200 cm h^{-1}, or 2 cm s^{-1} (Nelson, 1962). Although it seems that local and transit pathways are recognizable, the mechanism of sugar transport must be similar in all sieve tubes. A preferred sink cannot unload radioactive sucrose before all the unlabelled sucrose previously present in the sieve tube has been unloaded. Since unloading at a sink may have been proceeding rapidly before the label was applied, the labelled sucrose will proceed with higher velocity in these sections of the sieve tube system, and sections which do not lead to a sink will not be provided with labelled photoassimilates. The rate of unloading obviously has an influence on the rate of phloem loading.

When the root tips mentioned above grow rapidly, sucrose will be unloaded there preferentially, and partitioning appears to be directed mainly from source leaves to the growing root tips. However, since this transport is indicated – as in Fig. 6.2 – by labelling only a single source leaf, the contributions from other unlabelled source leaves are not visualized.

When the rate of photosynthesis was measured for a single continuously labelled source leaf, no alterations were recorded when the other source leaves were either darkened or detached from the plant (Geiger, 1976). In general, only after a few days will the new situation influence the rate of photosynthesis of the remaining single source leaf (King *et al.*, 1967; Thorne & Koller, 1974; review of literature, Neales & Incoll, 1968; compare also Sect. 6.7.2, Sink limitation).

It can be expected that such restrictions of source activities will mobilize additional sucrose pools and cause starch hydrolysis, until the rate of photosynthesis and the rates of phloem loading and unloading are balanced according to the demands for photoassimilates (Fondy & Geiger, 1985).

6.5.3 Conversion

When photosynthetically produced sucrose is not directly loaded into the phloem, it will be deposited in a source leaf pool. Limitation of inorganic phosphorus inhibits the efflux of triose phosphates out of the chloroplasts, and the fixed carbon is then deposited as starch inside the chloroplasts (see Ch. 1), which will be hydrolysed completely or in part during the night or in certain instances during the day (Fondy & Geiger, 1985). Alternatively, starch can be deposited when sucrose is provided to leaf fragments (Herold, 1978; Eschrich, 1984b). Deposition and hydrolysis of starch partly follows a diurnal rhythm in amylase activity (Pongratz & Beck, 1978). The starch cycle also occurs in detached leaves, and therefore seems not to be controlled by the activities of sources and sinks, starch–sugar interconversions appearing to be local affairs (Eschrich & Eschrich, 1987).

It has long been known that a leaf increases its sucrose content after being detached (Saproschnikoff, 1890; Ahrns, 1924; Köcher & Leonard, 1971). Obviously caused by water stress, this reaction is accompanied by hydrolysis of starch (Schroeder & Horn, 1922; Schroeder & Herrmann, 1931).

The occurrence of mannitol and other sugar alcohols in the sieve tube sap is restricted to certain plant families (see Ch. 4). The galactosides of sucrose which are oligosaccharides of the raffinose family (stachyose, verbascose) are constituents of the sieve tube sap of several plant families. They seem to be unloaded at the same rate as sucrose is unloaded. Their synthesis is based on the formation of galactinol (Tanner & Kandler, 1966), a reaction product of *myo*-inositol and UDP galactose, which can also be regarded as a photoassimilate. It has been reported that stachyose appears labelled in sieve tubes before sucrose when a leaf is exposed to $^{14}CO_2$ (Hendrix, 1968) indicating that galactose may be labelled prior to sucrose.

6.5.4 Apportionment

When the allocation of photoassimilates is reduced, the apportionment to the different sinks is also limited. This happens in complete darkness or at low light intensities, when the sucrose concentration in the sieve tube system is lowered. Within 48 h of darkness the osmotic values of the sieve tubes in mature maize leaves dropped from 600 to 200 mol m^{-3} (Evert *et al.*, 1978). Since wilting of darkened plants only occurs when they are not supplied with water, it can be assumed that the increased osmotic potential of the sieve tubes is still counteracting water potential, thus maintaining the turgor for translocation.

In senescing leaves the supply of photoassimilates is curbed. In nearly all cases, senescing organs are eventually shed by completion of an abscission layer. This can happen in different stages of development. Ripe fruits fall from the tree, flower parts are shed after pollination, single leaves can suddenly turn yellow and drop to the ground. Twigs can also be shed by completion of a preformed abscission layer (*Prunus domestica*). There are reports of the shedding of masses of young fruits, which can cover the ground at the bottom of the tree (*Acer* spec. div.; *Malus sylvestris*). This phenomenon is usually called 'physiological fruit fall'. The plant organizes this kind of natural sink limitation, and it is inferred that the plant protects its branches from breaking under the weight of the ripening fruits by this process. The most common abscission is the shedding of leaves in autumn.

Nothing is known on the regulation of this reduced apportionment of photoassimilates. Most publications mention the involvement of senescence hormones, ABA (abscisic acid) and ethylene, but it is far from clear how and where these hormones act. Abscission layers are mostly preformed, long before any sign of senescence appears.

6.6 Sinks

Any place in a plant which consumes or metabolizes photoassimilates is a sink. The amount of photoassimilates taken up by a sink describes its capacity.

6.6.1 *Irreversible sinks*

6.6.1.1 Permanent sinks

Growth depends on meristematic cells or tissues, and some meristems remain permanent sinks during the life of the plant. To these meristems belong the apices of the shoots and the roots – the apical meristems. They provide new elements to the sieve tube system by differentiation of procambial precursors. Apical growth proceeds during day and night, and thus continued unloading requires continuous loading at potential source sites.

Another permanent sink is the vascular cambium, which is a characteristic of gymnosperms and dicotyledons. This meristem also provides new sieve elements periodically, and in addition takes care of the expansion of the ray system in perennial plants which have secondary growth.

Thus, permanent sinks are responsible for the continuity of the transport system of photoassimilates, the phloem and the ray system. The strength of the permanent sinks can be decreased or they can shut down temporarily, depending on the climatic conditions in winter.

6.6.1.2 Temporary sinks

Some meristems are active only once for a limited period of time, or are reactivated periodically; they belong to the periodic or temporary sinks. The intercalary meristems of the shoot, of the leaves and of the flower organs are all examples of temporary sinks. They stop dividing as soon as the part of the plant which is developing from such a meristem has reached maturity (internodes, leaf blades, petioles, pedicels).

In active intercalary meristems, photoassimilates are consumed for division and growth of cells. Their sink strength vanishes when the processes of differentiation are finished. External influences can reactivate such meristems; for example, when a trampled down blade of grass comes into a vertical position again. Probably, the orientation to the light of a leaf blade is also managed by intercalary growth. Intercalary meristems are restricted to plant parts above the ground. Although some roots show intercalary contractions these are restricted to the zone between root and shoot (*Lilium* spec. div.).

Apical and laminar meristems of leaves, including the meristemoids, are also meristems of temporary activity. The period of activity of an apical leaf meristem can be very short (leaves with petioles) or very long (fern fronds). The marginal meristems of the leaf blade stop dividing when the leaf has reached its ultimate shape. The meristemoids of the lamina, which are responsible for the differentiation of hairs, stomata and studded areoles (*Gunnera*), seem to set the limits of their activity independent of the growth of the blade.

Flowers, fruits and seeds are also temporary sinks. It is not known whether those parts of the flower which are shed when seeds ripen are regularly emptied of material which had been invested during flower formation (Schumacher, 1932).

Another group of temporary sinks is represented by organs which serve for the vegetative propagation of the plant. Some of these organs, like potato tubers and fructan-storing tubers of *Helianthus tuberosus*, accumulate enormous amounts of photoassimilates. Other types are bud bulbs of *Allium ascalonicum* or bulbils of *Dentaria bulbifera*.

6.6.2 Reversible sinks

6.6.2.1 Depots

Important reversible sinks are the true storage tissues, which are missing only in annual plants. These are storage roots, bark and pith. The stored matter is utilized in periods of growth, flower formation or for the decrease of osmotic potential to resist drought and frost. The formation of dormancy callose in the secondary phloem of perennials can be regarded as a depot for photoassimilates, although this sink's capacity is small, and callose may have physiological functions related to unloading (Eschrich, 1965).

6.6.2.2 Functional sinks

6.6.2.2.1 Movement-fuelling sinks The living plant shows not only growth and reproduction; as a stationary organism it must perceive environmental stimuli, and, if necessary, react to protect itself against damage. In most cases, such reactions which can be observed or recorded are movements. Unlike dehiscence movements, those based on turgor pressure or growth must be fuelled with metabolic energy.

Movements which orient leaf blades to the changing position of the sun are – if not guided by intercalary growth – a reflection of functional sinks, although unloading of photoassimilates in such sinks has not yet been demonstrated.

Another still unexplained movement is the bending of stems for reorientation and gravitropic alignment. Many stems of herbaceous plants have plastids in their starch sheath, which obviously serve as statoliths.

Organs which show externally stimulated movements not based on growth are potentially reversible sinks. The best known is *Mimosa pudica* with its seismonastic (in part also nyctinastic) leaf movements. Since these movements are unusually rapid, *Mimosa* can be regarded as a demonstrative example of very fast unloading processes, as described below.

It has been shown by micro-autoradiography that ^{14}C-labelled photoassimilates stay in the sieve tubes of the pulvini as long as the leaf remains untouched (Fig. 6.12). Phloem transport does not differ from transport in irreversible sinks (growth). As soon as the *Mimosa* leaf is touched or otherwise stimulated, within a fraction of a second the labelled photoassimilates are unloaded in the pulvini, the label accumulating in the apoplastic free space (Fig. 6.13). The leaflets fold together, and the four pinnae and eventually the petiole bend down (Fromm, 1986).

The interpretation of these events fits into a proposed chain of mechanisms (Satter & Galston, 1973; Satter, 1979), but the missing link between stimulation and the shift of K^+ and Cl^- is now shown to be the phloem unloading of sucrose. The water potential in the apoplast of the pulvini decreases by the sudden accumulation of sucrose in the free space, and its influence on the breakdown of turgor in the cells of the motor cortex is therefore plasmolytic rather then electric.

Investigation of assimilate transport in the *Mimosa* leaf has shown that sucrose can be unloaded very quickly and can be limited to a small section of the phloem path. The time lapse between stimulation and unloading

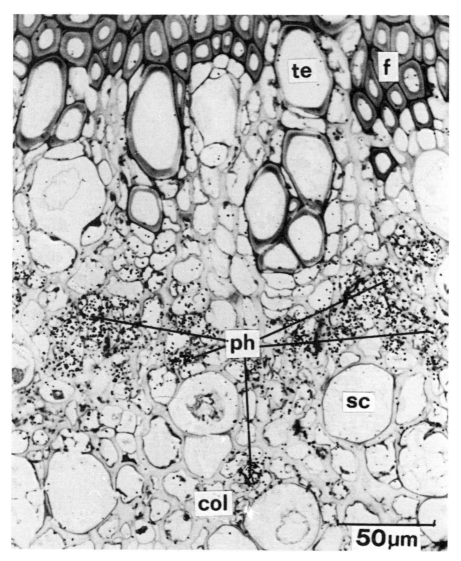

Fig. 6.12 Micro-autoradiography, cross-section of the primary pulvinus of non-stimulated *Mimosa pudica*. ^{14}C label is restricted to sieve tubes and companion cells. col, collenchymatous protophloem; f, libriform fibre; ph, phloem; sc, secretory cell; te, tracheary element (Fromm, 1986).

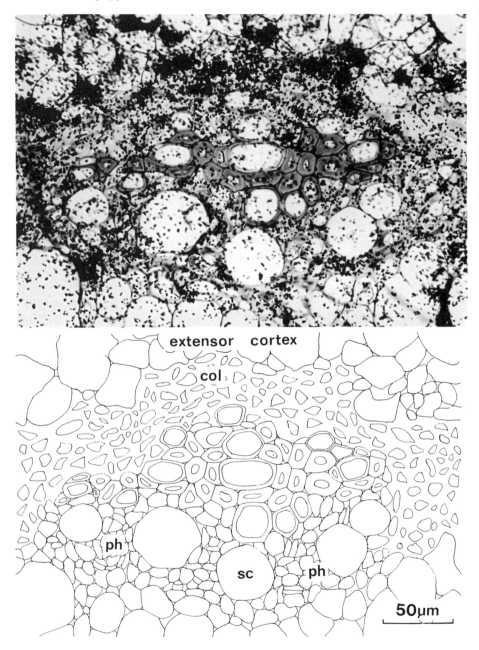

Fig. 6.13 Micro-autoradiograph, cross-section of a tertiary pulvinus of stimulated *Mimosa pudica*. ^{14}C label spread from the phloem mainly in the direction of the extensor cortex via the apoplast. Because of the dense labelling, a line-drawing below is used to explain the different tissue elements. col, collenchymatous tissue; ph, phloem strands; sc, secretory cells (Fromm, 1986).

plus movement leads to an average velocity of 10 cm s^{-1}, which is comparable to rate of movement of an action potential in slow-reacting nerves (Fig. 6.14).

There is no reason why sieve tubes should not be suited for the transmittance of action potentials. Such signals may move along the sieve tube plasmalemma. With regard to the unloading process, it can be assumed that an influx of positively charged ions from the free space will cause depolarization of the membrane potential of the sieve tube plasmalemma. Efflux of negative charges or an ATP-energized proton efflux will restore the original membrane potential (Sibaoka, 1962). With the proton efflux a sucrose-proton co-transport can be combined. If the unloaded sucrose is not removed from the free space, it will be reloaded, unless it is hydrolysed. This happens only in the pulvini where free space invertase is active. Interestingly, in the vascular tissue which extends between the pulvini active invertase is missing. It seems that the accumulation of hexoses in the free space causes the turgor decrease in the cortex of the extensor.

This hypothesis presumes a close relationship between the unloading of sucrose at sink sites, and the fast, ATP-fuelled unloading in reacting pulvini of *Mimosa pudica*. The ATP/ADP ratio decreases in all three types of pulvini when stimulated, but the increase during the phase of regeneration was recorded only in secondary pulvini, which perform nyctinastic movements in addition to the seismonastic reactions (Fromm, 1986).

Another hypothetical relationship exists between autonomous leaf movements (*Albizia julibrissin; Samanea saman*) and influences on the circadian clock by either phytochrome (P_{fr}) or sucrose (Satter *et al.*, 1975, 1977).

6.6.2.2.2 *Hormone-regulated unloading*

Unloading processes such as those found in *Mimosa* pulvini can, in principle, occur everywhere along the transport path of photoassimilates. Turgor pressure in sieve tubes is maintained by their retention of sucrose within the differentially permeable plasmalemma. With any kind of permeability increase of the sieve tube plasmalemma sucrose efflux occurs, because thermodynamically sucrose unloading is a spontaneous process of dilution. Therefore, it seems plausible that factors which cause an increase of membrane permeability may influence sucrose unloading.

During the sudden sucrose unloading in the stimulated *Mimosa* pulvinus described above it was assumed – because of the decrease in the ATP/ADP ratio – that sucrose left the sieve tube cytoplasm in protonated form with a consumption of energy. For this type of membrane transport, Tanner (1980) proposed a mechanism based on the observed increase in abscisic acid (ABA) content in sieve tubes. Increase of ABA concentration causes depolarization of the membrane potential and reduces the proton motive force. That means that ABA reduces the pH-potential (ΔpH)

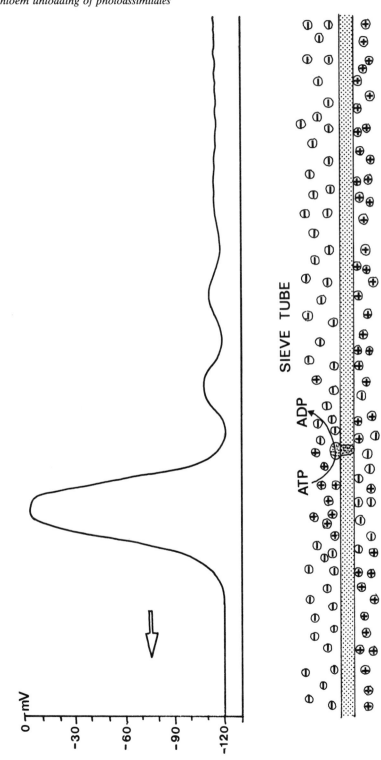

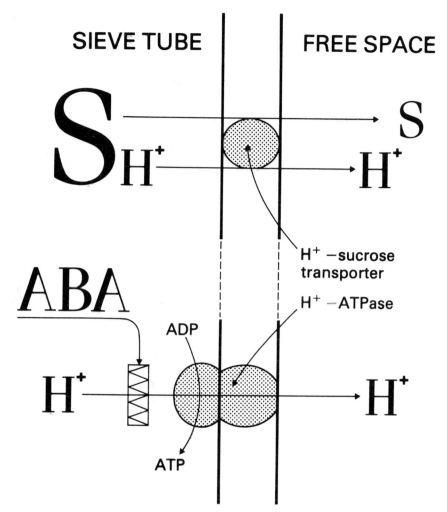

Fig. 6.15 Model for the depolarizing effect that abscisic acid (ABA) exerts on the membrane potential in sieve tubes (after Tanner, 1980).

Fig. 6.14 An action potential moving along a sieve tube. Its depolarizing effect on the membrane potential causes exchange of positive (free space) and negative (sieve tube) charges. The repolarization is managed by an ATP-fuelled cation pump. Protons (the positive charges), having entered the sieve tube, can be utilized for sucrose-proton co-transport into the free space.

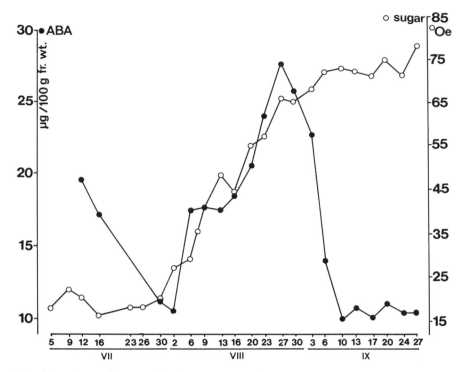

Fig. 6.16 Sugar increase in ripening grape berries in relation to ABA content. °Oe, degree Oechsle = specific gravity, as measured with an areometer at 20°C, and expressed in weight % (hexose). (Düring & Alleweldt, 1980).

across the sieve tube plasmalemma. This lowers the H^+-ATPase activity and permits a H^+-sucrose co-transport into the free space (Fig. 6.15).

When in addition free space invertase is present, the increasing hexose concentration will steepen the $sucrose_i$: $sucrose_o$ concentration gradient. When the hexoses are taken up into parenchyma cells, the ABA content of the sieve tubes must be higher than in the parenchyma cells. This model is one version of hormone-regulated unloading, which may operate in hexose-storing fruits like the grape berries.

It has been shown by Schussler *et al.*, (1984) that the seed coat of a growing soybean contains more ABA than the embryo, and that ABA stimulates sucrose uptake into the cotyledons.

Results which fit the proposed ABA hypothesis were obtained with grape berries. It is known that grape berries have a two-phasic exponential growth-ripening process (Coombe, 1976). Endogenous auxin, cytokinin and gibberellin are present only in the first phase, the growth phase. Endogenous ABA content does not increase until the beginning of the second, the ripening phase. Düring & Alleweldt (1980) have reported that the increasing ABA content runs parallel with the increasing sugar content

(Fig. 6.16). As soon as the sugar concentration reaches its final value, the ABA concentration drops back rapidly to the initial low level. This correlation between ABA increase and sugar deposition indicates that ABA has something to do with the regulation of sugar deposition. However, it need not be directly concerned with phloem unloading.

It has been confirmed recently (Offler & Patrick, 1986) that seeds are supplied by a combination of symplastic and apoplastic sucrose. Ovules of *Phaseolus vulgaris* and *Glycine max* have been investigated frequently. The ovules of the Fabaceae in general develop neither perisperm nor endosperm. The seed coat has a dense network of veins (Lush & Evans, 1980; Thorne, 1981; Offler & Patrick, 1984). In growing ovules, photoassimilates imported through the single bundle of the funiculus, are attracted by certain layers of the seed coat with considerable strength. Micro-autoradiographs have shown that in the young ovule of *Phaseolus* all tissues of the seed coat can periodically act as sinks. When the embryo is detached from the suspensor, the Malpighi cells (outer epidermis of the seed coat) start to elongate radially, and they incorporate great amounts of photoassimilates (Fig. 6.17). The inner epidermis of the seed coat, which has the appearance of a secretory epithelium in the early developmental stages (Fig. 6.18), disintegrates later. Thus, in the young ovule the seed coat seems to be the primary sink.

In many experiments, the embryo has been removed through a window cut into the seed coat, and replaced by a bathing solution or an agar filling (Thorne, 1982; Patrick, 1983; Thorne & Rainbird, 1983; Wolswinkel & Ammerlaan, 1983, 1985; Gifford & Thorne, 1985; Van Bel & Patrick, 1985). By analysing the bathing solution (or the agar filling) after some time in the cup formed by the seed coat, it has been shown that sucrose is secreted into the apoplast. It occurs there in concentrations of 90–200 mol m^{-3}, which is higher than the sucrose concentration in both seed coat and embryo (Gifford & Thorne, 1985). The rate of secretion is similar during day and night. This result shows that secretion can continue without the embryo, the designated sink. Besides amino acids and a small amount of glucose, mainly sucrose was obtained, showing that free space invertase activity was not present. The secretion was stimulated by the addition of EDTA, K$^+$, or 400 mol m^{-3} mannitol to the bathing solution.

In experiments with excised seed coat halves of ^{14}C-pre-loaded ovules of *Phaseolus vulgaris*, one half was used as the control. If 6-benzyl aminopurine (BAP) or ABA (each 10^{-5} molar) was added to the bathing solution in the other seed coat half, more ^{14}C was secreted than in the control cup. The difference in ^{14}C-secretion between control and hormone treatment was significant when statistical methods were applied (Clifford *et al.*, 1986). Other growth regulators (IAA, 1-naphthalene acetic acid, GA, gibberellin 4/7, 1-aminocyclopropane carboxyclic acid) had no effect. There are additional reports on hormone-regulated storage processes, which will be considered in Sect. 6.6.4, Grain filling.

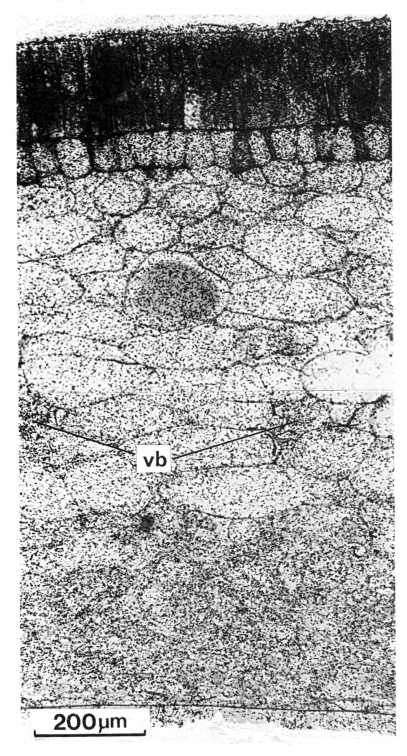

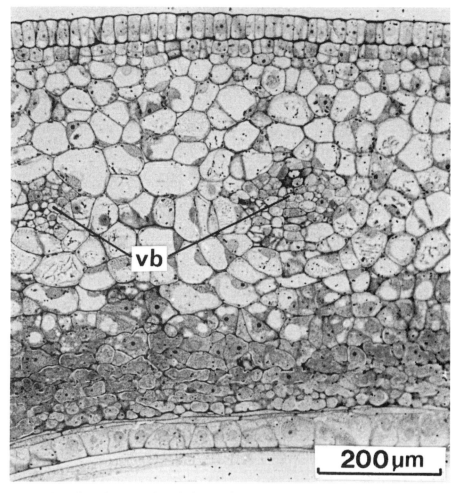

Fig. 6.18 Microphotography of the seed coat cross-section of a young ovule of *Phaseolus vulgaris* cv. Red Kidney. Note that the outer epidermis (the later Malpighi cells) is still undeveloped. vb, vascular bundles.

Fig. 6.17 Micro-autoradiography of a cross-section of the developing seed coat of a growing ovule of *Phaseolus vulgaris* cv. Red Kidney. The subtending leaf of the fruit was exposed to $^{14}CO_2$. Label mostly concentrated in the outer epidermis (Malpighi cells). vb, vascular bundles.

6.6.3 Seed-fruit relations

The method of studying unloading in embryo-free seed coats has been developed independently at three different places. This method is possible because the embryo has a different genome from the seed coat, and with the disruption of the suspensor the symplastic connection to the mother plant is missing. Since no vascular bundle extends into the embryo, water and nutrients must be provided by the seed coat. This situation seems to be similar in embryos of all angiosperms, although a perisperm may be interspersed and the storage can be taken over by endosperm.

Since the seed coat grows with the growing embryo, it constitutes its own sink. In single-seeded fruits like wheat, the fruit wall grows in contact with the seed. It uses photoassimilates, but functions solely as a substitute for the disintegrated seed coat. In big fruits like the pumpkins, conspicuous amounts of photoassimilates must be invested, mainly for the synthesis of cytoplasm and cell walls. Many fruits store sucrose at very high concentrations, for example: date (*Phoenix dactylifera*), 40–50% sucrose on a fresh weight basis; carob (*Ceratonia siliqua*), 40%; tamarind (*Tamarindus indica*), 30%; banana (*Musa sapientium*), 20% (Coombe, 1976).

There are several cultivars with parthenocarpic fruits (banana, seedless grapes, seedless *Citrus* fruits), which ripen and store considerable amounts of sugar. This shows that the fruit constitutes its own sink and cannot be regarded solely as the mediator for the supply of the ovules. The seed coat itself can be the only sink in an ovule if self-pollination has prevented fertilization and embryo development. Such empty seeds can occur also in wild forms of plants. The seed coat constitutes the terminal of the vascular tissue. Water, nutrients and photoassimilates have to traverse parenchymatic tissue and the inner epidermis of the seed coat before the embryo can get access to them. Since the apoplastic interface can accumulate fairly high concentrations of sucrose, secretion and active uptake processes must be involved in the nutrition of the embryo.

6.6.4 Grain filling

In the breeding of cereals grain yield dominates over flour quality: it is primarily a question of productiveness. The goal is to increase the grain:straw ratio by cultural and breeding processes (Jenner, 1982). Physiological aspects of crop seed filling have been reviewed by Thorne (1985).

In temperate zones it takes about 50 days from anthesis to ripeness. Results of numerous investigations have been reported (see Michael, 1984). Hormone concentrations have been related to illumination, water supply and fertilization. In such experiments, the movement of externally applied ^{14}C-labelled hormones does not necessarily mirror a natural process, even when the transport is via the phloem. However, as with the results described above, a marked change of ABA content was found during grain ripening. During development of the grain the endosperm had the higher ABA content, whilst at ripening the ABA content of the embryo was highest (Goldbach & Michael, 1976). Whether or not ABA stimulates or regulates sucrose unloading in the grain is only a guess, because an increased ABA concentration in the embryo could also maintain dormancy and protect the seed against untimely germination and outgrowth on the culm.

Another hormone, a cytokinin, was indicated by bioassays; its content decreased sharply during the ripening process (Michael & Seiler-Kelbitsch, 1972).

Determinations of the absolute concentrations of ABA in sieve tube sap have been obtained from fruit stalks of *Yucca* (195 ng ml^{-1}) and *Cocos* (90 ng ml^{-1}) (Hoad & Gaskin, 1980).

With regard to most of the proposed hormone-triggered unloading reactions, it must be noted that all biotests have shown that optimal activity occurs at very low hormone concentrations. Higher concentrations have negative or no influence. Therefore, it cannot be generalized that 'high hormone concentrations' suit their proposed effect on phloem unloading better than low ones. However, to determine the optimal hormone concentration, the bioassay employed should be an 'unloading test', which has not yet been described.

The pathway of photoassimilates into the grain has been investigated by Oparka & Gates (1981a,b) on caryopses of rice where the phloem ends inside the pericarp (Fig. 6.19). The sieve elements there are becoming small, whilst the companion cells have cytoplasmic contact with the adjacent parenchyma cells by pit connections. Of the seed coat only a cutin layer remains, forming a complete envelope around the nucellus, except at the pigment strand which runs along the xylem of the vascular bundle in the pericarp. All photoassimilates have to traverse this pigment strand to reach the aleurone layer and the inner endosperm. This process can include apoplastic steps of transport, as in corn (Felker & Shannon, 1980). Recent results have shown that photoassimilates move from the pedicel of the corn fruit into an apoplastic compartment, which arises by disintegration of the chalazal placenta cells 7 d after pollination. From there photoassimilates are taken up by transfer cells of the endosperm, which have developed their typical wall ingrowths 20 d after pollination (Shannon *et al.*, 1986).

In wheat endosperm slices it has been shown that externally applied sucrose is more readily transformed into starch than glucose or fructose. Starch synthesis is accelerated by fusicoccin (FC); therefore it is postulated that proton movements accompany the polymerization and accumulation of starch (Rijven & Gifford, 1983). Asymmetrically labelled [^{14}C](glucosyl)-sucrose is transported unaltered into the wheat fruit, randomization occurring only after the unloading of the sucrose (Robinson & Hendrix, 1983).

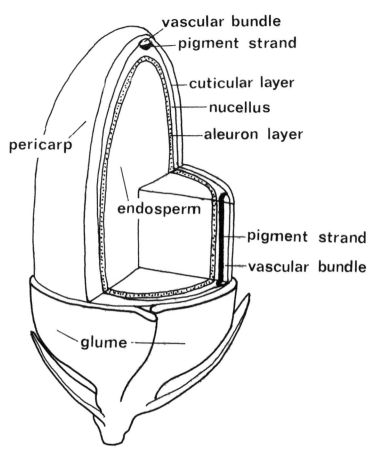

Fig. 6.19 Vascularization of a rice grain; three-dimensional representation; inner and outer glumes partially removed. All photoassimilates to be unloaded for starch deposition in the endosperm must traverse the pigment strand, because at any other place the nucellus and the endosperm are protected by the cuticular layer (Oparka & Gates, 1981).

It seems that sucrose reaches the endosperm unaffected by invertase, and thus the driving force for unloading in the wheat grain is the deposition of starch, which takes away osmotically effective sucrose, and at the same time inhibits its reloading.

6.6.5 *Fleshy fruits*

When one envisages a peach fruit, it is hard to imagine that the ovule could be nourished through the stony endocarp, although cavities with vascular bundles can be discerned. The fleshy exocarp of the Prunoideae is obviously a separate sink with a role in the ecology of seed dispersal. A wild cherry has a thin but sweet-tasting exocarp, which has probably contributed to the survival of the species because its seeds are dispersed by birds.

In a tomato fruit, the relation between fruit and seed growth appears to be closer. This fruit has been carefully investigated, and it was found that each single fruit of a truss reacts differently, if – for example – the leaves of the plant are detached. The differences are manifested in the ratios of C:N and of $K^+:Ca^{2+}$ (Ho, 1979, 1980). These results indicate that phloem unloading is selective and probably dependent on the water and cation supply.

6.7 Sink strength

6.7.1 *Influence of environmental factors*

Sink strength may be defined as the product of sink capacity and unloading velocity. The term sink strength is not an absolute entity, its relativity being demonstrated by the influences of environmental factors, for example light conditions.

6.7.1.1 Light

When the middle of a leaf strip of a pre-darkened (48 h) maize plant is illuminated in a CO_2-free atmosphere, starch can be recognized after 6–7 h. When a small region of the leaf strip is darkened with aluminium foil, starch synthesis does not occur in that region (Fig. 6.20). Photoassimilates, synthesized outside the CO_2-free cuvette, are transported via the phloem into the CO_2-free sink region where sucrose is unloaded. However, starch is synthesized in the chloroplasts of the bundle sheath only under illumination. Obviously, the sink strength is increased in the light and decreased in the dark. The driving force for unloading – usually described as 'demand' – is based on the velocity of the sucrose–starch conversion.

Another example shows the inhibitory influence of the environment on phloem unloading. A mature leaf pair of a pre-darkened *Gomphrena globosa* plant is arranged as a source–sink system as indicated in Fig. 6.21. When the source leaf is illuminated and exposed to $^{14}CO_2$, and the sink leaf is kept in a CO_2-free atmosphere, unloading in the sink will occur only when the sink leaf is protected against light. Fig. 6.22 shows the autoradiographs after 6 h photosynthesis in the $^{14}CO_2$-exposed source (Eschrich & Eschrich, 1987).

The two examples – *Zea mays* and *Gomphrena globosa*, both C_4 plants – exhibit sink strengths which oppose each other in relation to light or dark. This leads to the assumption that the methods of unloading have to be appraised differently, making the term 'sink strength' nearly obsolete, because it is not applicable to the physiological problems of phloem unloading.

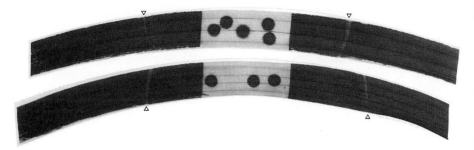

Fig. 6.20 Two leaf strips of mature maize leaves (*Zea mays*), originally 25 cm long, were taken from pre-darkened plants. The middle sections of the leaf strips were sealed in a cuvette (petri-dish halves, CO_2 removed with 20% KOH-soaked filter paper in the bottom) as indicated by triangles. Then the strips were exposed to white light (260 μmol m^{-2} s^{-1}). After 7 h the leaf strips were briefly boiled in water, destained in ethanol and stained with I_2KI. Photosynthates from outside the cuvette moved via the phloem to the inner section devoid of CO_2, and, after unloading, starch was synthesized there as it was in the other sections of the leaf strips. The inner section in part was darkened with aluminium foil (provided with perforations). In the darkened areas of the leaf no starch was synthesized.

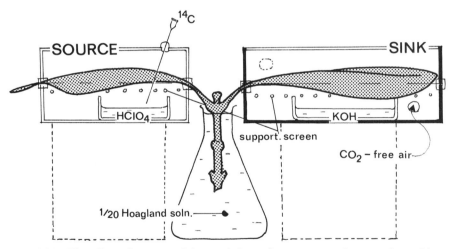

Fig. 6.21 The *Gomphrena* model: consisting of two mature leaves, about 14 cm long, connected to a leafless length of stem. The sink cuvette is impervious to light except for the upper lid (Eschrich & Eschrich, 1987).

6.7.1.2 Temperature

Since sucrose unloading is usually combined with metabolic processes, it can be accepted that – as it is everywhere in living organisms – cooling will slow down reactions and movements including the process of phloem unloading.

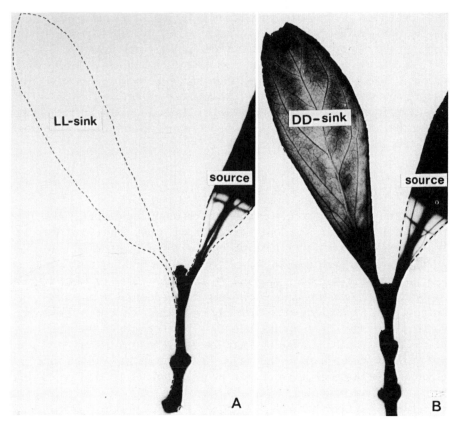

Fig. 6.22 Macro-autoradiographs of source–sink systems of *Gomphrena globosa*. The source leaves (only partly shown) were exposed to $^{14}CO_2$ for 30 min, and after 6 h translocation the systems were frozen, freeze-dried, and autoradiographed. *A*: Sink leaf illuminated with white light (LL, 180 μmol m^{-2} s^{-1}. *B*: Sink leaf kept in darkness (DD). (Eschrich & Eschrich, 1987)

6.7.1.3 Water supply

Water stress is naturally counteracted by a decrease of the osmotic potential in the corresponding part of the symplast. People may have experienced balcony plants, forgotten during a vacation period, which had not only dried out but also were often heavily infested with aphids. Interestingly, honey dew is still secreted by these aphids feeding on dry and brittle leaves. This honey dew appears to have an extremely high sugar content.

With increasing water stress, sieve tubes concentrate their sucrose solution, and as long as the osmotic potential is lower than the water potential of the apoplast, water can be drawn in for translocation. Water stress must hinder apoplastic phloem unloading, because less sucrose will leave the sieve tube. Accordingly, parenchyma cells have to obtain

247

additional sources of sugars, which may help protect against water stress. In detached leaves, water stress causes starch hydrolysis, obviously to decrease the osmotic potential of the mesophyll symplast.

There are several reports that proline content increases when water supply is decreased. It is postulated that this amino acid occurs as a 'stress factor', and can further decrease the osmotic potential (Hsiao, 1973). However, translocation experiments with detached maize leaf blades have shown that in sink regions of such blades proline accumulated only when they were kept dark (Eschrich, 1984a).

6.7.2 Sink limitation

Sink limitation is commonly employed for obtaining higher yields. For example, in viticulture the long shoots are cut when fruits start to grow. With this manipulation photoassimilates from the early matured leaves will move into the grapes, instead of being wasted on building new leaves.

Sink limitation has been applied by Thrower (1974) to force phloem transport into a *mature* leaf, which originally operated as a source. The purpose of such experiments is to obtain a sink without 'appendages', which allows the study of the unloading process proper, without being followed by growth, storage or secretion. Only mature leaves are suited for such purposes, because they have limited growth. Usually, when all other sinks of the plant are removed, photoassimilates are transported into such leaves and unloaded, especially when the plant was starved during a previous 48 h dark period.

A simple model for the forced phloem unloading is a mature maize leaf strip from a pre-darkened plant. It has been shown (Fig. 6.20) that sucrose unloaded in the sink region (CO_2-free atmosphere) is incorporated into starch, as visualized by staining with iodine reagent (Eschrich, 1975b).

By using ^{14}C-labelled photoassimilates it has been shown in *Amaranthus caudatus* and *Gomphrena globosa*, both dicotyledonous C_4 plants, that label was transported into mature leaves and unloaded, as shown by micro-autoradiography (Blechschmidt-Schneider & Eschrich, 1985). In contrast to the maize leaf strip, unloading occurred only when the sink leaf was darkened.

For elucidation of the mechanism of phloem unloading, sink limitation has been combined with pruning procedures to obtain a 'standard' source to be labelled under reproducible conditions, which makes the application of external sucrose via the apoplast unnecessary. With such a system, the influence a sink has on its source and vice versa can be investigated. One such system is the *Gomphrena* leaf pair system (Fig. 6.21), which will be mentioned again when sink/source relations are described.

Amongst varieties of *Coleus blumei* (Lamiaceae) some have variegated leaves, which have chlorenchyma in a marginal position, while the centre

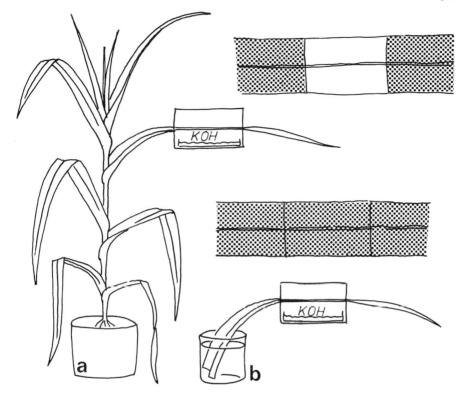

Fig. 6.23 A mature leaf of an illuminated maize plant, about 1 m high, partly enclosed in a translucent cuvette flushed with CO_2-free air. After 18 h the leaf section, kept devoid of CO_2, was free of starch (a). When this leaf was detached without removing the cuvette and put into water, 7 h later the starch in the bundle-sheath had been replenished and could be stained with I_2KI (b).

of the lamina is not green. The latter is a sink and the green rim of the leaf blade is the source. When all green tissue is cut away, and the remaining albino part of the blade is exposed to $^{14}CO_2$, no labelled carbon is fixed (although the guard cells of the stomata possess chloroplasts). When only one half of the green rim is cut away, the labelled photoassimilates transported in the veins spread beyond the midrib into the pruned albino part of the blade, up to the very ends of the veins on the cut edge. Even a small piece of green lamina will provide its labelled photoassimilates to most regions of the albino blade.

Extracts of the albino blade analysed by thin-layer chromatography and autoradiography show several labelled compounds in addition to sucrose, and labelled sucrose must therefore have been unloaded in the albino blade for metabolism. The variegated *Coleus* leaf is thus an operable source–sink system (Fisher & Eschrich, 1985).

6.7.3 Sink competition

When part of the mature leaf blade of a normally grown maize plant is encased in a translucent cuvette which is continuously flushed with CO_2-free air, the starch in this part of the leaf is hydrolysed despite the fact that this part of the leaf is illuminated as well as the remaining parts. Starch is completely absent after about 24 h (Fig. 6.23a). When the same leaf is then detached from the plant without removing the cuvette, and the cut end put into water, starch is again deposited after 7 h of illumination (Fig. 6.23b). Since CO_2 for photosynthesis was withheld, the starch could be synthesized only from the sucrose which was unloaded.

This experiment demonstrates the competition of sinks: as long as the mature leaf was attached to the whole plant, no sucrose was spared to feed the enclosed part of the blade. On the contrary, the CO_2-withheld part had to mobilize its reserves to fulfil its function as a source. Afterwards, when all sinks were removed, this part of the source leaf was supplied, because it was now the only sink left (Eschrich, 1984b). With this experiment it is also shown that switching from phloem loading (starch hydrolysis, sucrose loading) to phloem unloading (sucrose unloading, resynthesis of starch) can occur in the same section of the phloem pathway.

6.8 Sink/source relationships

When the *Gomphrena* system (Fig. 6.21) is used as a sourch/sink system (Fig. 6.20), the unloading process proper can be studied independently of growth and storage. Only respiration in the sink could require the import of some sucrose. (There are no data available on possible differences in

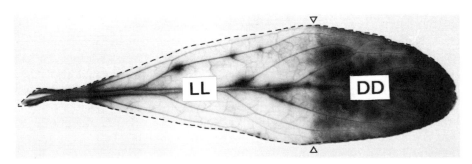

Fig. 6.24 Macro-autoradiograph of a *Gomphrena* model sink leaf, which was partially illuminated with photosynthetic active radiation (LL), and partially darkened (DD). The darkened part of the sink leaf shows phloem unloading (extracts contain metabolities of sucrose!). It triggers transport of labelled photoassimilates through the illuminated part of the sink leaf, although LL-conditions prevent unloading as shown in Fig. 6.22A (Eschrich & Eschrich, 1987).

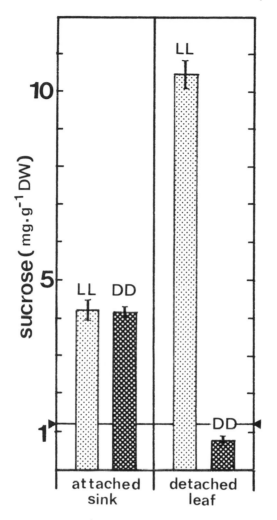

Fig. 6.25 Concentration of sucrose in attached sinks and in detached leaves of *Gomphrena* systems after 18 h white light (LL) illumination and darkness (DD). Sucrose level in the leaves of the pre-darkened plant indicated between arrowheads. (Eschrich & Eschrich, 1987).

respiration rates in source and sink.) In the *Gomphrena* system, it has been shown that unloading in the sink occurs only when the sink leaf is kept dark (Fig. 6.22). If only one part of the sink leaf is darkened (Fig. 6.24), the ^{14}C-labelled photoassimilates move through the veins of the illuminated part, but are not unloaded until they have reached the darkened part of the sink leaf. Import and unloading have to be regarded as separate processes. In such experiments it has been shown that contributions of the source are not limited to the supply of photoassimilates. It also appeared

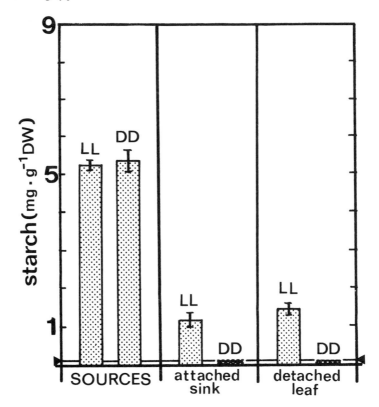

Fig. 6.26 Concentration of starch in sources and their illuminated (LL) or darkened (DD) sinks in *Gomphrena* systems and in detached leaves after 18 h treatments. The starch level of the leaves of the pre-darkened plant is very low; it is indicated between arrowheads. (Eschrich & Eschrich, 1987).

that the source has a strong influence on the regulation of the export-import potentials of the sink in relation to light-dark changes.

When measuring the amounts of sucrose and starch it was found that the mature *Gomphrena* leaf behaves differently as a sink in symplastic connection to the source than it does as a detached leaf, in both cases lacking CO_2. The most striking differences are: (a) sucrose and starch concentrations are the same in both the light part and the dark part of the sink when the leaf is attached to the source; but (b) in the detached leaf the sucrose content increases to a fairly high level when it is illuminated, and consequently the starch content increases too; and (c) in the dark neither sucrose nor starch can be found in the detached leaf (Figs. 6.25, 6.26).

The symplastic connection of a sink to its source stabilizes the sucrose content during the day and night. In the dark sucrose is unloaded; in the light no unloading occurs (cf. Fig. 6.22) but photorespiration provides

CO_2, obviously in sufficient amounts for photosynthetic production of sucrose (Eschrich & Eschrich, 1987).

As mentioned in Sect. 6.7.2, Sink limitation, extensive pruning of a plant may influence source activities, but only after a prolonged period of time (Neales & Incoll, 1968). This matter is still debated, and it must be assumed that photosynthesis is not directly controlled by the rate of consumption of sucrose. As long as a photosynthetic source is symplastically connected to a sink, export of sucrose is possible. If the latter is inhibited any surplus of photoassimilates will be stored in sucrose pools or as starch.

Results which may be important for further research were obtained by Wyse and Saftner (1982), who used tap-root slices of sugar-beet taken from plants exposed to different photoperiods (1–10 h) as sinks. Short photoperiods reduced the sink capacity of the root tissue for externally supplied [^{14}C] sucrose; in the leaf starch content increased when the root's sink capacity decreased. It is concluded that, depending on the sink capacity, the flux of photoassimilates from an active source can be limiting.

6.9 Biophysical and biochemical aspects of phloem unloading

From the biochemical point of view, the barrier between source and sink is the chloroplast envelope (Herold, 1980; see Ch. 1). This means that all compartments of a plant which are unable to carry out photosynthesis are sinks. Factors and effectors which have an influence on sucrose synthesis and phloem loading also have an influence on photosynthesis. According to the biochemical pathways in the photosynthetic cell, five locations can be attacked to interfere with the normal course of reactions (see Ch. 1). Sucrose itself inhibits sucrose-phosphate phosphatase and sucrose-phosphate synthetase by product accumulation. The latter enzyme is also inhibited by orthophosphate, and by nucleotide triphosphate. The fifth way of interference is fructose-6-phosphate, which inhibits the dephosphorylation of fructose-1,6-bisphosphate.

Phloem loading can also be influenced, for example inhibited, when the proton concentration is not sufficient for H^+-sucrose co-transport, or when H^+-ATPase or the sucrose transporter are inhibited or blocked (see Ch. 5). The long-distance transport of photoassimilates, on the other hand, is scarcely influenced, unless water potential in the apoplast decreases to values below the osmotic potential in the sieve tubes (see Ch. 7).

Phloem unloading is dependent on sink strength, which can be influenced by external factors. In the *Gomphrena* model it is the light which inhibits unloading. Since the sink in the *Gomphrena* model (Fig. 6.21) is a mature leaf which is normally concerned with the export of photoassimilates, it could be that light 'switches' the leaf to 'loading', although

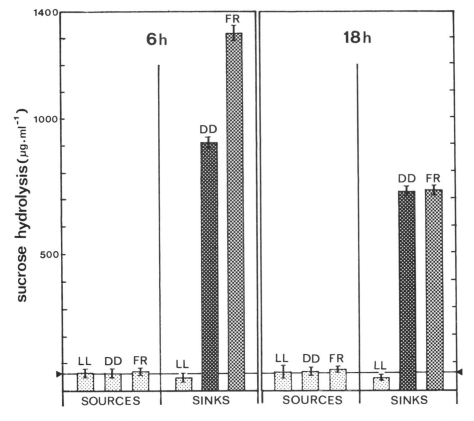

Fig. 6.27 Activity of soluble invertase in sources and their sink leaves; obtained in *Gomphrena* systems after 6 and 18 h treatments. The sink leaves were either illuminated with white light (LL), kept in darkness (DD) or irradiated with far-red (FR: above 690 nm). The level of invertase activity in leaves of the pre-darkened plants is indicated between arrowheads. Invertase assay conditions: 230–255 µg protein ml^{-1}, pH 5.0, 60 min incubation at room temperature. (Eschrich & Eschrich, 1987)

external CO_2 is lacking. It has been checked whether photoassimilates synthesized from photorespirational CO_2 are exported. For this, the source leaf and the length of stem were covered with aluminium foil, while the sink leaf was illuminated without a supply of external CO_2. There was no significant increase in sugar content in the darkened source or stem and thus it must be concluded that light inhibits phloem unloading in the mature sink leaf of the *Gomphrena* model.

Concomitant with the lack of sucrose export was the lack of invertase activity in the illuminated sink leaf. However, when the sink leaf was darkened, invertase level was high (Fig. 6.27). Since darkness is not a factor which stimulates enzyme activation, different light qualities were tested for their ability or inability to influence invertase activity. It was found that blue and red light did not increase invertase activity; far-red, on the contrary, stimulated the increase of invertase activity much more than complete darkness. In agreement with this, far-red also stimulated phloem unloading in the sink leaf.

These results indicate that a pigment similar to phytochrome could play a role in the regulation of phloem unloading and invertase activation. The influence of phytochrome on transport processes has not yet been investigated extensively (Britz *et al.*, 1985). Despite the technical difficulty of co-ordinating light quality and radiation energy, it is obvious that only the inactive form of the proposed phytochrome (P_r) would be produced by far-red irradiation. Thus, active phytochrome (P_{fr}) would inhibit both unloading and invertase activation. Since far-red activation of invertase is superior to its activation in darkness, it seems to be an effector-triggered process. In addition, the invertase activation takes some hours, during which the synthesis of the enzyme would be possible.

Phloem unloading must be considered as occurring from a supply system, which can be tapped everywhere to satisfy the demands of the adjacent cells; in some places on a small scale and in others on a large scale. Biophysically, unloading into the free space depends on the permeability of the sieve tube plasmalemma, which in turn is dependent on the number of ATPase molecules available for the exchange of charges (H^+, K^+, Cl^-). According to this picture, stimulations cause action potentials (as observed in *Mimosa*) which move in sieve tubes. They are fuelled by H^+-ATPase, as proposed for plant cells in general by Komor and Tanner (1974).

Tissue cultures of phloem tissue develop conspicuous amounts of sieve elements under certain conditions. Freeze-etch electron micrographs of such sieve elements have shown that the inner face of the plasmalemma is densely interspersed with intramembrane particles (Sjölund & Shih, 1983). If these particles are identical with ATPase molecules, their density of distribution would indicate how quickly charges can be exchanged.

The permanent occurrence of sucrose in the free space of vascular bundles points to the possibility that sucrose leakage is regulated. It could be called an eddy current of sucrose across the sieve element plasmalemma. The efflux of sucrose into the free space can be generated either by H^+-sucrose co-transport, ABA-stimulated increase of plasmalemma permeability, a combination of both mechanisms, or some other effectors not yet discovered.

Investigations with sucrose analogues which are not affected by invertase have shown that probably the glucose moiety is attached to the sucrose transporter in the membrane (Hitz *et al.*, 1985; Hitz, 1986; see Ch. 5). This result makes it difficult to imagine how the galactosides of the sucrose, which have their galactose moieties attached to the glucose, can penetrate the sieve tube plasmalemma. These galactosides, raffinose, stachyose and verbascose, have the fructose moiety sterically free. It also cannot be excluded that two sucrose transporters may exist; one for entry into the sieve tube and one for exit (Hitz *et al.*, 1985).

It has been shown that K^+ concentrations in sieve tubes are relatively high (about 1% in sieve tube exudates; see Ch. 4). The role of the potassium ion in relation to membrane transport processes is not yet fully understood. It is not only the depolarization of membrane potentials which is caused by external increase of K^+ concentration (Wright & Fisher, 1981). The singular thing is that cell types known for their high K^+ content (sieve tubes, motor cells of *Mimosa* pulvini, guard cells) always seem to have a high affinity for sucrose (Fromm, 1986).

Theoretically, the depolarizing effect of external K^+ would be compatible with the increase of membrane permeability. Thus, external K^+ could have functions like neutralization of the surplus of protons co-transported with sucrose during its loading (Heyser, 1980), or increase of the permeability of the sieve element plasmalemma to facilitate the unloading of H^+-sucrose complexes (Saftner & Wyse, 1980). The possible movement of cytoplasmic K^+ into the vascular free space has not yet been considered functionally.

As significant as the relatively high K^+ concentration in the sieve tubes is the almost complete absence of calcium in the sieve tube cytoplasm (Hall & Baker, 1972; Penot, 1979). The catchphrase 'potassium-calcium antagonism' is not applicable in relation to phloem unloading. Ca^{2+} in very low concentrations mobilizes liquid water out of its colloidally bound state (Eschrich & Eschrich, 1964). Furthermore, Ca^{2+} is necessary for the biosynthesis of β-1,3-glucans, namely callose (Eschrich, 1961; Fink *et al.*, 1987). Putting both facts together, it has been proposed that fast unloading of photoassimilates could occur on the callose-covered sieve plates, when changes of the water status of callose is triggered by Ca^{2+} (Eschrich, 1965).

Acid free-space invertase is widely distributed in vascular tissues of plants (Eschrich, 1980). The diverse functions of invertase and their forms have been described (Chin & Weston, 1973; Jones & Kaufman, 1975). This enzyme, which hydrolyses sucrose in an acid milieu, seems to be a glycoprotein which is bound like a salt to acidic groups of the cell wall (Lehle *et al.*, 1979). However, it is not known whether the relatively high energy of -6.6 kcal mol^{-1} (Edelmann, 1973) which is liberated during hydrolysis of sucrose is used in any way.

The products of sucrose hydrolysis, glucose and fructose, are not taken up into the sieve tube system. However, if these hexoses are taken up into vascular parenchyma they can accumulate, or they can be phosphorylated and, on a detour, can be reunited to form sucrose.

The second mechanism of sucrose unloading, which uses the symplastic route through plasmodesmata (Fig. 6.1b,c), must be kept moving either by metabolizing the sucrose in the cytoplasm of the vascular parenchyma cells (Fig. 6.1b), or by accumulating sucrose in the vacuoles (Fig. 6.1c).

In both cases, cytoplasmic invertase can catalyse the hydrolysis of sucrose. Histochemical invertase assays are not suitable to discriminate between cytoplasmic and vacuolar invertase (Krishnan *et al.*, 1985; Eschrich & Eschrich, 1987). Vacuolar invertase has been described by Boller & Kende (1979). When the vacuolar sucrose pool is a temporary depot, it may be that the vacuolar invertase is restricted to the tonoplast, and then hydrolyses sucrose only when re-entering the cytoplasm.

When sucrose is delivered to the meristems, invertase seems to play only a minor role. Instead, sucrose synthase (EC 2.4.1.13) is active in meristematic tissues (Giaquinta, 1979; Claussen *et al.*, 1985). For sucrose unloading in this case UDP and sucrose are converted to UDP-glucose and fructose. This reaction is catalysed optimally at pH 7.6, and since UDP may be restricted to cytoplasmic compartments, unloading accordingly must be symplastic. However, when UDP-glucose is used for further metabolism, its glucose is mainly transformed to G1P, leaving UMP as a by-product. This reaction is catalysed by pyrophosphatase (phosphodiesterase, EC 3.1.4.1), which is particle-bound and operates optimally at pH 5. Probably, unloading in meristems is much more compartmented. Sucrose, once outside the sieve element, can be split phosphorolytically, and UDP-glucose, as well as fructose, is used in many ways. In addition, it is not known how strictly the cytoplasm and cell wall in meristems are separated as metabolic compartments.

Acknowledgements

I am grateful to Dr Berthilde Eschrich, my wife, who undertook the enormous task of searching for the pertinent literature, and who spent countless hours in discussing details of the topic.

References

Ahrns, W. (1924). Weitere Untersuchungen über die Abhängigkeit des gegenseitigen Mengenverhältnisses der Kohlenhydrate im Laubblatt vom Wassergehalt. *Archiv für Botanik* **5**, 234–59.

Blechschmidt-Schneider, S. & Eschrich, W. (1985). Microautoradiographic localization of imported ^{14}C-photosynthate in induced sink leaves of two dicotyledenous C_4 plants in relation to phloem unloading. *Planta* **163**, 439–47.

Boller, Th. & Kende, H. (1979). Hydrolytic enzymes in the central vacuole of plant cells. *Plant Physiology* **63**, 1123–32.

Britz, S. J., Hungerford, W & Lee, D. R. (1985). Photoperiodic regulation of photosynthate partitioning in leaves of *Digitaria procumbens* Stent. *Plant Physiology* **78**, 710–14.

Chin, C. K. & Weston, G. D. (1973). Distribution in excised *Lycopersicon esculentum* roots of the principal enzymes involved in sucrose metabolism. *Phytochemistry* **12**, 1229–35.

Claussen, W., Loveys, B. R. & Hawker, J. S. (1985). Comparative investigations on the distribution of sucrose synthase activity and invertase activity within growing, mature and old leaves of some C_3 and C_4 plant species. *Physiologia Plantarum* **65**, 275–80.

Clifford, P. E., Offler, C. E. & Patrick, J. W. (1986). Growth regulators have rapid effects on photosynthate unloading from seed coats of *Phaseolus vulgaris* L. *Plant Physiology* **80**, 635–7.

Coombe, B. G. (1976). The development of fleshy fruits. *Annual Review of Plant Physiology* **27**, 507–28.

Dick, P. S. & ap Rees, T. (1975). The pathway of sugar transport in roots of *Pisum sativum*. *Journal of Experimental Botany* **26**, 305–14.

Düring, H. & Alleweldt, G. (1980). Effects of plant hormones on phloem transport in grapevines. In Eschrich, W. & Lorenzen, H. (eds.). *Phloem Loading and Related Processes*, pp. 339–47. Stuttgart, New York, Gustav Fisher Verlag.

Edelman, J. (1973). The role of sucrose in green plants. In Yudkin, J., Edelman, J. & Hough, L. (eds.). *Chemical, Biological and Nutritional Aspects of Sucrose*, pp. 95–102. London, Butterworths.

Esau, K. (1945). Vascularization of the vegetative shoots of *Helianthus* and *Sambucus*. *American Journal of Botany* **32**, 18–29.

Esau, K. (1965). *Vascular Differentiation in Plants*. New York, Holt, Rinehart and Winston.

Esau, K. (1969). *The Phloem. Encyclopedia of Plant Anatomy*, 2nd ed, Vol. V, part 2. Berlin, Stuttgart, Gebrüder Borntraeger.

Eschrich, W. (1953). Beiträge zur Kenntnis der Wundsiebröhrenentwicklung bei *Impatiens holsti. Planta* **43**, 27–74.

Eschrich, W. (1961). Untersuchungen über den Ab- und Aufbau der Kallose. *Zeitschrift fur Botanik* **49**, 153–218.

Eschrich, W. (1965). Physiologie der Siebröhrencallose. *Planta* **65**, 280–300.

Eschrich, W. (1970). Biochemistry and fine structure of phloem in relation to transport. *Annual Review of Plant Physiology* **21**, 193–214.

Eschrich, W. (1975a). Bidirectional transport. In Aronoff, S., Dainty, J., Gorham, P. R., Srivastava, L. M. & Swanson, C. A. (eds.). *Phloem Transport*, pp. 401–16. New York, London, Plenum Press.

Eschrich, W. (1975b). Das Experiment: Transport im Phloem. *Biologie in unserer Zeit* **5**, 26–8.

Eschrich, W. (1980). Free space invertase, its possible role in phloem unloading. In Eschrich, W. & Lorenzen, H. (eds.). *Phloem Loading and Related Processes*, pp. 363–78. Stuttgart, New York, Gustav Fischer Verlag.

Eschrich, W. (1983). Phloem unloading in aerial roots of *Monstera deliciosa*. *Planta* **157**, 540–7.

Eschrich, W. (1984a). Phloem unloading following reactivation in predarkened mature maize leaves. *Planta* **161**, 113–19.

Eschrich, W. (1984b). Untersuchungen zur Regulation des Assimilattransportes. *Berichte der Deutschen botanischen Gesellschaft* **97**, 5–14.

Eschrich, W. & Burchardt, R. (1982). Reactivation of phloem export in mature maize leaves after a dark period. *Planta* **155**, 444–8.

Eschrich, W. & Eschrich, B. (1964). Das Verhalten isolierter Callose gegenüber wässrigen Lösungen. *Berichte der Deutschen botanischen Gesellschaft* **77**, 329–31.

Eschrich, W. & Eschrich, B. (1987). Control of phloem unloading by source activities and light. *Plant Physiology and Biochemistry* **25**, 625–34.

Essiamah, S. K. & Eschrich, W. (1985). Changes of starch content in the storage tissues of deciduous trees during winter and spring. *IAWA Bulletin* **6**, 97–106.

Evert, R. F., Eschrich, W. & Heyser, W. (1978). Leaf structure in relation to solute transport and phloem loading in *Zea mays* L. *Planta* **130**, 279–94.

Felker, F. C. & Shannon, J. C. (1980). Movement of ^{14}C-labelled assimilates into kernels of *Zea mays* L. III. An anatomical examination and microautoradiographic study of assimilate transfer. *Plant Physiology* **65**, 864–70.

Fink, J., Jeblick, W., Blaschek, W. & Kauss, H. (1987). Calcium ions and polyamines activate the plasmamembrane-located 1.3.β-glucan synthase. *Planta* **171**, 130–5.

Fisher, D. G. & Eschrich, W. (1985). Import and unloading of ^{14}C assimilate into nonphotosynthetic portions of variegated *Coleus blumei* leaves. *Canadian Journal of Botany* **63**, 1708–12.

Fondy, B. R. & Geiger, D. R. (1985). Diurnal changes in allocation of newly fixed carbon in exporting sugar beet leaves. *Plant Physiology* **78**, 753–7.

Fritz, E., Evert, R. F. & Heyser, W. (1983). Microautoradiographic studies of (^{14}C) assimilate transport in the leaf of *Zea mays* L. *Planta* **159**, 193–206.

Fromm, J. (1986). Assimilattransport in ungereizten und gereizten Blattgelenken von *Mimosa pudica* L. Dissertation Forstl. Fakultät, Universität Göttingen.

Geiger, D. R. (1976). Effects of translocation and assimilate demand on photosynthesis. *Canadian Journal of Botany* **54**, 2337–45.

Geiger, D. R., Ploeger, B. J., Fox, R. C. & Fondy, B. R. (1983). Sources of sucrose translocated from illuminated sugar beet source leaves. *Plant Physiology* **72**, 964–70.

Giaquinta, R. T. (1979). Sucrose translocation and storage in the sugar beet. *Plant Physiology* **63**, 828–32.

Gifford, R. M. & Thorne, J. H. (1985). Sucrose concentration at the apoplastic interface between seed coat and cotyledons of developing soybean seeds. *Plant Physiology* **77**, 863–8.

Goldbach, H. & Michael, G. (1976). Abscisic acid content of barley grains during ripening as affected by temperature and variety. *Crop Science* **16**, 797–9.

Hall, S. M. & Baker, D. A. (1972). The chemical composition of *Ricinus* phloem exudate. *Planta* **106**, 131–40.

Hendrix, J. E. (1968). Labelling pattern of translocated stachyose in squash. *Plant Physiology* **43**, 1631–6.

Herold, A. (1978). Starch synthesis from exogenous sugars in tobacco leaf discs. *Journal of Experimental Botany* **29**, 1391–401.

Herold, A. (1980). Regulation of photosynthesis by sink activity – The missing link. *New Phytologist* **86**, 131–44.

Heyser, W. (1980). Phloem loading in the maize leaf. In Eschrich, W. & Lorenzen, H. (eds.). *Phloem Loading and Related Processes*, pp. 221–8. New York, Stuttgart, Gustav Fischer Verlag.

Heyser, W., Evert, R. F., Fritz, E. & Eschrich, W. (1978). Sucrose in the free space of translocating maize leaf bundles. *Plant Physiology* **62**, 491–4.

Heyser, W., Heyser, R., Eschrich, W., Leonard, O. A. & Rautenberg, M. (1976). The influence of externally applied organic substances on phloem translocation in detached maize leaves. *Planta* **132**, 269–77.

Hitz, W. D. (1986). Molecular determinants of sugar carrier specificity. In Cronshaw, J., Lucas, W. J. & Giaquinta, R. T. (eds.). *Phloem Transport*, pp. 27–39. New York, Alan R. Liss, Inc.

Hitz, W. D., Schmitt, M. R., Card, P. J. & Giaquinta, R. T. (1985). Transport and metabolism of 1′-fluorosucrose, a sucrose analog not subject to invertase hydrolysis. *Plant Physiology* **77**, 291–5.

Ho, L. C. (1979). Regulation of assimilate translocation between leaves and fruits in the tomato. *Annals of Botany* **43**, 437–48.

Ho, L. C. (1980). Control of import into tomato fruits. In Eschrich, W. & Lorenzen, H. (eds.). *Phloem Loading and Related Processes*, pp. 315–25. New York, Stuttgart, Gustav Fischer Verlag.

Hoad, G. V. & Gaskin, P. (1980). Abscisic acid and related compounds in phloem exudate of *Yucca flaccida* Haw. and coconut (*Cocos nucifera* L.). *Planta* **150**, 347–8.

Hsiao, T. C. (1973). Plant responses to water stress. *Annual Review of Plant Physiology* **24**, 519–70.

Jacobsen, K. R. (1986). Assimilattransport in Wundsiebröhren. Diss, Math. Natw. Fakultät, Universität Göttingen.

Jenner, C. F. (1982). Storage of starch. In Loewus, F. A. & Tanner, W. (eds.). *Plant Carbohydrates I. Encyclopedia of Plant Physiology 13A*, pp. 700–47. Berlin, Heidelberg, New York, Springer Verlag.

Jones, R. A. & Kaufman, P. B. (1975). Multiple forms of invertase in developing oat internodes. *Plant Physiology* **55**, 114–19.

King, R. W., Wardlaw, I. F. & Evans, L. T. (1967). Effect of assimilate utilization on photosynthetic rate in wheat. *Planta* **77**, 261–76.

Köcher, H. & Leonard, O. A. (1971). Translocation and metabolic conversion of ^{14}C-labelled assimilates in detached and attached leaves of *Phaseolus vulgaris* L. in different phases of leaf expansion. *Plant Physiology* **47**, 212–16.

Komor, E. & Tanner, W. (1974). The hexose-proton cotransport system of *Chlorella*. pH-dependent change in K_M values and translocation constants of the uptake system. *Journal of General Physiology* **64**, 568–81.

Krishnan, H. B., Blanchette, J. T. & Okita, T. W. (1985). Wheat invertases. Characterization of cell wall-bound and soluble forms. *Plant Physiology* **78**, 241–5.

Lang, A., Thorpe, M. R. & Edwards, W. R. N. (1986). Plant water potential and translocation. In Cronshaw, J., Lucas, W. J. & Giaquinta, R. T. (eds.). *Phloem Transport*, pp. 193–4. New York, Alan R. Liss.

Langenfeld-Heyser, R. (1987). Distribution of leaf assimilates in the stem of *Picea abies* L. *Trees* **1**, 102–9.

Larson, P. R. (1975). Development and organization of the primary vascular system in *Populus deltoides* according to phyllotaxy. *American Journal of Botany* **62**, 1084–99.

Lehle, L., Cohen, R. E. & Ballou, C. E. (1979). Carbohydrate structure of yeast invertase. Demonstration of a form with only core oligosaccharides and a form with completed polysaccharide chains. *Journal of Biological Chemistry* **254**, 12209–18.

Lush, W. M. & Evans, L. T. (1980). The seed coat of cowpeas and other grain legumes: structure in relation to function. *Field Crops Research* **3**, 267–86.

Michael, G. (1984). Speicherungsprozesse und ihre Regulation in Kulturpflanzen. *Berichte der Deutschen botanischen Gesellschaft* **97**, 1–4.

Michael, G. & Seiler-Kelbitsch, H. (1972). Cytokinin content and kernel size of barley grains as affected by environmental and genetic factors. *Crop Science* **12**, 162–5.

Neales, T. F. & Incoll, L. D. (1968). The control of leaf photosynthesis rate by the level of assimilate concentration in the leaf: a review of the hypothesis. *Botanical Review* **34**, 107–25.

Nelson, C. D. (1962). The translocation of organic compounds in plants. *Canadian Journal of Botany* **40**, 757–70.

Offler, C. E. & Patrick, J. W. (1984). Cellular structures, plasma membrane surface areas and plasmodesmatal frequencies of seed coats of *Phaseolus vulgaris* L. in relation to photosynthate transfer. *Australian Journal of Plant Physiology* 11, 79–99.

Offler, C. E. & Patrick, J. W. (1986). Cellular pathway and hormone control of short-distance transfer in sink regions. In Cronshaw, J., Lucas, W. J. & Giaquinta, R. T. (eds.). *Phloem Transport*, pp. 295–306. New York, Alan R. Liss.

Oparka, K. J. & Gates, P. (1981a). Transport of assimilates in the developing caryopsis of rice (*Oryza sativa* L.). Ultrastructure of the pericarp vascular bundle and its connection with the aleurone layer. *Planta* 151, 561–73.

Oparka, K. J. & Gates, P. (1981b). Transport of assimilates in the developing caryopsis of rice (*Oryza sativa* L.). The pathways of water and assimilated carbon. *Planta* 1452, 388–96.

Parthasarathy, M. V. (1980). Mature phloem of perennial monocotyledons. In Eschrich, W. & Lorenzen, H. (eds.). *Phloem Loading and Related Processes*, pp. 57–70. New York, Stuttgart, Gustav Fischer Verlag.

Patrick, J. W. (1983). Photosynthate unloading from seed coats of *Phaseolus vulgaris* L. General characteristics and facilitated transfer. *Zeitschrift für Pflanzenphysiologie* 111, 9–18.

Patrick, J. W. & McDonald, R. (1980). Pathway of carbon transport within developing ovules of *Phaseolus vulgaris* L. *Australian Journal of Plant Physiology* 7, 671–84.

Penot, M. (1979). Transport a longue distance des element minéraux dans le phlome. Relation entre organes, importance du controle hormonal. In Gagnaire-Michard, J. & Riedacker, A. (eds.). *Les Correlations entre les Racines et les Parties Aeriennes*. Comptes Rendu du Groupe d'Etudes des Racines, Vol. 9, pp. 94–133.

Pongratz, P. & Beck, E. (1978). Diurnal oscillation of amylolytic activity in spinach chloroplasts. *Plant Physiology* 62, 687–9.

Rijven, A. H, & Gifford, R. M. (1983). Accumulation and conversion of sugars by developing wheat grains. 3. Non-diffusional uptake of sucrose, the substrate preferred by endosperm slices. *Plant, Cell and Environment* 6, 417–25.

Robinson, N. L. & Hendrix, J. E. (1983). Translocation of [^{14}C] sucrose in wheat. *Plant Physiology* 71, 701–2.

Saftner, R. A. & Wyse, R. E. (1980). Alkali cation/sucrose co-transport in the root sink of sugar beet. *Plant Physiology* 66, 884–9.

Saproschnikoff, W. (1890). Bildung und Wanderung der Kohlenhydrate in den Laubblättern. *Berichte der Deutschen botanischen Gesellschaft* 8, 233–42.

Satter, R. L. (1979). Leaf movements and tendril curling. In Haupt, W. & Feinleib, M. E. (eds.). *Physiology of movements. Encyclopedia of Plant Physiology* 7, pp. 442–84. Berlin, Heidelberg, New York, Springer-Verlag.

Satter, R. L., Applewhite, P. B., Chaudhri, J. & Galston, A. W. (1975). P. phytochrome and sucrose requirement for rhythmic leaflet movement in *Albizzia*. *Phytochemistry and Photobiology* 23, 107–12.

Satter, R. L. & Galston, A. W. (1973). Leaf movements: rosetta stone of plant behaviour? *Bioscience* 23, 407–16.

Satter, R. L., Schrempf, M., Chaudhri, J. & Galston, A. W. (1977). Phytochrome and circadian clocks in *Samanea*. Rhythmic redistribution of potassium and chloride within the pulvinus during long dark periods. *Plant Physiology* 59, 231–5.

Schroeder, H. & Herrmann, F. (1931). Über die Kohlenhydrate und den Kohlen-

hydrat-Stoffwechsel der Laubblätter. I. Die Zunahme des Saccharosege-
haltes beim Welken. *Biochemische Zeitschrift* **235**, 407–24.

Schroeder, H. & Horn, T. (1922). Das gegenseitige Mengenverhältnis der Kohlenhydrate im Laubblatt in seiner Abhängigkeit vom Wassergehalt. *Biochemische Zeitschrift* **130**, 165–98.

Schumacher, W. (1932). Über Eiweissumsetzungen in Blütenblättern. *Jahrbuch für
Wissenschaftliche Botanik* **75**, 581–608.

Schussler, J. R., Brenner, M. L. & Brun, W. A. (1984). Abscisic acid and its
relationship to seed filling in soybeans. *Plant Physiology* **76**, 301–6.

Shannon, J. C., Porter, G. A. & Knievel, D. P. (1986). Phloem unloading and
transfer of sugars into developing corn endosperm. In Cronshaw, J., Lucas,
W. J. & Giaquinta, R. T. (eds.). *Phloem Transport*, pp. 265–77. New
York, Alan R. Liss.

Sibaoka, T. (1962). Excitable cells in *Mimosa*. *Science* **137**, 226.

Sjölund, R. D. & Shih, C. Y. (1983). Freeze-fracture analysis of phloem structure
in plant tissue cultures. II. The sieve element plasma membrane. *Journal
of Ultrastructure Research* **82**, 189–97.

Tanner, W. (1980). On the possible role of ABA on phloem unloading. In
Eschrich, W. & Lorenzen, H. (eds.). *Phloem Loading and Related
Processes*, pp. 349–51. New York, Stuttgart, Gustav Fischer Verlag.

Tanner, W. & Kandler, O. (1966). Biosynthesis of stachyose in *Phaseolus vulgaris*.
Plant Physiology **41**, 1540–2.

Thorne, J. H. (1981). Morphology and ultrastructure of maternal seed tissues of
soybean in relation to the import of photosynthate. *Plant Physiology* **67**,
1016–25.

Thorne, J. H. (1982). Characterization of the active sucrose transport system of
immature soybean embryos. *Plant Physiology* **70**, 953–8.

Thorne, J. H. (1985). Phloem unloading of C and N assimilates in developing
seeds. *Annual Review of Plant Physiology* **36**, 317–43.

Thorne, J. H. & Koller, H. R. (1974). Influence of assimilate demand on photosynthesis, diffusive resistances, translocation and carbohydrate levels of
soybean leaves. *Plant Physiology* **54**, 201–7.

Thorne, J. H. & Rainbird, R. M. (1983). An *in vivo* technique for the study of
phloem unloading in seed coats of developing soybean seeds. *Plant Physiology* **72**, 268–71.

Thrower, S. L. (1974). Sink limitation and import of assimilate into mature leaves.
New Phytologist **73**, 685–7.

Turgeon, R. (1980). The import to export transition: experiments on *Coleus
blumei*. In Eschrich, W. & Lorenzen, H. (eds.). *Phloem Loading and
Related Processes*, pp. 91–7. New York, Stuttgart, Gustav Fischer Verlag.

Turgeon, R. & Webb, J. A. (1973). Leaf development and phloem transport in
Cucurbita pepo: transition from import to export. *Planta* **113**, 179–91.

Turgeon, R. & Webb, J. A. (1975). Physiological and structural ontogeny of the
source leaf. In Aronoff, S., Dainty, J., Gorham, P. R., Srivastava, L. M.
& Swanson, C. A. (eds.). *Phloem Transport*, pp. 297–326. New York,
London, Plenum Press.

Van Bel, A. J. E. & Patrick, J. W. (1985). Proton extrusion in seed coats of *Phaseolus vulgaris* L. *Plant, Cell and Environment* **8**, 1–6.

Wolswinkel, P. & Ammerlaan, A. (1983). Phloem unloading in developing seeds
of *Vicia faba* L. (The effect of several inhibitors on the release of sucrose
and amino acids by the seed coat.) *Planta* **158**, 205–15.

Wolswinkel, P. & Ammerlaan, A. (1985). Effect of potassium on sucrose and
amino acid release from the seed coat of developing seeds of *Pisum sativum*.
Annals of Botany **56**, 35–43.

Wright, J. P. & Fisher, D. B. (1981). Measurement of the sieve tube membrane potential. *Plant Physiology* **67**, 845–8.

Wyse, R. E. & Saftner, R. A. (1982). Reduction in sink-mobilizing ability following periods of high carbon flux. *Plant Physiology* **69**, 226–8.

Zimmermann, M. H. (1973). The monocotyledons: their evolution and comparative biology. IV. Transport problems in arborescent monocotyledons. *Quarterly Review of Biology* **48**, 314–21.

7 Physiological aspects of phloem translocation

J. A. Milburn and J. Kallarackal

7.1 Introduction

This chapter will concentrate on the interpretation of evidence for the mechanism of long-distance transport of photoassimilates and the effects of environmental factors on both structural and physiological aspects of this process. An understanding of the physiology of phloem transport requires a knowledge of phloem structure and readers are referred to Ch. 3 and various review publications (e.g. Parthasarathy, 1975; Cronshaw, 1981; Kursanov, 1984). We concentrate here on the physiological aspects of sieve tube transport (for earlier reviews see Milburn, 1975; Baker, 1978). Pressure driven mass-flow as described by Münch (1930) is today the most widely accepted mechanism for long-distance transport. Weatherley (1975) constructed a theoretical model sieve tube, consistent with the Münch hypothesis, which had the following physiological and anatomical characteristics to enable it to act as a functional unit: (a) a semi-permeable membrane for osmosis leading to a state of turgor and also actively able to transfer sugars in and out of the sieve tubes; and (b) continuity for mass-flow along the sieve elements with minimum drag from static structures, especially the arguably porous sieve plates.

The semi-permeable nature of sieve tubes has been unequivocally demonstrated, except when older sieve tubes have become non-functional (see Parthasarathy, 1975). The presence of mitochondria and plastids, seen on the inner periphery of the outermost plasmalemma membrane, is indicative of the presence of a parietal cytoplasm maintaining the functional viability of this membrane. The osmotic capacity of the sieve tube membrane is demonstrable by its ability to take up water (see Hall & Milburn, 1973), and the membrane hydraulic conductivity has been measured (e.g. Wright & Fisher, 1983). The capacity of the sieve tubes to take up water at loading or unloading sites and along the pathway is indisputable (see below). Studies have shown the ability of sieve tubes to load sugars against a concentration gradient, not only at the usual loading sites (Komor *et al.*, 1977), but also along the pathway (Malek & Baker, 1977) by a proton co-transport system (see Ch. 5).

The major objection to the acceptance of mass-flow as a model is the possible magnitude of the resistance to flow created by the sieve plates if they are blocked structurally. The porosity of sieve plates has therefore been a subject of much controversy among phloem physiologists (see Ch. 3 and review by Spanner, 1978).

Few related publications have appeared in the literature since the publication of the above review. One was the freeze-fracture analysis of phloem from tissue cultures where the tissue samples were fixed without any mechanical injury (Sjölund *et al.*, 1983). 'Open' sieve plate pores were demonstrated in this study. Another investigation was the observation of open sieve plates in exuding phloem of *Ricinus* fruit peduncle (Kallarackal & Milburn, 1983). The peduncles were fixed in acrolein for electron microscopy. Non-exuding phloem processed in the same way showed pores blocked by P-protein. Thus recent studies, although not totally immune from criticism, add more evidence that the sieve plates are 'open' and that the displaced P-protein helps in the initial sealing, followed by the deposition of callose (Parthasarathy, 1975). The absence of P-protein from the sieve plates of several monocotyledons is probably compensated by plastids and other inclusions which block the pores upon injury (Walsh & Melaragno, 1981). Some members of the Cucurbitaceae have sieve pores with large diameters and have evolved the ability to seal by gellation of the phloem sap (McEuen & Hill, 1982; Read & Northcote, 1983). The dependence upon air for gellation was demonstrated clearly by Milburn (1971), who showed that freshly collected exudate sealed in capillaries remained liquid for several weeks, but once the capillaries were broken to admit oxygen the contents gelled. Clearly, for translocation to take place in the normal plant, either the protein is chemically protected or the concentration of oxygen must be minimal.

7.2 Environmental effects

The environment of sieve tubes can be regarded as: (a) internal, e.g. hormonal or solute levels; or (b) external, e.g. light, temperature, mechanical pressure. In this section we examine selected key areas.

7.2.1 *Factors influencing sieve plate permeability*

Two materials seem to be especially important in regulating the permeability of sieve plates: callose and P-protein.

7.2.1.1 Callose synthesis and degradation

Callose, a β-(1–3)-glucan, first separated from the sieve plates of *Vitis vinifera* by Kessler (1958), yields on hydrolysis mainly glucose and traces of uronic acids. Callose is synthesized in response to wounding and chemical stimulation (Englemann, 1965), making the callose content of sieve tubes in the undisturbed tissue difficult to determine. Quantitative extraction methods for callose are not yet available; estimations are usually based on somewhat subjective microscopic evaluations. Despite these difficulties reasonable estimates of callose content have been made by the freeze-fixation method using light (Eschrich & Currier, 1964; Eschrich *et al.*, 1965) and electron (Cronshaw & Anderson, 1969) microscopy.

Is callose universally present in sieve tubes? Conflicting reports make this question difficult to answer. There are reports of detectable callose in conducting sieve tubes of trees during summer months (Zimmermann, 1960; Evert & Derr, 1964). There is good evidence that callose deposition fluctuates on a seasonal basis in tree bark (see Zimmermann, 1971), appearing to act as a semi-permanent sealing agent. There is also some evidence that actively functioning sieve tubes have little or no callose when compared to old or degenerating sieve tubes. Milburn & Kallarackal (1984) observed fully callose-plugged sieve tubes in the older outermost layers of phloem cells in *Ricinus*, whereas the younger sieve tubes near the cambium had only a lining of callose on the cell wall regions of the sieve plate.

Temperature has a marked effect on callose deposition on sieve plates. McNairn & Currier (1968) showed that ^{14}C-labelled assimilate transport was blocked when a 4 cm portion of the hypocotyl of a young cotton plant was heated to 45°C. Electron microscopy of similarly treated plants showed an extra deposition of callose in the heated region (Shih & Currier, 1969). McNairn (1972) observed diurnal changes in the translocation rate of cotton plants growing in the Californian valley, which showed a decline in phloem transport towards the evening when the temperature rose beyond 44°C. Callose deposition on sieve plates monitored simultaneously by microscopy was inversely related to the translocation rate. When the stem temperature of *Ricinus* was raised to 45 ± 2°C, the ability to collect exudate was lost (Milburn, unpublished). Ultrasound (Currier & Webster, 1964) and wounding (Thomas & Hall, 1975) have also been shown to cause callose formation.

In general, the blockage or slowing down of transport by callose is argued from anatomical observations. Physiological experiments by Eschrich (1957) do not support this view unequivocally. He proposed that callose, because of its inherent capacity to swell by absorbing water, might regulate flow mechanically in response to changes in water balance. Theoretically one might expect freely exuding phloem sap of a fully hydrated plant to become diluted until virtually pure water in composition. In fact, it seems to seal before this point is reached, which may support the hypothesis proposed by Eschrich. However, more work is needed to establish the hypothesis.

7.2.1.2 P-protein deposition and removal

Even less firm information seems to be available on the factors affecting the P-protein lodgement in pores of sieve plates. It is generally agreed that it constitutes the main sealing mechanism in most plants when phloem is wounded. In numerous studies it seems to be deposited more rapidly than callose.

If P-protein is to be deposited rapidly when sieve tubes are punctured, then a rush of pressurized sap towards the wound would provide an ideal trigger. There is excellent evidence of such a surge in that the so-called slime plugs are always orientated towards a wound site, as indicated in Fig. 7.1.

Exudation of phloem sap from incisions in *Ricinus* stems can be stopped instantly by compressing the stem near the wound with the fingers. Such compression (massage) can be interpreted as inducing an artificial pressure surge which triggers deposition (Milburn, 1970, 1971; Kallarackal & Milburn, 1983). There is an interesting corollary to these experiments because the effect of stem compression is only temporary. In periods varying from a half to two hours, the permeability of the sieve tubes is normally restored. Whereas incisions made immediately after squeezing fail to exude phloem sap, after a recovery period fresh incisions in the *same area* do so and even incisions made previously sometimes begin to exude spontaneously in a similar manner.

Observations of this kind, along with evidence on the effect of temperature (Sect. 7.2.2), support the idea that P-protein can be both deposited and removed from sieve plate pores over periods up to a few hours. Only in this way could the restoration of sieve tube permeability be reconciled with direct experimental evidence.

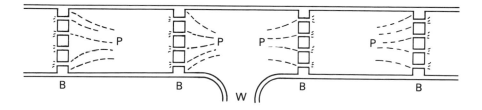

Fig. 7.1 Diagram showing the orientation in a sieve tube of slime plugs suspended on sieve plates B composed of P-protein (P) fixed after wounding at W.

7.2.2 *Temperature*

The effects of low or high temperature and changes around ambient were reviewed by Geiger & Sovonick (1975). The general consensus is that when an entire plant is subjected to temperature treatments, the optimal range for translocation is between 20°C and 30°C. Lang (1978) attempted to study the temperature dependence (Q_{10}) of the translocation rate when the source, sink and the pathway were subjected to temperature changes. His results suggest that translocation rate is normally under the control of all these three components but that the pathway predominates in controlling the rate. Marowitch *et al.* (1986) studied the effect of temperature variation on translocation in whole plants. Loading, unloading and long-distance transport were all affected by extremes of temperature. However, this was not connected with photosynthesis in source leaves. They suggested that occlusion of sieve tubes could be induced by temperature extremes. Loading and unloading are definitely under metabolic control and affected directly by temperature.

A number of contradictory reports have resulted when the effect of extremes of temperature (low or high) on translocation rate are measured. In fact, it is not difficult to explain many of these contradictions on the basis of chilling and heat sensitivity. For example, *Cucurbita melopepo*, a chilling-sensitive species, shows a marked reversible inhibition of transport below 10°C (Webb, 1970). However, in certain other plants, e.g. sugar-beet and willow, despite localized path-chilling, translocation continued undiminished after a short period of inhibition (Swanson & Geiger, 1967). Such plants can be considered as chilling-insensitive plants. Similarly, inflorescence stalks of *Yucca* continued to exude for up to 24 h at 0°C (Tammes *et al.*, 1969).

Is there a threshold for each plant below which it becomes chilling-sensitive? The available investigations show that at least in chilling-sensitive plants, this threshold exists. Thus in *Phaseolus vulgaris* (Giaquinta & Geiger, 1973) it is at about 10°C. However, in chilling-insensitive plants this threshold is difficult to visualize because there is no inhibition till the temperature reaches the sub-zero range. The difference in response of chilling sensitivity even occurs between ecotypes of the same species (Geiger, 1969) and sometimes between young and old shoots of the same plant. In *Sorghum*, a temperature block of 5°C had no apparent effect on the mass transfer of photosynthates when the plants were maintained at 21°C, but when the plants were kept at 30°C, the 5°C block stopped the mass transfer (Wardlaw & Bagnall, 1981).

The existence of two different response patterns in the above instances suggests that there may be two different mechanisms for producing translocation inhibition. Giese (1973) suggested that a Q_{10} of 1.2 to 1.5 indicated a mechanism limited by a physical process, such as diffusion or viscosity, while a Q_{10} of 2 to 4 indicated a thermochemical process as the basis of the mechanism. Thus, *Phaseolus* showed a Q_{10} of approximately 6 below 10°C, whereas sugar-beet showed a Q_{10} of 1.3 throughout. In plants like *Yucca* (Tammes *et al.*, 1969), *Ricinus* (Giaquinta & Geiger, 1973) and *Beta* (Geiger & Sovonick, 1970), the increase in phloem sap viscosity is suggested as causing inhibition of translocation through chilling. Experiments on *Salix* by Weatherley & Watson (1969) indicated that energy metabolism is only indirectly responsible for maintaining translocation. However, complete inhibition of translocation in both chilling-sensitive and chilling-insensitive plants can be due to physical disruption and blockage of the sieve tubes. As already mentioned, temperature changes can cause callose deposition in the sieve plates. Low temperature may also increase the number of callose plugs (Majumder & Leopold, 1967; Giaquinta & Gieger, 1973). EM studies on the chilled zone indicate that low temperature inhibition of translocation is due to cytoplasmic changes (Nobel, 1974). Nevertheless, unequivocal evidence for physical blockage is still lacking.

Based on a number of physical and physiological parameters Christy & Ferrier (1973) used a mathematical model for sieve element function and subsequently applied their model to the cold-inhibited recovery phenomena in sugar-beet (Ferrier & Christy, 1974). Their model suggests that an increase in pathway resistance could only be brought about by sieve plate blockage and not by a change in viscosity of the phloem sap. A build-up of pressure on the source side of the cold block promotes a flow of concentrated sap through this zone. Just below the cold block, water is loaded into the sieve elements, thereby restoring the velocity. The Christy and Ferrier model is interesting because it takes into account the sink unloading rate, the sucrose concentration in the pathway, the length and position of the cold block and the hydraulic conductivity of the sieve tube and its plasma membrane.

However, some investigations show that the time of recovery is the same irrespective of the lengths of petiole cooled (Geiger, 1969; Swanson, 1969; Grusak & Lucas, 1984). The prediction that sink unloading rate also influences the recovery rate has been questioned (Grusak & Lucas, 1984). Christy and Ferrier's model envisages a passive role for the pathway, as proposed by Münch's pressure flow. This passive role has also been questioned by some investigators (Geiger & Sovonick, 1970; Minchin *et al.*, 1983), suggesting that a pathway reloading of sugars occurs, especially in cases where recovery of cold-induced inhibition occurs. Grusak & Lucas (1984) have demonstrated a re-routing of phloem transport during cold recovery in sugar-beet plants via sieve tube anastamoses. In brief, there seems to be a general agreement that increased viscosity cannot be the critical factor in cold-induced inhibition, but it could play a major role when the cooled path is comparatively long (e.g. 6 cm in sugar-beet).

Faucher *et al.* (1982) and Faucher & Bonnemain (1986) claimed that the low-temperature inhibition of phloem translocation affects species with P-protein (*Vicia, Plantago*) but not those without P-protein (*Zea* and *Triticum*). Similarly, an exhaustive study by Lang & Minchin (1986) on cold-temperature effects on elated translocation relates the chilling sensitivity of plants to their phylogenetic position. Dicotyledons are, as a rule, chilling-sensitive, but the monocotyledons, gymnosperms and ferns tested proved variable. This suggests an apparent correlation between chilling sensitivity and the occurrence of P-protein (P-protein is unusual in monocotyledons, and is totally absent in gymnosperms and ferns; see Ch. 3). The results do not always correlate well and Lang and Minchin have explained chilling sensitivity in terms of changes in symplast:apoplast concentrations and perhaps in membrane properties. Calcium has also been proposed as the primary physiological transducer of chilling injury (Minorsky, 1985).

Chilling sensitivity to translocation has been used recently for studies on assimilate partitioning control (Pickard *et al.*, 1978a, b; Minchin *et al.*, 1983; Thorpe *et al.*, 1983; Thorpe & Lang, 1983). This has the advantage of temporarily stopping translocation to a particular sink without affecting other sinks; later it is possible to reverse the effect by warming. Moreover, the hormonal imbalances caused by the conventional method, i.e. sink removal, can be overcome by this method. On the basis of a study on the effect of chilling treatment on source/sink relationship, Grusak & Lucas (1986) suggested that the cooling treatments induced the transmission of a rapid, physical signal along the phloem pathway to the crown region of the plant, which in turn caused alterations in the pathway. The observation by Fisher (1983) of stylet exudation from field-grown willow at -11 to $-13°C$ shows that sieve tubes can indeed transport in functional winter phloem. The solute potential of the phloem sap was double that of summer phloem, and whether the sap normally moves in sieve tubes at this temperature, or is merely induced to do so by a release of turgor via the stylet, is unknown. However, the answer may be inferred from the fact that many plants like *Salix* can flower in the cold conditions of early spring.

Inhibition of translocation due to high temperature treatments ($40–60°C$) probably results from changes in the structural integrity of sieve tubes. For example, at $40°C$, phloem translocation was inhibited by both constriction of plasmodesmata and sieve pores (Shih & Currier, 1969). As already discussed, high temperatures can also induce the formation of sieve plate callose.

Comparatively minor changes in temperature around ambient have also been shown to affect the translocation rate. Momentary disruption of translocation was observed in *Ipomoea* when an 80 mm stem segment was brought down from $27.1°C$ to $21.4°C$ (Pickard *et al*, 1978b) although stepwise *increases* in temperature caused neither inhibition nor acceleration. Similar results have been obtained for *Ricinus* (see Fig. 7.2).

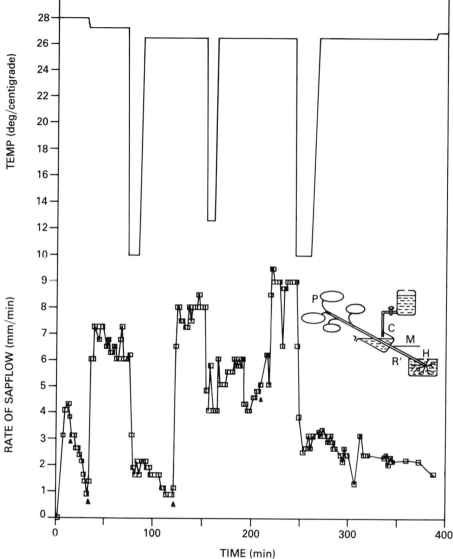

Fig. 7.2 Effect of stem temperature changes on phloem sap exudation from a *Ricinus* plant (P). The plant was (H) hydroponically grown and 82 cm tall. Incisions were made in a previously prepared strip of bark 5 mm wide, which was re-cut periodically; sap was collected in calibrated microcapillaries (M) above a ring (R). It can be seen that lowering the water temperature to 10°C in cold bath (C) over a stem length of 3 cm, produced a sharp reduction in exudation which could not be restored by rewarming; re-cutting a few mm of bark (▲) from the exudation site was an essential additional tratment. Apparently the cooling checks sieve tube transport which allows the incision wound to become partially sealed. (Ambient temperature 20°C.)

271

7.2.3 Water status

Several investigators have reported an inhibition of assimilate translocation by water stress (Hartt, 1967; Brevedan & Hodges, 1973). However, it has been claimed that photosynthesis is more sensitive to water stress than translocation (Munns & Pearson, 1974; Johnson & Moss, 1976; McPherson & Boyer, 1977; Sung & Krieg, 1979). Thus it is clear that the main problem in studying the effect of water stress on translocation is in separating the process of photosynthesis from phloem loading and long-distance transport.

The interaction between water and translocation flow in plants has received less attention, although the likelihood of such an interaction has been well recognized (Lang, 1973; Milburn, 1975; Grange & Peel, 1978; Jenner, 1982) and clear evidence was provided by studies of *Ricinus* exudation (Hall & Milburn, 1973). Milburn (1975) emphasized that *Ricinus* plants exuded from the bark even when the plants were subjected to severe wilting. Smith & Milburn (1980) made a more detailed study on the effect of water stress on phloem exudation. Recently, Lang & Thorpe (1986) found that when different parts of the root system of a plant were subjected to different water potentials, translocation was favoured towards the more negative apoplastic water potentials. Several previous workers have noted that roots may actually elongate faster when soil water is depleted (e.g. Malik *et al.*, 1979). Further experimentation is needed to see if this effect could be manipulated to change assimilate partitioning to the advantage of a crop.

It remains to be explained why the water movement in the phloem is apparently affected so little by differences in water stress, according to reports on some plants. This is especially true in the supply to some fruit sinks (Tromp, 1977). As Wiersum (1966) suggested, the supply of water to the fruits through the xylem pathway is considerable only in the early stages; thereafter, water mainly reaches the fruits via the phloem pathway.

The effect of transpiration has also been studied to some extent. The increase in transpiration rate of a sink leaf has been shown to increase the rate of translocation (Swanson *et al.*, 1976). Using developing leaf sinks of *Phaseolus vulgaris*, different translocation rates were demonstrated for leaves transpiring differently in a range of humidities. Sink manipulation studies (girdling or removal of sinks) have been shown to cause stomatal closure. This is apparently due to the accumulation of abscisic acid (ABA) in the leaves. ABA concentration has been shown to be dependent on the extent of phloem export from leaves (Hocking *et al.*, 1972). Pickard *et al.* (1979) have reported a decreased export from the source leaf when transpiration was increased. Generally, it appears from previous work that transpiration in a source leaf reduces the rate of export, whereas high transpiration from a sink leaf favours importation.

As already pointed out, not only photosynthesis, but also transpiration seems to interfere with studies on the effect of water status on translocation. Wardlaw (1969) indicated that the major effect of water stress on translocation is to delay and reduce the rate of transfer of sugars from the assimilating tissue to the conducting tissue and thus phloem loading is brought into picture. Investigations aimed at studying the effect of water status on phloem translocation should therefore take into account the complications introduced by photosynthesis, transpiration and phloem loading.

It has been shown that during water stress a greater proportion of carbon is channelled into amino acids (Lawlor, 1976). Carbon isotopic studies show that water stress affects both stomatal closure and also translocation (Watson & Wardlaw, 1981).

7.2.4 Photosynthesis

Photosynthesis is the major sugar manufacturing process in the source leaves of plants (see Ch. 1), hence its possible influence on the translocation rate of sap has been studied by many investigators.

Christy & Swanson (1976) used both steady-state labelling and pulse labelling on sugar-beet leaves enclosed in two different concentrations of CO_2 to generate and track different rates of photosynthesis. Their results indicate that the translocation rate and velocity are dependent on leaf sucrose concentration rather than any direct dependence on photosynthetic rate. However, above a sucrose concentration of 15 μg cm^{-3}, translocation appears to be uncoupled from sucrose concentration. Other investigators have also published evidence indicating that translocation is dependent on the source leaf sucrose concentration (Husain & Spanner, 1966; Geiger & Batey, 1967).

Concurrent photosynthesis is probably necessary to maintain the concentration of sugars in the pool (Ho, 1976) since Husain & Spanner (1966) showed that a darkened leaf can continue to promote translocation if supplied with external sucrose. The same can occur if stored starch is mobilized (see Ch. 8).

There are very few reliable methods available to study the immediate effects of current photosynthesis on translocation. A method developed by Geiger & Fondy (1979), although cumbersome, is promising. Our experience with the exuding system suggests that this could possibly be exploited to advantage. For instance, it would be worth investigating the changes in exudation rate and sucrose concentration while subjecting the source leaves to different rates of photosynthesis by controlling the amount of light or CO_2 concentrations.

7.2.5 Respiration

The results of early experiments on the effect of respiration on translocation were quite contradictory. Some workers found an inhibitory effect on translocation when oxygen was excluded for a sufficient length of stem tissue (Curtis, 1929; Mason & Phillis, 1936), whilst others failed to observe any significant effect of anoxia in the path of translocation (Bauer, 1953; Willenbrink, 1957; Ullrich, 1961). Most later experiments centred around the application of respiratory inhibitors and stimulators to the transporting system. In one such study, Willenbrink (1966) applied KCN to an isolated bundle system of *Pelargonium* still transporting to the leaf and found that the transporting capacity of the phloem was greatly reduced. With the removal of the respiratory inhibitor, translocation resumed within 90 minutes. The difficulty of isolating and exposing undamaged vascular bundles in other systems makes it difficult to repeat, and hence confirm, these experiments.

In experiments with respiratory stimulators like ATP, Kursanov & Brovcenko (1961) claimed a stimulation of labelled assimilate translocation in sugar-beet. However, Ullrich (1962) concluded that ATP may stimulate uptake into the phloem but not the transport along it. Later Sij & Swanson (1972) and Qureshi & Spanner (1973) observed transient and marked translocation inhibition due to an anaerobic atmosphere around the phloem.

Regarding the respiratory activity of the phloem tissue, one of the early studies (Kursanov & Turkina, 1952a,b) indicated an exceptionally high oxygen uptake for the vascular bundles. A re-investigation of this result by Ziegler (1958), however, made it clear that the respiration of the phloem was no greater than that of the xylem. Canny & Markus (1960) also found that respiration values for phloem were very low, unless phloem slices were cultured in sugar solution. Coulson *et al.* (1972) also found no stoichiometric relationship between sucrose transport rates and ATP turnover. Likewise, no anoxia response was detected in moonflower stems (Pickard *et al.*, 1979).

Contrary to many previous studies, Fensom *et al.* (1984) observed complete inhibition of ^{11}C translocation when a 5 cm portion of the sunflower petiole was treated with gaseous nitrogen for five minutes or longer. Based on this and similar observations, they concluded that translocation is apparently under the control of a metabolic respiratory mechanism (Fensom, 1981; Fensom *et al.*, 1984). It has been found, using ^{11}C, that leaves could not export assimilates when the available oxygen was $< 2\%$ even when the same leaf fixed CO_2 (Pickard *et al.*, 1978; 1979; Jahnke *et al.*, 1981; Grodzinski *et al.*, 1984).

7.3 Tracing the pathway: dyes, viruses and isotopes

7.3.1 Dye tracers

Ever since the pioneering experiments by Dixon & Ball (1922) on the movement of assimilates from source to sink in potato, investigators have been interested in tracing visually the movement of assimilates in plants. Fluorescent dyes were, and still are, one of the favourite tracers in the phloem. Thus, fluorescein provided much information about phloem translocation (Schumacher, 1930, 1933, 1967). Other workers using this method have failed to trace phloem translocation (Bauer, 1953). Dyes are usually applied to the abraded leaf tissues, which could allow xylem uptake. Another major problem in using dye tracers in phloem research is the extreme fragility of the tissue to manipulation. Recently it has been found that fluorescein can be fed through the exposed cotyledons of 5-day-old *Ricinus* seedlings (Kallarackal & Komor, unpublished). A cuticle is absent so there is no need for abrasions and the fed dye has been detected in phloem sap collected from the hypocotyl.

Recently, investigators have micro-injected fluorescent dyes, e.g. lucifer yellow, into the sieve tubes mainly to study the loading of assimilates into the phloem (see Chs. 2 and 5). Komor & Orlich (1986) have successfully micro-injected lucifer yellow into the sieve tubes of *Ricinus* cotyledons by ionophoresis. Enclosing the fluorescent dyes in liposomes and pressure-injecting them into sieve tubes is also very promising. Madore *et al.* (1986) have used this method to inject the mesophyll cells of *Ipomoea* (see Ch. 2).

7.3.2 Virus particles

Viruses are mainly transmitted from one plant to another via insect vectors and then move in the same general pattern as the sugars in the phloem (Bennet, 1937). Viruses have been extracted from phloem exudates, demonstrating conclusively that they can move within sieve tubes. Esau *et al.* (1967) detected virus particles in the sieve plate pores of *Beta vulgaris*, using electron microscopy.

7.3.3 Radioisotope tracers

The use of radioisotopes in the study of phloem translocation was pioneered by O. Biddulph in the 1940s. No other tool has contributed so much to our knowledge of translocation patterns in *intact* plants as radio-isotopes in their various forms. The application of $^{14}CO_2$ via the photo-synthetic pathway has been used routinely to follow the path of labelled carbohydrates.

Radioactive tracers are used in phloem translocation studies in two different ways. One of them is destructive sampling, where the translocation pathway is cut into small segments after feeding tracer, to obtain a *spatial profile*. This method is mostly used in ^{14}C investigations because of the low energy radiation given out by this isotope. A second method uses collimated detectors positioned along the translocation pathway to monitor the movement of radioactivity as a function of time, a *temporal profile*. Temporal profiles can also be obtained conveniently by using isotopes which can be monitored *in vivo*, like [11]C. One can obtain temporal profiles with ^{14}C, but this necessitates destructive measurements on several plants. Minchin (1979) has analysed the relationship between spatial and temporal tracer profiles in detail.

Translocation studies using isotopes were reviewed by Canny (1973). General trends can be summarized here as follows:

1. The movement of tracer from application at a source to the sink is longitudinally confined. Lateral spread of label is restricted.
2. Mature leaves, having completed growth, do not import tracer. For a comparison of the import, export and growth rate of an expanding leaf, see Thrower (1962).
3. The contribution of upper leaves to an apical sink is greater than that of lower leaves. The contribution of lower leaves to a root sink is greater than that of upper leaves.
4. Fruits derive their phloem assimilates almost entirely from adjacent leaves or bracts, as suspected by Münch (1930).
5. Removal of source leaves diverts the burden of sink supply to the remaining sources, altering the patterns of distribution.

Although ^{14}C is the most widely used isotope in the study of phloem translocation, some studies have been made with ^{11}C, the short-lived isotope. Moorby *et al.* (1963) were probably the first to use ^{11}C, studying assimilate movement in soybean. This isotopic method is attractive in that measurements on equivalent pulses are repeatable because the method is non-destructive, scintillation counters positioned near the stem in which the tracer is moving allowing non-invasive counting (Moorby *et al.*, 1963; Minchin, 1979). However, the limited half-life (21 min) and costly equipment limit its use as a routine tool in most physiological investigations (see More & Troughton, 1973 and Minchin, 1986, for details on the use of short-lived isotopes).

Using ^{11}C as a tracer, Minchin and his co-workers have shown that there is a tendency for dispersive flow in the phloem pathway, i.e. a passive unloading of sugars and an active reloading occurs into and from the sieve tubes (Minchin & Thorpe, 1984; Minchin *et al.*, 1984). Bishop *et al.* (1986) have compared the speed of translocation by two methods, namely the method of 'advancing intercept'. [11]C and ^{13}N have been used to calculate the speed of translocation (Moorby, 1977; Troughton *et al.*, 1977). In contrast, Ross & Tyree (1980) claimed that it is impossible to make any meaningful estimates of the velocity using tracers because of exchange of tracers with 'cold' non-tracer isotopes in cells surrounding the sieve tubes. Comparison of these two techniques indicates much higher velocities by the latter method, in which the radioactive detectors are placed very close together, so that the movements of small pulses can be detected. Bishop *et al.* (1986) claim their procedure reduces error because there is minimal exchange with ^{12}C isotopes along the path.

7.4 Sap sampling and measurement: manipulative techniques

In this section we will consider the measurement of phloem sap turgor pressure by different methods, exudation studies, insect feeding techniques, chemical induction of exudation and also methods to determine pressure gradients in sieve tubes.

7.4.1 *Water relations of the phloem: indirect methods of measurement*

The water relations parameters of the phloem are important in explaining the different mechanisms of phloem translocation (Kaufmann & Kramer, 1967). Although comparatively few studies are available, successful attempts have been made in measuring the turgor pressure (P), hydraulic conductance (L_p) and elasticity (ϵ) of the phloem.

Since the phloem is sensitive to manipulations, direct measurements of turgor pressure have been difficult. Table 7.1 summarizes the various attempts made by different investigators to measure the sieve tube P. As with most other plant cells, P estimates are calculated from the difference between water and solute potentials using the equation:

$$P = \psi_p = \psi_s - \psi_w \qquad (7.1)$$

where ψ_w is the water potential of the tissue, usually measured by a pressure chamber or psychrometer, ψ_p is the P and ψ_s is the solute potential of the phloem sap. When applied to the phloem, the above equation would mean that when xylem tension is subtracted from the solute potential of the phloem sap, we get the P of the phloem sieve tubes. However, this assumption requires the ready availability of water to the sieve tubes so that equilibrium conditions can be achieved. Practical measurements in the field have shown that phloem sap concentrations decrease at night when xylem sap tensions fall, but increase by day when xylem sap tensions increase (Huber *et al.*, 1937; Ziegler, 1956; Milburn, 1975). Milburn (1975) has explained this in terms of the increased loading of storage sugars at night and the increased demand of the root sink at night due to higher water availability.

Table 7.1 Phloem turgor pressure (*P*) measurements

Method	Range in value (MPa)	Reference
1. Calculation	0.1–1.5	Huber *et al.*, (1937) Kaufmann & Kramer (1967); Zimmermann (1971); Milburn (1975); Rogers and Peel (1975); Fisher (1978); Vreugdenhil (1985).
2. Aphid stylet exudation rate	0.8–4.0	Mittler (1957); Weatherley *et al.* (1959); Rogers & Peel (1975); Barlow & Randolph (1978).
3. Manometric methods	0.3–1.0	Tammes (1952); Hammel (1968); Milburn (1974); Sheikholeslam & Currier (1977); Wright & Fisher (1980).
4. Transducer methods	0.5–1.2	Lee (1981a,b); Sovonick-Dunford *et al.* (1981); Wright & Fisher (1983).

Using the above procedure, phloem sap solute potentials have been measured at different points in the pathway to determine gradients, which are important in driving a mass-flow of solution. Thus, a solute potential gradient of 0.02–0.035 MPa has been measured for tree species (Huber *et al.*, 1937; Hammel, 1968; Zimmermann, 1971). In a small plant like *Ricinus*, this gradient was much steeper, giving an overall gradient of 1.0 MPa for a plant 0.5 m tall (Milburn, 1974, 1980). Comparable gradients have been found by cytochemical measurements of the sugar concentrations in the source and sink of soybean (Fisher, 1978). Solute concentrations in the sieve tubes have also been measured using a plasmolytic method (Fisher & Evert, 1982; Warmbrodt, 1987). K^+ concentration measurements have also shown similar gradients in the bark of *Mannihot* and the phloem sap of *Ricinus* (Vreugdenhil, 1985). However, in *Salix* a gradient twice as great as that of *Ricinus* was found (Rogers & Peel, 1975; Grange & Peel, 1978). It is worth noting that no such gradient could be detected in some cucurbits (Richardson *et al.*, 1984). Measurable pressure gradients can be expected to develop only when hydraulic resistances hinder flow.

Although the above indirect method of measuring *P* is simple in its application, the validity can be checked only by direct measurements. This is because the xylem water potentials ψ_w at the source and sink and the hydraulic conductivity (L_p) of the sieve tube system can exert a direct influence on *P*. This has been verified recently by Wright & Fisher (1983), who by comparing direct measurements with calculated values arrived at an experimental error of + 0.04 MPa for willow sieve tubes.

A direct approach has been to calculate *P* from the exudation rate from severed aphid stylets using Poiseuille's equation. This method has yielded results ranging from 0.8 to 4.0 MPa, which differs from values determined by other methods (Table 7.1). Since aphid stylets are tapered and Poiseuille's equation indicates that volume flow rates are proportional to the fourth power of its radius, the above method is less dependable.

Hammel (1968) introduced a direct method for measuring P in sieve tubes of *Quercus*. A hypodermic needle glued to a glass capillary with a sealed end punctures the phloem tissue and acts as a manometer. The ratio of air column length at ambient pressure to the length of air column at pressure $(P) \times$ barometric pressure in mmHg/760 mmHg can be plotted as a function of the gauge pressure P. Although considerable modifications have been made to the above device, the principle behind measurements cited in Table 7.1 is the same. For example, Wright & Fisher (1980) used an aphid stylet instead of a steel needle to attach a manometer. Apart from using a version of Hammel's method, Milburn (1975) also used a pressure-cuff, resembling the device used to measure human blood pressure, on exuding bark. However, the former method is liable to suffer from internal leakage, and the latter method also suffers from drawbacks, e.g. phloem fibres around the sieve tubes, which make it difficult to estimate applied pressure reliably. Moreover, it is not known if phloem tissues can withstand any external pressure during exudation because the resistance of the sieve tubes to externally applied stimulus is rather poor.

The advent of the pressure transducer has led several investigators to make sieve tube P measurements using a device from Hammel's (1968) method and the pressure probe of Zimmerman *et al.* (1969). Theoretically, this method should be ideal for measuring the sieve tube P and estimated values using this method were very close to theoretically calculated values (Wright & Fisher, 1983).

At present it is difficult to argue for high accuracy by the indirect method of calculating sieve tube P from water potential and solute potential measurements. For routine measurements of water potentials leaf samples must be used, which may not indicate the exact water potentials in the stem or the root at the desired location (except during predawn when flow is nearly zero and complete equilibrium can be assumed). Hence, an extrapolation of sieve tube turgor in stem bark of a plant from leaf water potentials can only provide a rough estimate, especially during vigorous transpiration.

The range of net water movement into or out of a plant cell is determined by the solute potential of the sap (ψ_s), the hydraulic conductance (L_p) of its plasma membrane, and the volumetric elastic modulus (ε) of its surrounding cell wall. The ψ_s of the phloem sap has been dealt with above and a few measurements have been made on the L_p and ε of the sieve tubes (see Table 7.2).

It is apparent that these few measurements render comparisons difficult. Values obtained for ε fall into the low range for plant cell walls (Dainty, 1976), i.e. sieve tube walls are generally more elastic than other plant cell walls. Wright & Fisher (1983) noticed a sluggishness of the ψ_s response and

279

Table 7.2 Membrane hydraulic conductance (L_p) and volumetric elastic modulus (ε) measurements on sieve tubes

Genera	L_p (cm s^{-1} bar^{-1})	ε (bar)	Reference
Fraxinus americana	—	68–75	Lee (1981)
Quercus borealis	9.6×10^{-8}	58	Sovonick-Dunford *et al.*, (1982)
Salix exigua	$5.5 \pm 2.1 \times 10^{-8}$	27 ± 21	Wright & Fisher (1983)
Ricinus communis	—	16 ± 4	Kallarackal & Milburn (1985a)

suggested that this could be caused by complexity in the elastic properties of the cell wall, possibly a time-dependent or viscoelastic component in its stretching response.

7.4.2 Exudation methods

The earliest known record of the collection of sugar-rich saps from the palms goes back to the first century AD in Sanskrit texts. An alcoholic beverage made from the sap of *Phoenix* is mentioned in one such book (Sutrasthana – Section, Chapter 45, Verse 174). Phloem sap is collected from different palms like *Arenga, Borassus, Caryota, Cocos, Corypha, Nypa* and *Elaeis* in Asia, *Raphis* in Africa, and *Copernicia, Jubaea* and *Mauritia* in South America (Van Die & Tammes, 1975). Apart from these palms, *Fraxinus, Agave* and *Yucca* also yield commercial quantities of phloem sap.

Much useful information about phloem translocation has been derived from exudation data (Milburn, 1975; Pate *et al.*, 1975; Van Die & Tammes, 1975) but like many other techniques requiring manipulation it has been subjected to criticism. One objection faced by early investigators on phloem exudation was that the sap collected might be found in xylem rather than phloem. The analysis of physical and chemical factors controlling exudation has largely resolved this problem. Although phloem is usually positively pressurized (in contrast with xylem which is normally negatively pressurized), it seals when injured but the sealing rate, if slow, allows some exudation. This has been unequivocally demonstrated in exuding plants (e.g. Van Die & Tammes, 1966; Milburn & Zimmermann, 1977). Exuding sap is derived from current photosynthates as well as stored sugars in the leaves (Van Die, 1975; Hall *et al.*, 1977; Milburn & Zimmermann, 1977).

Chemical characteristics which distinguish phloem sap from xylem sap are the alkaline pH of the former (*c.* 8.0), predominance of sugars especially sucrose in most plants (10–20%), high K$^+$ content (*c.*

100 mol m^{-3}), very low Ca^{2+} content, and high amino acid content. The type of sugar present in the phloem can vary depending on the species (see Ziegler, 1975; Komor, 1982).

Some deciduous dicotyledons such as *Populus, Carpinus, Alnus, Betula* and *Acer* bleed xylem sap in the early spring; the latter two support small industries. Their sap, which is slightly acidic, contains a relatively high concentration of sugars not normally expected in the xylem (Lohr, 1953). In *Betula* the phenomenon is a manifestation of root pressure, whilst in *Acer* the pressurization arises from the wood following freezing and thawing cycles (Milburn & O'Malley, 1984) and although transport via ray cells is involved, this phenomenon should not be confused with phloem transport.

Another system which can be easily mistaken for phloem exudation is the latex exudation from plants like *Hevea* or *Musa*. It is not known if the collected latex in *Hevea* contains any phloem sap, laticifers being closely adjacent to sieve tubes. In *Musa*, latex is often a clear sap devoid of sugar (Baker *et al.*, 1989). When latex exudes the concentration of the solutes decreases rapidly as exudation proceeds (Buttery & Boatman, 1976; Milburn *et al.*, 1988), unlike phloem exudation where the concentration usually remains steady. This is because the phloem system has sugar being loaded at the source region or along the pathway as a continuous process, whereas the latex seems to be a rather static secretory system, at least in the intact condition.

One of the biggest problems in exploiting phloem exudation to study translocation physiology is that most plants which exude freely, e.g. palms, live in tropical regions of the world. Palms are tall and difficult to handle for physiological experimentation. These handicaps were to a large extent overcome by Milburn (1970) who found that stems and petioles of *Ricinus*, the castor bean, could yield considerable quantities of sap (*c.* 1 cm^3 h^{-1}). Exudation levels could be enhanced by massaging the bark regularly for a few days before slicing for exudation (Milburn, 1971) or by repeated re-cutting over a period of time. Later, the manipulation studies were extended to *Cocos* to explain the induction of exudation in palms (Milburn & Zimmermann, 1977). Similarly, Pate *et al.* (1974, 1975, 1979) have studied in detail the exuding system of *Lupinus albus* (see Ch. 4). Good exudation rates were obtained from the distal ends of fruits. A cryopuncture technique, which gives enhanced exudation rates (Pate *et al.*, 1984) unfortunately seems successful for this species only.

Although sieve tubes seem ideally designed to seal upon wounding, their inability in certain genera to perform this sealing function efficiently is not yet understood. The general tendency of the palms to exude considerable quantities of phloem sap could be hypothesized to result from the scant amount of P-protein filaments present in certain of their organs or tissues (cf. Parthasarathy, 1975). However, the collection of phloem sap from some non-palmaceous plants opposes this view. It has been found in

Ricinus fruit stalks that the sieve plates of exuding phloem are *not* plugged by P-protein, whereas in non-exuding phloem sieve plates *are* sealed (Kallarackal & Milburn, 1983). No such differences were detectable in palms by Parthasarathy and Zimmermann (reported in Milburn, 1975). Future research on this problem must be directed towards a better under-standing of P-protein. Some controversy exists regarding its synthesis, with Weber *et al.* (1974) proposing that P-protein is synthesized in differen-tiating sieve elements, whereas Nuske & Eschrich (1976) cite evidence for its *de novo* synthesis in companion cells. Another approach to exudation is to argue that plants which seal inefficiently have been spared predation pressure through having toxins (e.g. ricin and ricinine in *Ricinus*) or unsweetened sap as in the Cucurbitaceae. It seems that plants which exude have plentiful photosynthates and there is a correlation between insolation and capacity to support exudation.

The question of the effectiveness of pre-treatments practised on palms to enhance exudation deserves mention, the inflorescence being beaten with a mallet or massaged between the fingers before slicing for exudation (Bose, 1947; Kallarackal, 1975). It has been found that massage pre-treat-ment of part of a stem in *Ricinus* (Milburn, 1971) and bending of the inflorescence in *Cocos* (Milburn & Zimmermann, 1977) can also enhance the exudation. Enhanced exudation from massaged plants may be caused by a loss of sieve tube elasticity in response to mechanically overstressing the cellulose fibrillar nets within the sieve tube walls (Lee, 1981b). This hypothesis was examined for the phloem of *Ricinus* by measuring the volumetric elastic modulus (ε), which was not significantly different for massaged and unmassaged bark (Kallarackal & Milburn, 1985a). However, the thickness of phloem tissues in the bark increased by 160% in the massaged plants and the number of sieve tubes was greater. Based on this observation, Kallarackal and Milburn (1985a) proposed that massage gives a response similar to wounding (Jacobs, 1970), which in turn promotes the regeneration of cambial cells to produce more phloem. Since phloem sap is usually under positive pressure, any increase in cross-sectional area will increase the output of sap when cut and probably decrease the pressure gradient, so reducing the tendency to seal.

One interesting aspect of phloem exudation is that in most cases exam-ined, the sap is collected from inflorescences of fruits. Do the young ovules attract assimilates eventually required for fruit growth? In *Arenga* and *Elaeis* it is the *male* inflorescence which gives the exudate (Tammes, 1933; Van Die, 1975). Hence it is difficult to ascribe exudation to the presence of ovules on the inflorescence, though nectaries in male inflorescences may be equivalent sinks. Smith & Milburn (1980a,b) found that the rate of exudation was dependent on the number of source leaves present in *Ricinus* plants. Later Kallarackal & Milburn (1984) observed a specific mass transfer (SMT) rate nearly 20 times greater for an exuding peduncle of *Ricinus* when compared with the rate towards an intact fruit, which is similar to palms (Van Die & Tammes, 1975). However, the comparative values were much less for exudation from stem tissue (Smith & Milburn, 1980a,b). In *Ricinus* seedlings the SMT values were almost identical in exuding and intact systems (Kallarackal & Komor, unpublished).

From the above investigations, it seems that the major determinant for tissue differences in exudation is probably not hormonal but the availability of a number of sources to supply a wound. It is found that during the reproductive growth of a plant the root sink receives less assimilates than the reproductive sink (Kallarackal & Milburn, 1985). Hence it is not surprising that exudate can be collected more readily from reproductive sinks than from vegetative sinks.

Several investigators have attempted to answer the question of why some plants exude while others do not. Van Die & Tammes (1966) proposed callose formation as the main determining factor, but electron microscopic observations of exuding peduncles of *Ricinus* showed callose on sieve plates, but no P-protein plugging the pores (Kallarackal & Milburn, 1983). The general consensus regarding the function of P-protein and callose seems to favour P-protein for the immediate sealing of sieve tubes, whereas callose completes it later (Parthasarathy, 1975), prolonged exudation being brought about if neither functions. It is difficult to understand why if pre-treatment is required to produce good exudation, it is certainly not necessary in many cases; however, repeated incisions which act like an artificial sink once exudation begins have a similar effect. Future research should be directed towards a closer analysis of pre-treatment effects so that eventually methods can be applied to the majority of plants to enable phloem sap to be sampled readily. Some success is presently being achieved using methods described below (Sects. 7.4.3 and 7.4.4).

7.4.3 Exudation via aphid stylets

Several investigators followed the pioneering work of Kennedy & Mittler (1953), who devised the technique of using aphids to collect phloem sap exudate. Aphids (Hemiptera) feed on sieve tube sap of plants using fine mouth parts called stylets (Auclair, 1963; Evert *et al.*, 1968). Pectic enzymes secreted by the stylets help penetration between parenchyma cells by breaking the bonding between them (McAllan & Adams, 1961; Miles, 1968). Aphids consume large quantities of phloem sap (Dixon & Logan, 1973; Barlow & Randolph, 1978), although much of it is excreted as honey-dew. Most aphids occur on angiosperms; however, they are also reported on other groups of plants (see Dixon, 1975).

The feeding habits of aphids have been exploited in studying phloem physiology. By anaesthetizing a feeding aphid, then severing its stylet, it is possible to collect virtually pure phloem sap from the cut end of the stylet. It has been found that such a stylet in willow can exude up to

4 μl h^{-1} for 5 d (Peel, 1975). Some investigators have used the honey-dew secreted by the aphids for their studies on phloem sap (Peel, 1963, 1965; Ford & Peel, 1967). This is useful in plants which do not exude from stylets and has the added advantage of collecting more sap than from a stylet. However, the phloem sap undergoes compositional changes while passing through the insect's gut and the honey-dew flow is apparently controlled by activity of the cibarial pump.

The velocity of sap flow and SMT have been calculated using the aphid stylet technique, yielding values of 120 cm h^{-1} and 12 g cm^{-2} h^{-1} respectively (Weatherley *et al.*, 1959). These are close to those obtained for intact systems, showing that the characteristics of stylet exudation are essentially similar. The composition of sieve tube sap from stylet exudation is also in good agreement with the overall grain composition of wheat (Fisher, 1987). Fisher & Frame (1984) made a detailed study of stylet exudates from 41 plant species and 35 insect species, and Munns & Fisher (1987) used stylets to study wheat phloem sap composition. The rate of exudation depended on the plant species, physiological condition of the plant and size of the insect species used. They have also devised a radio frequency microcautery probe for cutting the stylets more effectively.

7.4.4 *Pressure gradients and pressure flow*

Münch's original hypothesis (1930) would account for phloem transport when the pressure potential (derived from solute concentration differences) became converted into a flux (see Milburn, 1975). Such transport would have to be unidirectional within single sieve tubes but could be in opposite directions within separate sieve tubes which could be adjacent. No good evidence has been produced to support the idea that transport could occur in opposite directions in the same sieve tube.

Despite the strong evidence from exudation experiments, which all indicate that flow is generated from a highly pressurized liquid system escaping through pressure released towards atmospheric pressure, the operation of pressure flow has still been questioned. A further point is that an exuding plant, e.g *Ricinus*, not only drives sap from an incision (which operates like an artificial sink), but it also loads new sap to the extent that the entire volume of all sieve tubes in a plant must be refilled up to three times each hour. There can be no doubting the capacity of the system to operate as Münch envisaged; indeed, the rate of flow in such circumstances is often many times faster than normally found in an intact plant.

It is not easy actually to demonstrate pressure flow, because the sieve plate sealing system will not allow sieve tubes to be used like simple mechanical pipes. Furthermore, if the sieve pores are open there is very little drop in pressure along a sieve tube; it is only when flow is obstructed that a pressure difference across the obstruction builds up to an easily measurable value. Nevertheless, two experimental approaches have given strong direct evidence of pressure flow in operation. Small manometers were used to study phloem sap transport in the manna ash (*Fraxinus ornus*) in Sicily, where the trees are tapped annually. Milburn (unpublished results) was able to plot the pressure profile immediately above an exuding wound. These manometers showed that the pressure in the sieve tubes of the water-deficient trees was about 4 bar at the time. This value fell more or less exponentially towards zero (i.e. atmospheric pressure) at a large incision in the bark, over a distance of 50 mm.

Another approach used by Minchin & Thorpe (1987) has been to modify the flow of labelled assimilates using a pressure chamber. The upper shoot of a *Phaseolus* plant was enclosed in a pressure chamber while ^{14}C photoassimilate was monitored. Transport towards the chamber stopped immediately when pressure was applied but tended to recover within about five minutes. When the pressure was released the flow increased again.

Along somewhat similar lines Munns *et al.* (1988) have used a pressure chamber to induce exudation of phloem sap from *Lupinus* plants in a manner analogous to that used commonly to extract sap from the xylem. The composition of the sap collected indicates that it is mainly derived from sieve tubes, but it is hard to be certain of the extent of xylem sap contamination.

All of these results support Münch's hypothesis that phloem sap can be driven by pressure differences through sieve tubes. Obviously the hydraulic resistance to flow across the sieve plates must be appropriately low.

7.4.5 Chemical induction of exudation

The enhancement of exudation by chelating agents like ethylenediaminetetracetic acid (EDTA), deserves special mention. EDTA was used by King & Zeevaart (1974) to enhance exudation of labelled solutes from the cut petioles of *Perilla*. A concentration of 20 mol m^{-1} EDTA applied 2–4 h after excision gave optimal exudation. Exudation continued after transferring the cut end from an EDTA solution to water. The authors suggested that EDTA probably binds Ca^{2+} at the cut end, which in turn prevents the sealing mechanism, especially the callose formation, because Ca^{2+} inhibited the EDTA response. Other possible modes of action for EDTA, like prevention of gelling by proteinaceous material in the phloem, were not ruled out.

Subsequently this method has been used by a number of workers to identify phloem-mobile metabolites in a number of plants (Dickson, 1977; Fellows *et al.*, 1978; Tully & Hanson, 1979; Urquhart & Joy, 1981). Simpson & Dalling (1981) used the method to collect phloem sap from senescing wheat leaves to identify the amino acids exported. They noticed unexpectedly high concentrations of phenylalanine and tyrosine in the

EDTA-collected exudate. These were considered to be artefacts. They also observed that considerable amounts of Ca^{2+} were released into the collection medium, probably from xylem and damaged cells, and suggested a pre-treatment for 20 min in EDTA followed by exudation into water alone (see Appendix).

Costello *et al.* (1982), while using EDTA to get exudates from *Fraxinus*, found that the concentration effect of EDTA on sap exudation was probably due to the EDTA entering the transpiration stream. Although they obtained good exudation rates of 2 mol m^{-3} EDTA concentration, continuous presence of EDTA in the medium was essential for prolonged exudation. Like King & Zeevaart (1974) they have also suggested a callose inhibition role for EDTA in enhancing exudation. However, the electron micrograph presented by Costello *et al.* (1982) shows very little P-protein on the sieve plate. It is worth re-examining if P-protein or callose is really affected by EDTA treatment, because P-protein is probably more involved in immediate sealing than callose (Kallarackal & Milburn, 1983). It is also suspected that EDTA may be simply making all the phloem parenchyma cells leaky, in which case the exudate would not be pure phloem sap.

Fellows *et al.* (1978) studied exudation from the pods of soybean using 20 mol m^{-3} EDTA. Their term 'leakage' indicates a progressive process, since the surge of sap commonly observed from incisions in exuding plants is absent here. The velocity of translocation calculated from this method was 0.3 cm min^{-1} (18 cm h^{-1}), which compares well with other velocity measurements in soybean. Exudation stopped when the leaking pod was transferred from EDTA to water. Tully & Hanson (1979), using the EDTA method to study phloem exudation from turgid or water-stressed barley leaves, found that even wilted leaves could be induced to exude by this method. An enriched supply of CO_2 around the experimental leaf reduced transpiration by reducing the stomatal aperture, and prevented the suction of EDTA into the leaf tissues. pH control of the EDTA-induced exudation medium proved critical for induction of prolonged exudation. Hanson & Cohen (1985) showed that at pH 7.3, the exudation was better than at a lower pH, ascribing this to a better chelation of EDTA with Ca^{2+} at this particular pH.

Fellows & Zeevaart (1983) compared the carbohydrate partitioning in a detached leaf with EDTA exudation from an attached leaf and found that, although the export rate of ^{14}C from a detached leaf was 10% less than from an attached leaf, differences in non-structural carbohydrate partitioning in the two leaves were insignificant.

Since the method of exudation using EDTA and other chemicals has good scope for work on conventionally non-exuding plants, it is worth further examining possible artefacts from treatment with EDTA. From examination of previous studies, the treatment schedule followed by Simpson & Dalling (1981) seems promising because it avoids the continued presence of EDTA in the exuding medium. Also the entry of EDTA into the transpiration stream must be restricted where possible; its effect has not yet been fully investigated.

7.4.6 Problems and interpretation of exudation data

Although the exudation data from severed tissues and aphid stylets have provided a lot of information, the conclusions are not uncontroversial. Exudate collected from incisions in bark or from aphid stylets is phloem sap samples resulting from current photosynthates or stored materials in the plant body (Peel & Weatherley, 1962; Hall *et al.*, 1971). A major practical problem in collecting exuded sap is the evaporation of the sample, especially when the exudation rate is low. This is all the more difficult in aphid stylet droplets, which are often 1–2 μl. In samples of phloem sap from incisions, contamination from the injured tissue could occur and also it is difficult to ascertain the number of sieve tubes that have been severed. The latter is especially important for studies on mass transfer rates and velocity of sap flow. In the aphid stylet technique, however, the stylet draws sap from one sieve tube only. Miles (1986) has suggested that IAA could be produced in response to aphid saliva after its injection into a plant. A number of studies indicate an effect of hormones in controlling phloem transport (see review by Patrick & Wareing, 1980). If this is so, the exudation studies using an aphid stylet could be influenced by insect-produced hormones, especially when aphids are concentrated in particular zones on the plant body (Dixon, 1975).

In spite of the above problems, exudation from incisions and stylets has provided substantial amounts of data on translocation. Exudation of phloem sap is indeed an artificial phenomenon and appropriate caution should be exercised in interpreting the results. An understanding of the following facts is important in using exudation techniques to study phloem physiology.

The sieve tube pathway in plants is under considerable hydrostatic pressure of nearly 2.0 MPa (Milburn, 1975). When severed, this pressure is immediately released at the incision, creating a surge of sap. The pressure is released to nearby atmospheric pressure and is more slowly transmitted along the rest of the pathway. When turgor pressure release occurs in the sieve tubes, the water potential decreases, and water from surrounding tissues enters the sieve tubes. This can cause a consequent dilution of the phloem sap, as reported in *Fraxinus americana* by Zimmermann (1961). However, later investigators have failed to observe any remarkable change in sap solute potentials in other plants (Van Die & Tammes, 1975; Milburn & Zimmermann, 1977; Kallarackal & Milburn, 1984). This means that the influx of water into sieve tubes is usually matched simultaneously by either loading at the sources or along the pathway. The aphid stylet technique probably creates a gradient of pressure potential similar to the incision technique but on a more localized scale. The incision technique is an extremely useful tool to investigate the effect of changing turgor pressures on several phenomena connected with phloem transport (Milburn, 1970, 1971; Van Die & Tammes, 1975; Smith & Milburn, 1980a, b; Kallarackal & Komor, unpublished). Milburn and his co-workers have used a model to explain differences between exuding and intact systems (Milburn & Zimmermann, 1977; Kallarackal & Milburn, 1984).

7.5 Whole plant, non-invasive studies

Calculation of translocation rates and speed of translocation based upon daily increments of dry weight developed with the pioneering experiments of Dixon (1922) on potato tubers. He estimated that the dry matter (carbohydrates) was moving through the stolons of potato at a rate of $4.5 \text{ g cm}^{-2} \text{ h}^{-1}$ (SMT) of the phloem cross-sectional area. He calculated also the velocity of transport as 50 cm h^{-1} and then favoured the xylem pathway, concluding that such fast movement could not possibly occur in the phloem with its small elements and viscous contents. We know that the SMT and velocity obtained by Dixon are well within the range measured by more modern experimental techniques.

Later workers repeated the experiments of Dixon and his co-workers on certain other plants, considering that the transfer of volume through the phloem was a good measure of the translocation capacity of this tissue. Among these, the work of Mason & Lewin (1926) on *Dioscorea*, Mason & Maskell (1928) on *Gossypium*, Crafts (1933) on *Solanum* and Clements (1950) on *Kigelia*, are worthy of mention. In most of these studies the translocation data were expressed as the volume transferred in unit time, given by the equation:

$$\text{volume transfer} = \text{area} \times \text{velocity} \qquad (7.2)$$
$$(\text{cm}^3 \text{ h}^{-1}) \qquad (\text{cm}^2) \quad (\text{cm h}^{-1})$$

The above expression envisages the movement of a solution of organic substances in water through the phloem. Since water content in the tissues is variable depending on the water status of the plant and the environment, Canny (1960) thought that it would be better to ignore the water in the transporting system and consider only the *dry weight* transfer. This SMT concept is expressed by Canny (1960, 1973) thus:

$$\text{SMT} = \text{velocity} \times \text{concentration} \qquad (7.3)$$
$$(\text{g cm}^{-2} \text{ h}^{-1}) \quad (\text{cm h}^{-1}) \quad (\text{g cm}^{-3})$$

As already mentioned, translocation is basically the transport of dry weight from the sources of its production to the various destinations like fruits, roots, developing leaves and flowers, storage tissues, etc., although the control of translocation can be much more complex. As seen from the above expression, the measurement of this dry weight transferred is the most important parameter in determining the magnitude of translocation. The disadvantage of dry weight measurement is that it can be assesed only by destructive sampling, leading to the need for replicate samplings and cumbersome experimental routines.

In the past, the SMT has been measured in three ways: the measurement of sink gains, the measurement of loss from sources, and the SMT of exuding systems.

7.5.1 Sink gain methods

The gain in dry weight of a sink can be estimated by plotting its growth curve as a function of time. For example, the growth of a fruit (increase in dry weight) against time is a typical sigmoid curve. In such a situation the gain in dry weight over a certain period of fruit growth is not always uniform during the different stages. Although previous investigators had either taken the dry weight accumulation from the exponential part of the sigmoid, or taken an average from the whole growth period, Canny (1975) preferred the former to the latter because this gave the maximum capacity of the phloem to transport dry matter. In his book on phloem transport, Canny (1973) concentrated on the topic of SMT and published detailed tables on the SMT values (his Table 1) in different plants. He finally arrived at a figure of 4 g cm^{-2} h^{-1} as a general value for most plants when the cross-sectional area of the phloem was taken into account. Later, he refined this value to 6 g cm^{-2} h^{-1} taking into calculation only the sieve tube cross-sectional area as the specific transport path.

The validity of the above value was questioned with the publication of a study on the seminal root of wheat, reporting a much higher SMT value of 131 g cm^{-2} h^{-1} for the intact unpruned wheat roots (Passioura & Ashford, 1974). In the developing fruits of *Lupinus* on the other hand, Pate *et al.* (1978) estimated an SMT value of 3.95 g cm^{-2} h^{-1}, which is very close to Canny's value. They also found that the phloem carried 97% of the dry matter and 27% of the water entering the fruit. Kallarackal & Milburn (1984) made a comparative study of SMT in intact, pruned and exuding fruit bunches of *Ricinus*. When the fruit bunch was thinned to a simple fruit-bearing system, it showed an SMT twice that of the unpruned intact system. In the latter system, however, the SMT value closely resembled that of Canny's general figure. The SMT measurements were made on 30–35 d old *Ricinus* fruits when their rate of dry weight accumulation capacity was maximum. Therefore, it seems that Canny's value for SMT is acceptable for many unmanipulated sinks, when growth rates are maximal. The importance of determining the SMT when the growth rate of sinks is maximal cannot be overstressed; otherwise lower values will be obtained which do not indicate the true conducting capacity of the phloem.

From a further study of the SMT in roots of chickpea (Kallarackal & Milburn, 1985b), it is clear that the SMT varies during the growth of a sink. The plant grew steadily, reaching a maximum 40 d after germination; then the SMT fell, coincident with the initiation of anthesis. Cross-sectional areas of sieve tubes were monitored and it could be seen that SMT values were not influenced by this factor. The effect of distance between source and sink has been also studied to some extent. Mason & Maskell (1928) indicated an effect of length of the pathway on translocation, although later studies have shown that under normal conditions, several metres seem to be necessary to increase appreciably the resistance to translocation (Canny, 1973; Hansen & Christensen, 1974; Hanson, 1975). There are reasons to believe that the control is either in the sources or in the sinks. In the single-fruit system of *Ricinus* (Kallarackal & Milburn, 1984), the SMT is only twice that of an unpruned fruit bunch although we would expect an SMT equivalent to the number of fruits removed from a bunch, which is usually 8–12 fruits. It appears that the turgor pressure in the sieve tubes of the fruit sink is not noticeably changed by fruit thinning and the greater mass transfer rate achieved in pruned fruit systems could be caused by other factors, e.g. phytohormones (see Patrick & Wareing, 1980).

7.5.2 Loss from source leaves

Very few attempts appear in the literature measuring SMT from data on dry weight loss from the leaves. Although photosynthesis and respiration can be quantified, the method has many practical problems. The water relations of the leaf and the variability in dry weight per unit area are technicalities difficult to circumvent.

Crafts (1931) made SMT estimations on *Phaseolus* leaves in light and dark conditions and arrived at figures of 0.65 and 0.15 g h^{-1} cm^{-2} phloem cross-sectional area respectively. Geiger *et al.* (1969) used the amount of ^{14}C translocated out of the leaf to estimate the SMT in beet leaves to yield a figure of 1.7–0.5 g h^{-1} cm^{-2} phloem area from 13 such measurements. Simultaneous monitoring of CO_2 uptake and dry weight accumulation by Terry & Mortimer (1972) in beet leaves gave translocation rate as the calculated difference between them. Since the phloem cross-sectional areas were not measured in this work, Canny (1975) recomputed the SMT from other estimates of the phloem cross-sectional area in beet leaves (Geiger *et al.*, 1969) and obtained values of 0.4 to 0.6 g h^{-1} cm^{-2} phloem area in the light, and 0.12 to 0.15 in the dark.

The above measurements on leaf export indicate that SMT values for translocation from leaves are lower than values obtained from sink imports. One reason for this discrepancy could be the possible error involved in the method (see Thoday, 1910). Secondly, the method does not take into account the controlling influence of the sink. According to Münch's pressure flow hypothesis, the turgor pressure in the sink should be a critical factor in determining the rate of translocation. Most SMT measurements cited by Canny (1973) are on unmodified plants, which means that several source leaves contribute to a single sink. This would certainly influence the pressure gradient between the sources and sink when compared to a single leaf acting as a source for a particular sink. Sink manipulation studies have already demonstrated an effect on SMT (Kallarackal & Milburn, 1984).

7.5.2.1 SMT from exuding systems

Two types of exuding systems have already been mentioned in this chapter, namely exudation from bark incisions and aphid stylets. Although exudation is viewed as an unusual feature of phloem translocation, it has contributed much to the understanding of SMT in plants. Tammes (1933, 1952), from a detailed study on the exuding inflorescence of *Arenga*, reported a mass transfer rate of 99 g h^{-1} cm^{-2} phloem area. Further studies on the exuding inflorescence of *Cocos* also revealed a high SMT value (Milburn & Zimmermann, 1977). Kallarackal & Milburn (1984) reported an SMT value for an exuding *Ricinus* peduncle which was 20 times that for the intact system. From the above SMT determinations it is clear that the rate of translocation in an exuding system is indeed faster than in an intact one, and therefore represents an artificial system. However, in a comparison of the exuding seedling with an intact seedling of *Ricinus*, the SMT in both situations was similar and near to Canny's values (Kallarackal & Komor, unpublished). It should be remembered that the SMT determinations in the exuding inflorescences of palms and *Ricinus* plants were made with all the source leaves intact on the plant. In such a situation, the reservoir of assimilates will be so large that on incision the surge will give rise to a high SMT value. In seedlings by contrast, the sources are two cotyledons only, and in the intact system itself dry weight of a similar magnitude is required for the active growth during this stage. Smith & Milburn (1980b) showed the effect of leaf removal on exudation in *Ricinus* plants. The high SMT values obtained in exuding systems throws light on the capacity of sieve tubes to convey higher amounts than they ordinarily do. In other words, the sieve tube cross-sectional area is seldom a limitation to phloem translocation.

The exuding aphid stylet represents another interesting system to study SMT. Canny (1961) studied such a system in the willow tree and arrived at an SMT value very near to his expected standard. However, more work is required to confirm that an exuding stylet is transporting similarly to an intact plant system. This is because the stylet exudation would partly depend on the current turgor of the sieve tube in question.

7.5.2.2 Problems in measuring SMT

Canny (1973) discussed in detail the problems faced by physiologists in determining SMT to various plant parts. Depending on the organ studied, several precautions are needed to get an exact estimate of this transport parameter. In most SMT measurements presented in Canny (1973) the method of sampling has been destructive. In fact, it is almost impossible to overcome this problem. When destructive sampling is adopted as a means of determining the dry weight increase in the sink organ, it should be remembered that most sinks, especially the fruits, follow a sigmoid pattern of growth curve. This means that the analysis of growth curves is important before deciding at what stage of the growth the SMT is to be determined. As mentioned above, Canny (1975) suggested the exponential part of the growth curve as the most ideal period to estimate SMT because this indicates the maximal capacity of the phloem to translocate. In the study by Kallarackal & Milburn (1985b) previously referred to, on the SMT into the roots of chickpea from very early stages to the flowering and fruiting period, they found that the SMT to the roots was drastically reduced after flowering. They interpreted this as a slowing down of root metabolism and growth, making the roots a less competitive sink than developing fruits. However, during the whole measurement period, the cross-sectional area of the sieve tubes increased steadily. This again implies that the cross-sectional area of the sieve tubes is not a limiting factor in translocation.

When determining the dry weight increases of sink organs, it is important to estimate the contribution of sink photosynthesis to carbohydrate accumulation. Developing fruits are largely heterotrophic and 95% or more of their carbon is imported from the leaves. However, in the developing fruits of several plants like barley, tomato, pears and grapes (Edwards & Walker, 1983), some CO_2 fixation occurs, probably as a mechanism to re-absorb the carbon respired by the fruits (Sambo *et al.*, 1977). In general, in green fruits, photosynthesis is a secondary process, where some respired CO_2 or CO_2 from malate decarboxylation may be re-assimilated.

Many fruits have a relatively high rate of respiration. Carbon imported from the leaves in the form of sugars and amino acids may be assimilated primarily into lipids, proteins and oligosaccharides utilizing the energy of respiration. For the tricarboxylic acid cycle to function there must be continuous formation of oxaloacetate to react with acetyl coenzyme. A. During this process CO_2 evolves last from the system, thus:

acetyl coenzyme A + oxaloacetate $\rightarrow$ $2CO_2$ + oxaloacetate

As already mentioned, part of the CO_2 is trapped in fruit photosynthesis. However, this loss of CO_2 can be very high in non-photosynthetic sinks, e.g. roots. For example, in the roots of chickpea, up to 80% of the imported carbon was lost through respiration (Kallarackal & Milburn, 1985b). The necessity of determining the respiratory loss in SMT estimations is obvious. The determination of the sieve tube cross-sectional area has proved to be the most difficult part in making SMT estimations (Canny, 1973, 1975). Estimates show such a wide range of figures that we find them difficult to document here. It is possible that the proportion of sieve tubes in the phloem could vary in different plants. However, correct identification of sieve tubes in the phloem tissue requires the laborious process of serial sectioning followed by fluorescence microscopy, as suggested by Canny (1973). Milburn & Kallarackal (1983) tried this method on different parts of the *Ricinus* plant and found variations in the percentage of sieve tubes in different regions. To get over this laborious method they used a novel method of mild maceration which separated the phloem tissue as a flat sheet suitable for quantifying the sieve tube cross-sectional area. Although sieve tubes can be identified structurally, their functional viability is more difficult to ascertain. Callose blocking of old sieve tubes (Milburn & Kallarackal, 1983) and lignification of the walls (Cartwright *et al.*, 1977) could make a sieve tube non-functional. It is also not known at what stage during differentiation they start becoming functional. The negative staining method suggested by Fisher (1976) might be used to determine the functional viability of sieve tubes, but this method is somewhat cumbersome and a simple, efficient test is required urgently.

7.5.3 *Bidirectional transport*

Simultaneous transport in two different directions occurs from the time a young leaf begins to export. Very young leaves are typically strong sinks, whereas mature leaves are often strong sources. Thus, the maturation of a leaf is not synchronous, it passes through a transition period during which it functions as both sink and source. The major vien system of a leaf differentiates acropetally, whereas the maturation of the minor vein system and lamina tissue proceeds basipetally, so that the leaf apex matures first (Larson *et al.*, 1980). Consequently, during the transition from sink to source, a leaf may export assimilates from its mature apical region and at the same time import photosynthatcs to its immature base.

Eschrich (1975) concluded that although bidirectional transport was possible through adjacent sieve tubes within a vascular bundle, it was unlikely to occur within the same sieve tube. Simultaneous bidirectional transport within a vascular bundle has been demonstrated in *Vicia faba* (Fritz, 1973). He suggested that the first-formed metaphloem sieve tubes were translocating acropetally while later-formed ones were transporting basipetally. Indirect evidence for a similar situation has been demonstrated in cottonwood (Isebrands & Larson, 1980). Fellows & Geiger (1974) have also demonstrated bidirectional transport in the same bundle when sugar accumulation and phloem loading capacity in the companion cell-sieve tube complex of the minor viens reaches the threshold necessary to induce mass flow.

Studies using ^{14}C labelling on *Populus* have shown that bidirectional transport might occur in different parts of the same sieve tube at the same time (Vogelmann *et al.*, 1982). Their labelling patterns suggest that as the lamina matures basipetally, export begins in sieve tubes of dorsal bundles and advances as a front down the midvein and petiole. At the same time, import ceases in the same sieve tubes of the dorsal bundles and retreats as a front down the midvein and petiole. Thus, during a brief transitional stage, advance of the export front coincides with retreat of the import front in different parts of the same sieve tubes. The above study was possible through a painstaking analysis of the vascular system of *Populus* (Fisher & Larson, 1983).

In several plants it has been shown that the maturing vascular and structural tissues are strong sinks for assimilates (Kocher & Leonard, 1971; Chopwick & Forward, 1974; Dickson & Larson, 1975; Isebrands *et al.*, 1976; Blechschmidt-Schneider & Eschrich, 1985). Therefore, it is possible that the bidirectional transport within a single sieve tube might serve a common sink created by differentiating vascular and structural tissues. This is analogous to the artificial sink created by an aphid stylet, in which the phloem sap moves from opposite directions toward the stylet (Peel, 1975). In this regard it should be borne in mind that bidirectional transport in sieve tubes in this manner poses no challenge to Münch's mass flow hypothesis (cf. Milburn, 1975).

References

Auclair, J. L. (1963). Aphid feeding and nutrition. *Annual Review of Entomology* **8**, 439–90.

Baker, D. A. (1978). *Transport Phenomena in Plants*. London, Chapman and Hall.

Baker, D. A., Kallarackal, J. & Milburn, J. A. (1989). Water relations of the banana. II. Physico-chemical aspects of the latex and other tissue fluids. *Australian Journal of Plant Physiology* (in press).

Barlow, C. A. & Randolph, P. A. (1978). Quality and quantity of plant sap available to the pea aphid. *Annals of the Entomological Society of America* **71**, 46–8.

Bauer, I. (1953). Zur Frage der Stoffbewegungen im der Pflanze mit besonderer-Berücksichtigung der Wanderung von Fluorochromen. *Planta* **42**, 367–451.

Bennett, C. W. (1937). Correlation between movement of the curly-top virus and translocation of food in tobacco and sugar beet. *Journal of Agricultural Research* **54**, 479–502.

Bishop, H. T., Thompson, R. G., Aikman, D. P. & Fensom, D. S. (1986). Fine structure aberrations in the movement of ^{11}C and ^{13}N in the stems of plants. *Journal of Experimental Botany* **37**, 1780–94.

Blechschmidt-Schneider, S. & Eschrich, W. (1985). Microautoradiographic local-isation of imported ^{14}C photosynthate in induced sink leaves of two dico-tyledonous C_4 plants in relation to phloem loading. *Planta* **163**, 439–47.

Bose, J. C. (1947) *Plants and their Autographs*. London, Longman, Green & Co.

Brevdan, E. R. & Hodges, H. F. (1973). Effect of moisture deficits on ^{14}C trans-location in corn (*Zea mays* L.). *Plant Physiology* **52**, 436–9.

Buttery, B. R. & Boatman, S. G. (1976). Water deficits and flow of latex. In Kozlowski, T. T. (ed.). *Water Deficits and Plant Growth*, Vol. IV, pp. 234–89. New York, London, Academic Press.

Canny, M. J. (1960). The rate of translocation. *Biological Reviews of the Cambridge Philosophical Society* **35**, 507–32.

Canny, M. J. (1961). Measurements of the velocity of translocation. *Annals of Botany* **25**, 152–67.

Canny, M. J. (1973). *Phloem Translocation*. Cambridge University Press.

Canny, M. J. (1975). Mass transfer. In Zimmermann, M. H. & Milburn, J. A. (eds.). *Transport in Plants. Encyclopedia of Plant Physiology*, New Series, Vol. 1, pp 139–53. Heidelberg, Springer-Verlag.

Canny, M. J. & Markus, K. (1960). Metabolism of phloem isolated from grapevine. *Australian Journal of Biological Sciences* **13**, 292–9.

Cartwright, S. C, Lush, W. M. & Canny, M. J. (1977). A comparison of trans-location of labelled assimilate by normal and lignified sieve elements in wheat leaves. *Planta* **134**, 207–8.

Chopwick, R. E. & Forward, D. F. (1974). Translocation of radioactive carbon after the application of ^{14}C-alanine and $^{14}CO_2$ to sunflower leaves. *Plant Physiology* **53**, 21–7.

Christy, A. L. & Ferrier, J.M. (1973). A mathematical treatment of Münch's pressure-flow hypothesis of phloem translocation. *Plant Physiology* **52**, 531–8.

Christy, A. L. & Swanson, C. A. (1976). Control of translocation by photosyn-thesis and carbohydrate concentrations of the source leaf. In Wardlaw, I. F. & Passioura, J. B. (eds.). *Transport and Transfer Processes in Plants*, pp. 329–38. New York, Academic Press.

Clements, H. F. (1940). Movement of organic solutes in the sausage tree, *Kigelia africana*. *Plant Physiology* **15**, 689–700.

Costello, L. R., Bassham, J. A. & Calvin, M. (1982). Enhancement of phloem exudation from *Fraxinus uhdei* Wenz. (evergreen ash) using ethylenedi-aminetetra-acetic acid. *Plant Physiology* **69**, 77–82.

Coulson, C. L., Christy, A. L., Cataldo, D. A. & Swanson, C. A. (1972). Carbo-hydrate translocation in sugar beet petiole in relation to petiolar respiration and adenosine-5'-triphosphate. *Plant Physiology* **49**, 919–23.

Crafts, A. S. (1931). Movement of organic material in plants. *Plant Physiology* **6**, 1–41.

Crafts, A. S. (1933). Sieve-tube structure and translocation in the potato. *Plant Physiology* **8**, 81–104.

Cronshaw, J. (1981). Phloem structure and function. *Annual Review of Plant Phys-iology* **32**, 465–84.

Cronshaw, J. & Anderson, R. (1969). Sieve plate pores of *Nicotiana*. *Journal of Ultrastructural Research* **27**, 134–48.

Currier, H. B. & Shih, C. Y. (1968). Sieve tubes and callose in *Elodea* leaves. *American Journal of Botany* **55**, 145–52.

Currier, H. B. & Webster, D. H. (1964). Callose formation and subsequent disappearance: ultrasound stimulation. *Plant Physiology* **39**, 843–7.

Curtis, O. F. (1929). Studies on solute translocation in plants. Experiments indicating that translocation is dependent on the activity of living cells. *American Journal of Botany* **6**, 154–68.

Dainty, J. (1976). Water relations of plant cells. In Lüttge, U. & Pitman, M. G. (eds.). *Transport in Plants. Encyclopedia of Plant Physiology*, New Series, Vol. IIa, pp. 12–35. Heidelberg, Springer-Verlag.

Dickson, R. E. (1977). EDTA-promoted exudation of ^{14}C-labelled compounds from detached cottonwood and bean leaves as related to translocation. *Canadian Journal of Forestry Research* **7**, 277–84.

Dickson, R. E. & Larson, P. R. (1975). Incorporation of ^{14}C-photosynthate into major chemical fractions of source and sink leaves of cottonwood. *Plant Physiology* **56**, 185–93.

Dixon, A. F. G. (1975). Aphids and translocation. In Zimmermann, M. H. & Milburn, J. A. (eds.). *Transport in Plants. Encyclopedia of Plant Physiology*, New Series, Vol. I, pp. 154–70. Heidelberg, Springer-Verlag.

Dixon, A. F.G. & Logan, M. (1973). Leaf size and availability of space to the sycamore aphid *Drepanosiphum platanoides*. *Oikos* **24**, 58–63.

Dixon, H. H. (1922). Transport of organic substances in plants. *Nature* **110**, 547–51.

Dixon, H. H. & Ball, N. G. (1922). Transport of organic substances in plants. *Nature* **109**, 236–7.

Edwards, G. & Walker, D. (1983). *C₃, C₄: Mechanisms and Cellular and Environmental Regulation of Photosynthesis*. London, Blackwell Scientific Publications.

Englemann, E. M. (1965). Sieve element of *Impatiens sultanii* wound reaction. *Annals of Botany* **29**, 83–101.

Esau, K., Cronshaw, J. & Hoefert, L. L. (1967). Relation of beet yellows to the phloem and to the movement in the sieve tube. *Journal of Cell Biology* **32**, 71–87.

Eschrich, W. (1957). Callosebildung in Plasmolysierten *Allium cepa*-Epidermen. *Planta* **48**, 578–86.

Eschrich, W. (1975). Bidirectional transport. In Zimmermann, M. H. & Milburn, J. A. (eds.). *Transport in Plants. Encyclopedia of Plant Physiology*, New Series, Vol. II, pp. 245–55. Heidelberg, Springer-Verlag.

Eschrich, W. & Currier, H. B. (1964). Identification of callose by its diachrome and fluorochrome relations. *Stain Technology* **39**, 303–7.

Eschrich, W., Currier, H. B., Yamaguchi, S. & McNairn, R. B. (1965). Der Einfluss Verstärkter Callosebildung auf den Stofftransport in Siebröhren. *Planta* **65**, 49–64.

Evert, R. F. & Derr, W. F. (1964). Callose substance in sieve elements. *American Journal of Botany* **51**, 552–9.

Evert, R. F., Eschrich, W., Medler, J. T. & Alfieri, F. J. (1968). Observations on penetration of linden branches by stylets of the aphid *Longistigma caryae*. *American Journal of Botany* **55**, 860–74.

Faucher, M. & Bonnemain, J. L. (1986). Remote effect of localised thermic treatments on the uptake of sucrose by the veins. In Cronshaw, J., Lucas, W. J. & Giaquinta, R. T. (eds.). *Phloem Transport*, pp. 55–7. New York, Alan R. Liss.

Faucher, M., Bonnemain, J. L. & Doffin, M. (1982). Effets de refroidissements localisés sur la circulation libérienne chez quelque espèces avec ou sans proteins-P et influence du mode de refroidissement. *Physiologie Végétale* **20**, 395–405.

Fellows, R. J., Egli, D. B. & Leggett, J. E. (1978). A pod leakage technique for phloem translocation studies in soybean (*Glycine max* L. merr.). *Plant Physiology* **62**, 812–14.

Fellows, R. J. & Geiger, D. R. (1974). Structural and physiological changes in sugar beet leaves during sink to source transition. *Plant Physiology* **54**, 877–85.

Fellows, R. J. & Zeevaart, J. A. D. (1983). Comparison of ethylenediamenetetra-acetate-enhanced exudation from detached and translocation from attached leaves. *Plant Physiology* **71**, 716–18.

Fensom, D. S. (1981). Problems arising from a Münch-type pressure flow mechanism of sugar transport in phloem. *Canadian Journal of Botany* **59**, 425–32.

Fensom, D. S., Thompson, R. G. & Alexander, K. G. (1984). Stem anoxia temporarily interrupts translocation of ¹¹C-photosynthate in sunflower. *Journal of Experimental Botany* **35**, 1582–94.

Ferrier, J. M. & Christy, A. L. (1974). Time-dependent behaviour of a mathematical model for Münch translocation. Application to recovery from cold inhibition. *Plant Physiology* **55**, 511–14.

Fisher, D. B. (1976). Histochemical approaches to water-soluble compounds and their application to problems in translocation. In Wardlaw, I. F. & Passioura, J. B. (eds.). *Tranport and Transfer Processes in Plants*, pp. 237–46. New York, Academic Press.

Fisher, D. B. (1978). An evaluation of the Münch hypothesis for phloem transport in soybean. *Planta* **139**, 25–8.

Fisher, D. B. (1983). Year-round collection of willow sieve tube exudate. *Planta* **159**, 529–33.

Fisher, D. B. (1987). Changes in the concentration and composition of peduncle sieve tube sap during grain filling in normal and phosphate-deficient wheat plants. *Australian Journal of Plant Physiology* **14**, 147–56.

Fisher, D. B. & Evert, R. F. (1982). Studies on the leaf of *Amaranthus retroflexus* (Amaranthaceae): quantitative aspects and solute concentration in the phloem. *American Journal of Botany* **69**, 1375–88.

Fisher, D. B. & Frame, J. M. (1984). A guide to the use of the exuding-stylet technique in phloem physiology. *Planta* **161**, 385–93.

Fisher, D. B. & Larson, P.R. (1983). Structure of leaf/branch gap parenchyma and associated vascular tissues in *Populus deltoides*. *Botanical Gazette* **144**, 73–85.

Ford, J. & Peel, A. J. (1967). Preliminary experiments on the effect of temperature on the movement of ¹⁴C-labelled assimilates through the phloem of willow. *Journal of Experimental Botany* **18**, 406–15.

Fritz, E. (1973). Microautoradiographic investigations on bidirectional translocation in the phloem of *Vicia faba*. *Planta* **112**, 169–79.

Geiger, D. R. (1969). Chilling and translocation inhibition. *Ohio Journal of Science* **69**, 356–66.

Geiger, D. R. & Batey, J. W. (1967). Translocation of ¹⁴C sucrose in sugar beet during darkness. *Plant Physiology* **64**, 1743–9.

Geiger, D. R. & Fondy, B. R. (1979). A method for continuous measurement of export from a leaf. *Plant Physiology* **64**, 361–5.

Geiger, D. R., Saunders, M. A. & Cataldo, D. A. (1969). Translocation and accumulation of translocate in the sugar beet petiole. *Plant Physiology* **44**, 1657–65.

Geiger, D. R. & Sovonick, S. A. (1970). Temporary inhibition of translocation velocity and mass transfer rate by petiole cooling. *Plant Physiology* **46**, 847–9.

Geiger, D. R. & Sovonick, S. A. (1975). Effects of temperature, anoxia and other metabolic inhibitors on translocation. In Zimmermann, M. H. & Milburn, J. A. (eds.). *Transport in Plants. Encyclopedia of Plant Physio-*

logy, New Series, Vol. I, pp. 256–86. Heidelberg, Springer-Verlag.

Giaquinta, R. T. & Geiger, D. R. (1973). Mechanism of inhibition of translocation by localized chilling. *Plant Physiology* **51**, 372–7.

Giese, A. C. (1973). *Cell Physiology*. Philadelphia, Saunders.

Grange, R. R. & Peel, A. J. (1978). Evidence for solution flow in the phloem of willow. *Planta* **138**, 15–23.

Grodzinski, B., Jahnke, S. & Thompson, R. (1984). Translocation profiles of ^{11}C and ^{13}N-labelled metabolites after assimilation of $^{11}CO_2$ and 13-labelled ammonia gas by leaves of *Helianthus annuus* L. and *Lupinus alba* L. *Journal of Experimental Botany* **35**, 678–90.

Grusak, M. A. & Lucas, W. J. (1984). Recovery of cold-inhibited phloem translocation in sugar beet. *Journal of Experimental Botany* **35**, 389–402.

Grusak, M. A. & Lucas, W. J. (1986). Cold-inhibited phloem translocation in sugar beet. III. The involvement of the phloem pathway in source-sink partitioning. *Journal of Experimental Botany* **37**, 277–88.

Hall, S. M., Baker, D. A. & Milburn, J. A. (1971). Phloem transport of ^{14}C-labelled assimilates in *Ricinus*. *Planta* **100**, 200–7.

Hall, S. M. & Milburn, J. A (1973). Phloem transport in *Ricinus*. Its dependence on the water balance of the tissues. *Planta* **109**, 1–10.

Hammel, H. T. (1968). Measurement of turgor pressure and its gradient in the phloem of oak. *Plant Physiology* **43**, 1042–8.

Hansen, P. (1975). Carbohydrate allocation. In Landsberg, J. J. & Cutting, C. V. (eds.). *Environmental Effects on Crop Physiology*, pp. 247–58. London, Academic Press.

Hansen, P. & Christensen, J. V. (1974). Fruit thinning. III. Translocation of ^{14}C-assimilates to fruit from near and distant leaves in the apple 'Golden Delicious'. *Horticultural Research* **14**, 41–5.

Hansen, S. D. & Cohen, J. D. (1985). A technique for collection of exudate from pea seedlings. *Plant Physiology* **78**, 734–8.

Hartt, C. E. (1967). Effect of moisture supply on translocation and storage of ^{14}C translocates in sugarcane. *Plant Physiology* **42**, 338–46.

Ho, L. C. (1976). The relationship between the rates of carbon transport and of photosynthesis in tomato leaves. *Journal of Experimental Botany* **27**, 87–97.

Hocking, T. J., Hillman, J. R. & Wilkins, M. B. (1972). Movement of abscisic acid in *Phaseolus vulgaris* plants. *Nature New Biology* **235**, 124–5.

Huber, B., Schmidt, E. & Jahnel, H. (1937). Untersuchungen über den assimilatström. *Tharandt. Forstl. Jahrb.* **88**, 1017–50.

Husain, A. & Spanner, D. C. (1966). The influence of varying concentrations of applied sugar on the transport of tracers in cereal leaves. *Annals of Botany* **30**, 549–61.

Isebrands, J. G., Dickson R. E. & Larson, P. R. (1976). Translocations and incorporation of ^{14}C into the petiole from different regions within developing cottonwood leaves. *Planta* **128**, 185–93.

Isebrands, J. G. & Larson, P. R. (1980). Ontogeny of major veins in the lamina of *Populus deltoides* Bart. *American Journal of Botany* **67**, 23–33.

Jacobs, W. P. (1970). Regeneration and differentiation of sieve tube elements. *International Review of Cytology* **28**, 239–73.

Jahnke, S., Stocklin, G. & Willenbrink, J. (1981). Translocation profiles of ^{11}C-assimilates in the petiole of *Marsilea quadrifolia* L. *Planta* **153**, 56–63.

Jenner, C. F. (1982). Movement of water and mass transfer into developing grains of wheat. *Australian Journal of Plant Physiology* **9**, 69–82.

Johnson, R. R. & Moss, D. N. (1976). Effect of water stress on $^{14}CO_2$ fixation and translocation in wheat during grain filling. *Crop Science* **16**, 697–701.

Kallarackal, J. (1975). Palm tapping – An art, science and industry. *Science Reporter* **12**, 172–4.

Kallarackal, J. & Milburn, J. A. (1983). Studies on the phloem sealing mechanism in *Ricinus* fruit stalks. *Australian Journal of Plant Physiology* **10**, 561–8.

Kallarackal, J. & Milburn, J. A. (1984). Specific mass transfer and sink-controlled phloem translocation in castor bean. *Australian Journal of Plant Physiology* **11**, 483–90.

Kallarackal, J. & Milburn, J. A. (1985a). Phloem sap exudation in *Ricinus communis*: elastic responses and anatomical implications. *Plant, Cell and Environment* **8**, 239–45.

Kallarackal, J. & Milburn, J. A. (1985b). Respiration and phloem translocation in the roots of chickpea (*Cicer arietinum*). *Annals of Botany* **56**, 211–18.

Kaufman, M. J. & Kramer, P. J. (1967). Phloem water relations and translocation. *Plant Physiology* **42**, 191–4.

Kennedy, J. S. & Mittler, T. E. (1953). A method for obtaining phloem sap via the mouth parts of aphids. *Nature* **171**, 528.

Kessler, G. (1958). Zur Charakterisierung der Siebröhrencallose. *Bericht der Schweizerischen botanischen Gesellschaft* **68**, 5–43.

Kessler, R. W. & Zeevaart, J. A. D. (1974). Enhancement of phloem exudation from cut petioles by chelating agents. *Plant Physiology* **53**, 96–103.

Kocher, H. & Leonard, O.A. (1971). Translocation and metabolic conversion of ^{14}C-labelled assimilates in detached and attached leaves of *Phaseolus vulgaris* L. in different phases of leaf expansion. *Plant Physiology* **47**, 212–16.

Komor, E. (1982). Transport of sugar. In Loewus, F. A. & Tanner, W. (eds.). *Plant Carbohydrates I. Enclyopedia of Plant Physiology*, New Series, Vol. 13A, pp. 635–676. Heidelberg, Springer-Verlag.

Komor, E. & Orlich, G. (1986). Sugar-proton symport: from single cells to phloem loading. In Cronshaw, J., Lucas, W. J. & Giaquinta, R. T. (eds.). *Phloem Transport*, pp. 53–7. New York, Alan R. Liss.

Komor, E., Rotter, M. & Tanner, W. (1977). A proton-cotransport system in a higher plant: sucrose transport in *Ricinus communis*. *Plant Science Letters* **9**, 153–62.

Kursanov, A. L. (1984). *Assimilate Transport in Plants*. Amsterdam, New York, Oxford, Elsevier.

Kursanov, A. L. & Brovcenko, M. I. (1961). The influence of ATP on the entry of assimilates into the translocation system of sugar beet. *Fiziologiya Rast* **8**, 270–8.

Kursanov, A. L. & Turkina, M. V. (1952a). The respiration of vascular bundles. *Doklady Academii nauk SSSR* **84**, 1073–6.

Kursanov, A. L. & Turkina, M. V. (1952b). The respiration of translocating tissues and the transport of sucrose. *Doklady Academii nauk SSSR* **85**, 649–52.

Lambers, H., Simpson, R. J., Beilharz, V. C. & Dalling, M. J. (1982). Translocation and utilization of carbon in wheat (*Triticum aestivum*). *Physiologia Plantarum* **56**, 18–22.

Lang, A. (1973). A working model of a sieve tube. *Journal of Experimental Botany* **24**, 896–904.

Lang, A. (1978). Interactions between source, path and sink in determining phloem translocation rate. *Australian Journal of Plant Physiology* **5**, 665–74.

Lang, A. & Minchin, P. E. H. (1986). Phylogenetic distribution and mechanism of translocation inhibition by chilling. *Journal of Experimental Botany* **37**, 389–98.

Lang, A. & Thorpe, M. R. (1986). Water potential, translocation and assimilate partitioning. *Journal of Experimental Botany* **37**, 389–98.

Larson, P. R., Isebrands, J. G. & Dickson, R. E. (1980). Sink to source transition of *Populus* leaves. *Bericht der Deutschen botanischen Gesellschaft* **93**, 79–90.

Lawlor, D. W. (1976). Assimilation of carbon into photosynthetic intermediates of water-stressed wheat. *Photosynthetica* **10**, 431–9.

Lee, D. R. (1981a). Synchronous pressure-potential changes in the phloem of *Fraxinus americana* L. *Planta* **151**, 304–8.

Lee, D. R. (1981b). Elasticity of phloem tissues. *Journal of Experimental Botany* **32**, 251–60.

Lohr, E. (1953). Die Zuckerarten im Blutungssaft von *Betula* und *Carpinus*. *Physiologia Plantarum* **6**, 529–32.

Madore, M. A., Oross, J. W. & Lucas, W. J. (1986). Symplastic transport in *Ipomoea tricolor* source leaves. *Plant Physiology* **82**, 432–42.

Majumder, S. K. & Leopold, A. C. (1967). Callose formation in response to low temperature. *Plant Cell Physiology* **8**, 775–8.

Malek, F. & Baker, D. A. (1977). Proton co-transport of sugars in phloem loading. *Planta* **135**, 279–99.

Malik, R. S., Dhankar, J. S. & Turner, N. C. (1979). Influence of soil water deficits on root growth of cotton seedlings. *Plant and Soil* **53**, 109–15.

Marowitch, J., Richter, C. & Hoddinott, J. (1986). The influence of plant temperature on photosynthesis and translocation rates in bean and soybean. *Canadian Journal of Botany* **64**, 2337–42.

Mason, T. G. & Lewis, C. J. (1926). On the rate of carbohydrate transport in greater yam, *Dioscorea alata* Linn. *Scientific Proceedings of the Royal Dublin Society* **18**, 203–5.

Mason, T. G. & Maskell, E. J. (1928). Studies on the transport of carbohydrates in the cotton plant. II. The factors determining the rate and direction of movement of sugars. *Annals of Botany* **42**, 571–636.

Mason, T. G. & Phillis, E. (1936). Oxygen supply and the inactivation of diffusion. *Annals of Botany* **50**, 455–99.

McAllan, J. W. & Adams, J. B. (1961). The significance of pectinase in plant penetration by aphids. *Canadian Journal of Zoology* **39**, 305–10.

McEuen, A. R. & Hill, H. A. O. (1982). Superoxide, hydroxen peroxide, and the gelling of phloem sap from *Cucurbita pepo*. *Planta* **154**, 295–7.

McNairn, R. B. (1972). Phloem translocation and heat induced callose formation in field-grown *Gossypium hirsutum* L. *Plant Physiology* **50**, 366–70.

McNairn, R. B. & Currier, H. B. (1968). Translocation blockage by sieve plate callose. *Planta* **82**, 369–80.

McPherson, H. G. & Boyer, J. S. (1977). Regulation of grain yield by photosynthesis in maize subjected to a water deficiency. *Agronomy Journal* **69**, 714–18.

Milburn, J. A. (1970). Phloem exudation from castor bean: induction by massage. *Planta* **95**, 272–6.

Milburn, J. A. (1971). An analysis of the response in phloem exudation on application of massage to *Ricinus*. *Planta* **100**, 143–54.

Milburn, J. A. (1974). Phloem transport in *Ricinus*: concentration gradients between source and sink. *Planta* **117**, 303–19.

Milburn, J. A. (1975). Pressure flow. In Zimmermann, M. H. & Milburn, J. A. (eds.). *Transport in Plants. Encyclopedia of Plant Physiology*, New Series, Vol. I, pp. 3–38. Heidelberg, Springer-Verlag.

Milburn, J. A. (1980). The measurement of turgor pressure in sieve tubes. *Bericht der Deutschen botanischen Gesellschaft* **93**, 153–66.

Milburn, J. A. & Kallarackal, J. (1984). Quantitative determination of the sieve tube dimensions in *Ricinus, Cucumis* and *Musa. New Phytologist* **96**, 383–95.

Milburn, J. A., Kallarackal, J. & Baker, D. A. (1989). Water relations of the banana. *Australian Journal of Plant Physiology* (in press).

Milburn, J. A & O'Malley, P. E. R. (1984). Freeze-induced sap absorption in *Acer pseudoplatanus*: a possible mechanism. *Canadian Journal of Botany* **62**, 2101–6.

Milburn, J. A. & Zimmermann, M. H. (1977). Preliminary studies on sap flow in *Cocos nucifera* L. II. Phloem transport. *New Phytologist* **96**, 383–95.

Miles, P. W. (1968). Studies on the salivary physiology of plant-bugs: experimental induction of galls. *Journal of Insect Physiology* **14**, 97–106.

Minchin, P. E. H. (1979). The relationship between spatial and temporal tracer profiles in transport studies. *Journal of Experimental Botany* **30**, 1171–8.

Minchin, P. E. H. (ed.). (1986). Short-lived isotopes in Biology. *Proceedings of an International Workshop on Biological Research with Short-lived Isotopes.*

Minchin, P. E. H., Lang, A. & Thorpe, M. R. (1983). Dynamics of cold-induced inhibition of phloem transport. *Journal of Experimental Botany* **34**, 156–62.

Minchin, P. E. H., Ryan, K. G. & Thorpe, M. R. (1984). Further evidence of apoplastic unloading into the stem of bean: identification of the phloem buffering pool. *Journal of Experimental Botany* **35**, 1744–53.

Minchin, P. E. H. & Thorpe, M. R. (1984). Apoplastic phloem unloading in the stem of bean. *Journal of Experimental Botany* **35**, 538–50.

Minchin, P. E. H. & Thorpe, M. R. (1987). Measurement of unloading and reloading of photoassimilates within the stem of bean. *Journal of Experimental Botany* **38**, 211–20.

Minorsky, P. V. (1985). An heuristic hypothesis of chilling injury in plants: a role for calcium as the primary physiological transducer of injury. *Plant, Cell and Environment* **8**, 75–94.

Mittler, T. E. (1957). Studies on the feeding and nutrition of *Tuberolachnus salignus* (Gmelin) (Homoptera Aphididae). *Journal of Experimental Botany* **34**, 334–41.

Moorby, J. (1977). Integration and regulation of translocation within the whole plant. In Jennings, D. H. (ed.). *Integration of Activity in the Higher Plant*, pp. 425–54. Cambridge, Cambridge University Press.

Moorby, J., Ebert, M. & Evans, N. T. S. (1963). The translocation of ¹¹C-labelled photosynthate in soybean. *Journal of Experimental Botany* **14**, 210–20.

Moore, R. D. & Troughton, J. H. (1973). Production of ¹¹CO₂ for use in plant translocation studies. *Photosynthetica* **7**, 271–4.

Munns, R. & Pearson, C. J. (1974). Effect of water deficit on translocation of carbohydrate in *Solanum tuberosum. Australian Journal of Plant Physiology* **1**, 529–37.

Munns, R., Tonnet, M. L., Shennan, C. & Gardner, P. A., (1988). Effect of high external NaCl concentration on ion transport within the shoot of *Lupinus Albus* II. Ions in phloem sap. *Plant Cell and Environment* **11**, 291–300.

Nobel, P. S. (1974). Temperature dependence of the permeability of chloroplasts from chilling-sensitive and chilling-resistant plants. *Planta* **115**, 369–72.

Nuske, J. & Eschrich, W. (1976). Synthesis of P-protein in mature phloem of *Cucurbita maxima. Planta* **132**, 109–18.

Parthasarathy, M. V. (1975). Sieve element structure. In Zimmermann, M. H. & Milburn, J. A. (eds.). *Transport in Plants. Encyclopedia of Plant Physiology*, New Series, Vol. 1, pp. 3–38. Heidelberg, Springer-Verlag.

Passioura, J. B. & Ashford, A. E. (1974). Rapid translocation in the phloem of wheat roots. *Australian Journal of Plant Physiology* **1**, 521–7.

Pate, J. S., Kuo, J. & Hocking, P. J. (1978). Functioning of conducting elements of phloem and xylem in the stalk of the developing fruit of *Lupinus albus* L. *Australian Journal of Plant Physiology* **5**, 321–36.

Pate, J. S., Layzell, D. B. & Atkins, C. A. (1979). Economy of carbon and nitrogen in a nodulated and non nodulated (NO_3-grown) legume. *Plant Physiology* **64**, 1083–8.

Pate, J. S., Peoples, M. B. & Atkins, C. A. (1984). Spontaneous phloem bleeding from cryopunctured fruits of a ureide-producing legume. *Plant Physiology* **74**, 499–505.

Pate, J. S., Sharkey, P. J. & Lewis, O. A. M. (1974). Phloem bleeding from legume fruits: a technique for study of fruit nutrition. *Planta* **120**, 229–43.

Pate, J. S., Sharkey, P. J. & Lewis, O. A. M. (1975). Xylem to phloem transfer of solutes in fruiting shoots of legume, studied by a phloem bleeding technique. *Planta* **122**, 11–26.

Patrick, J. W. & Wareing, P. F. (1980). Hormonal control of assimilate movement and distribution. In Jeffcoat, B. (ed.). *Aspects and Prospects of Plant Growth Regulators*, pp. 65–84, Proceedings Joint Symposium British Crop Protection Council and British Plant Growth Regulator Group.

Peel, A. J. (1963). The movement of ions from the xylem solution into the sieve tubes of willow. *Journal of Experimental Botany* **14**, 438–47.

Peel, A. J. (1965). The effect of changes in the diffusion potential of the xylem water on sieve-tube exudation from isolated stem segments. *Journal of Experimental Botany* **16**, 249–60.

Peel, A. J. (1975). Investigations with aphid stylets into the physiology of the sieve tube. In Zimmermann, M. H. & Milburn, J. A. (eds.). *Transport in Plants. Encyclopedia of Plant Physiology*, New Series, Vol. 1 pp. 171–95. Heidelberg, Springer-Verlag.

Peel, A. J. & Weatherley, P. E. (1962). Studies in sieve-tube exudation through aphid mouthparts. I. The effects of light and girdling. *Annals of Botany* **26**, 633–46.

Pickard, W. F., Minchin, P. E. H. & Troughton, J. H. (1978a). Real time studies of carbon-11 translocation in moonflower. I. The effect of cold blocks. *Journal of Experimental Botany* **29**, 993–1001.

Pickard, W. F., Minchin, P. E. H. & Troughton, J. H. (1978b). Transient inhibition of translocation in *Ipomoea alba* L. by small temperature reductions. *Australian Journal of Plant Physiology* **5**, 127–30.

Pickard, W. F., Minchin, P. E. H. & Troughton, J. H. (1979). Real time studies of carbon-11 translocation in moonflower. II. Further experiments on the effects of a nitrogen atmosphere, water stress and chilling, and a qualitative theory of stem translocation. *Journal of Experimental Botany* **30**, 307–18.

Qureshi, F. A. & Spanner, D. C. (1973). The effect of nitrogen on the movement of tracers down the stolon of *Saxifraga sarmentosa* L. with some observations on the influence of light. *Planta* **110**, 131–44.

Read, S. M. & Northcote, D. H. (1983). Chemical and immunological similarities between the phloem proteins of three genera of the *Cucurbitaceae*. *Planta* **158**, 119–27.

Richardson, P. T., Baker, D. A. & Ho, L. C. (1984). Assimilate transport in cucurbits. *Journal of Experimental Botany* **35**, 1575–81.

Rogers, S. & Peel, A. J. (1975). Some evidence for the existence of turgor pressure gradients in the sieve tubes of willow. *Planta* **126**, 259–67.

Ross, M. S. & Tyree, M. T. (1980). A reanalysis of the kinetics of ^{14}C photosynthate translocation in morning glory vines. *Annals of Botany* **46**, 727–38.

Sambo, E. Y., Moorby, J. & Milthorpe, F. L. (1977). Photosynthesis and respiration of developing soybean pods. *Australian Journal of Plant Physiology* **44**, 713–21.

Sauter, J. J. (1967). Der Einfluss verschiedener Temperaturen auf die Reservestärke in parenchymatischen Geweben von Baumsprossachsen. *Zeitschhrift für Pflanzenphysiologie* **56**, 340–52.

Schumacher, W. (1930). Untersuchungen über die Lokalisation der Stoffwanderung in den Leitbündeln höherer Pflanzen. *Jahrbuch für wissenschafttliche Botanik* **73**, 770–823.

Schumacher, W. (1933). Untersuchungen über die Wanderung des Fluoresceins in den Siebröhren. *Jahrbuch für wissenschaftliche Botanik* **77**, 685–732.

Schumacher, W. (1967). Die Fernleitung der Stoffe im Pflanzenkörper. In Ruhland, W. (ed.). *Encyclopedia of Plant Physiology*, Vol. XIII, pp. 60–177. Berlin, Heidelberg, New York, Springer.

Sheikholeslam, S. N. & Currier, H. B. (1977). Phloem pressure differences and ^{14}C assimilate translocation in *Ecballium elaterium*. *Plant Physiology* **59**, 376–80.

Shih, C. Y. & Currier, H. B. (1969). Fine structure of phloem cells in relation to translocation in the cotton seedling. *American Journal of Botany* **56**, 464–72.

Sij, J. & Swanson, C. A. (1972). Effect of petiole anoxia on phloem transport in squash. *Plant Physiology* **51**, 368–71.

Simpson, R. J. & Dalling, M. J. (1981). Nitrogen redistribution during grain growth in wheat (*Triticum aestivum* L.). III. Enzymology and transport of aminoacids from senescing flag leaves. *Planta* **151**, 447–56.

Sjölund, R. D., Shih, C. Y. & Jensen, K. G. (1983). Freeze-fracture analysis of phloem structure in plant tissue cultures. III. P-protein, sieve area pores and wounding. *Journal of Ultrastructure Research* **82**, 198–211.

Smith, J. A. C. & Milburn, J. A. (1980). Phloem transport, solute flux and the kinetics of sap exudation in *Ricinus communis* L. *Planta* **148**, 35–41.

Sovonick-Dunford, S., Ferrier, J. M. & Dainty, J. (1982). Water relation parameters of the phloem. Determinations of volumetric elastic modulus and membrane conductivity using an applied force method and shrinkage and swelling of tissues in solutions at different osmotic pressure. *Annals of Botany* **51**, 27–37.

Sovonick-Dunford, S., Lee, D. R. & Zimmerman, M. H. (1981). Direct and indirect measurements of phloem turgor pressure in white ash. *Plant Physiology* **68**, 121–6.

Spanner, D. C. (1978). Sieve-plate pores, open or occluded? A critical review. *Plant, Cell and Environment* **1**, 7–20.

Sung, F. J. M. & Kreig, D. R. (1979). Relative sensitivity of photosynthetic assimilation and translocation of ^{14}Carbon to water stress. *Plant Physiology* **64**, 852–6.

Swanson, C. A. (1969). Mechanism of translocation of plant metabolites. In Dinauer, E. C (ed.). *Physiological Aspects of Crop Yield*, p. 167. American Society of Agronomy, Crop Science Society of America, Madison, Wisconsin.

Swanson, C. A. & Geiger, D. R. (1967). Time course of low temperature inhibition of sucrose translocation in sugar beets. *Plant Physiology* **42**, 751–6.

Swanson, C. A., Hoddinott, J. & Sij, J. W. (1976). The effect of selected sink leaf parameters on translocation rates. In Wardlaw, I. F. & Passioura, J. B. (eds.). *Transport and Transfer Processes in Plants*, pp. 347–56. New York, Academic Press.

Tammes, P. M. L. (1933). Observations on the bleeding of palm trees. *Recueil des travaux botaniques néerlandica* **30**, 514–36.

Tammes, P. M. L (1952). On the rate of translocation of the bleeding-sap in the fruit stalk of *Arenga*. *Proceedings Koniklijke Nederlandse Akademie van Wittenschappen* **55**, 141–3.

Tammes, P. M. L., Vonk, C. R. & Van Die, J. (1969). Studies on phloem exudation from *Yucca flaccida* Haw. VII. The effect of cooling on exudation. *Acta Botanica Neerlandica* **18**, 224–9.

Terry, N. & Mortimer, D. C. (1972). Estimation of the rates of mass carbon transfer by leaves of sugar beet. *Canadian Journal of Botany* **50**, 1049–54.

Thoday, D. (1910). Experimental researches on vegetable assimilation. V. A critical examination of SACHS' method for using increase of dry weight as a measure of carbon dioxide assimilation in leaves. *Proceedings of the Royal Society (London) Series B.* **82**, 1–55.

Thorpe, M. R. & Lang, A. (1983). Control of import and export of photosynthate in leaves. *Journal of Experimental Botany* **34**, 231–9.

Thorpe, M. R., Lang, A. & Minchin, P. E. H. (1983). Short-term interactions between flows of photosynthate. *Journal of Experimental Botany* **34**, 10–19.

Thrower, S. L. (1962). Translocation of labelled assimilates in the soybean. *Australian Journal of Biological Science* **15**, 629–49.

Tromp, J. (1977). Growth and mineral nutrition of apple fruits as affected by temperature and relative air humidity. In *Environmental Effects on Crop Physiology*, pp. 101–21. London, Academic Press.

Troughton, J. H., Currie, B. G. & Chang, F. H. (1977). Relation between light level, sucrose concentration and translocation of carbon 11 in *Zea mays* leaves. *Plant Physiology* **59**, 808–20.

Tully, R. E. & Hanson, A. D. (1979). Amino acids translocated from turgid and water-stressed barley leaves. *Plant Physiology* **64**, 460–6.

Ullrich, W. (1961). Zur Sauerstoffabhängigkeit des Transportes in den Siebröhren. *Planta* **57**, 402–29.

Ullrich, W. (1962). Zur Wirkung von Adenosintriphosphat auf den Fluorescein-transport in den Siebröhren. *Planta* **57**, 713–7.

Urquhart, A. A. & Joy, K. W. (1981). Use of phloem exudate technique in the study of amino acid transport in pea plants. *Plant Physiology* **68**, 7504.

Van Die, J. (1975). The use of phloem exudates from several representatives of the Agavaceae and Palmae in the study of translocation of assimilates. In Aronoff, S., Dainty, J., Gorham, P. R., Srivastava, L. M. & Swanson, C. A. *Phloem Transport*, NATO Advanced Study Institute Series. New York, London, Plenum Press.

Van Die, J. & Tammes, P. M. L. (1966). III. Prolonged bleeding from isolated parts of the young inflorescence. *Proceedings Koniklijke Nederlandse akademie van Wetenschappen* **C69**, 648–54.

Van Die, J. & Tammes, P. M. L (1975). Phloem exudation from monocotyledonous axes. In Zimmermann, M. H. & Milburn, J. A. *Phloem Transport*. *Encyclopedia of Plant Physiology*, New Series, Vol. I, pp. 196–222. Heidelberg, Springer-Verlag.

Vogelmann, T. C., Larson, P. R. & Dickson, R. E. (1982). Translocation pathways in the petioles and stem between source and sink leaves of *Populus deltoides* Bartr. ex Marsh. *Planta* **156**, 345–58.

Vreugdenhil, D. (1985). Source-to-sink gradient of potassium in the phloem. *Planta* **163**, 238–40.

Walsh, M. A. & Melaragno, J. E. (1981). Structural evidence of plastid inclusions as a possible sealing mechanism in the phloem monocotyledons. *Journal of Experimental Botany* **32**, 311–20.

Wardlaw, I.F. (1969). The effect of water stress on translocation in relation to photosynthesis and growth. II. Effect during leaf development in *Lolium temulentum* L. *Australian Journal of Biological Science* **22**, 1–16.

Wardlaw, I. F. & Bagnall, D. (1981). Phloem transport and the regulation of growth of *Sorghum bicolor* (Moench) at low temperature. *Plant Physiology* **68**, 311–4.

Warmbrodt, R. D. (1987). Solute concentrations in the phloem and apex of the root of *Zea mays*. *American Journal of Botany* **74**, 394–402.

Watson, B. T. & Wardlaw, I. F. (1981). Metabolism and export of ^{14}C-labelled photosynthate from water-stressed leaves. *Australian Journal of Plant Physiology* **8**, 143–53.

Weatherley, P. E. (1975). Some aspects of the Münch hypothesis. In Aronoff, S., Dainty, J., Gorham, P. R., Srivastava, L. M. & Swanson, C. A. *Phloem Transport*, pp. 535–55. New York, Plenum Press.

Weatherley, P. E., Peel, A. J. & Hill, G. P. (1959). The physiology of the sieve tube. Preliminary investigations using aphid mouth parts. *Journal of Experimental Botany* **10**, 1–16.

Weatherley, P. E., Watson, B. T. (1969). Some low-temperature effects on sieve tube translocation in *Salix viminalis*. *Annals of Botany*, **33**, 845–53.

Webb, J. A. (1970). Translocation of sugars in *Cucurbita melopepo* V. The effect of leaf-blade temperature on assimilation and transport. *Canadian Journal of Botany* **48**, 935–42.

Weber, C., Franke, W. W. & Kartenbeck, J. (1974). Structure and biochemistry of phloem proteins isolated from *Cucurbita maxima*. *Experimental Cell Research* **87**, 79–106.

Wiersum, L. K. (1966). Calcium content of fruits and storage tissues in relation to the mode of water supply. *Acta botanica neerlandica* **15**, 406–18.

Willenbrink, J. (1957). Über die Hemmung des Stofftransports in den Siebröhren durch lokale Inaktivierung verschiedener Atmungsenzyme. *Planta* **48**, 269–342.

Willenbrink, J. (1966). Zur lokale Hemmung des Assimilattransports durch Blausäure. *Zeitschrift für Pflanzenphysiologie* **55**, 119–30.

Wright, J. P. & Fisher, D. B. (1980). Direct measurement of sieve tube turgor pressure using severed aphid stylets. *Plant Physiology* **65**, 1133–5.

Wright, J. P. & Fisher, D. B. (1983). Estimation of the volumetric elastic modulus and membrane hydraulic conductivity of willow sieve tubes. *Plant Physiology* **73**, 1042–7.

Ziegler, H. (1956). Untersuchungen über die Leitung und Sekretion der Assimilate. *Planta* **47**, 447–500.

Ziegler, H. (1958). Über die Atmung und den Stofftransport in den isolierten Leitbündeln der Blattstiele von *Heracleum mantegazzianum* Somm. et Lev. *Planta* **51**, 186–200.

Ziegler, H. (1975). Nature of transport substances. In Zimmermann, M. H. & Milburn, J. A. (eds.). *Transport in Plants. Encyclopedia of Plant Physiology*, New Series, Vol. 1, pp. 59–100. Heidelberg, Springer-Verlag.

Zimmermann, M. H. (1957). Translocation of organic substances in trees. II. On the translocation mechanism in the phloem of white ash (*Fraxinus americana* L.). *Plant Physiology* **32**, 399–90.

Zimmermann, M. H. (1960). Absorption and Translocation: transport in the phloem. *Annual Review of Plant Physiology* **11**, 167–9.

Zimmermann, M. H. (1961). Movement of organic substances in trees. *Science* **133**, 73–9.

Zimmermann, M. H. (1971). Transport in the phloem. In Zimmermann, M. H. & Brown, C. L. (eds.). *Trees, Structure and Function*, pp. 221–79. Berlin, Heidelberg, New York, Springer.

Zimmermann, U., Rade, H. & Steudle, E. (1969). Kontinuierliche Druckmessung in Pflanzenzellen. *Naturwissenschaften* **56**, 634.

8 Source/sink regulation

L. C. Ho, R. I. Grange and A. F. Shaw

8.1 Introduction

Previous chapters dealt with the anatomical, biochemical and biophysical characteristics of photoassimilate transport. The advances made in these areas have improved understanding of the mechanisms regulating transport, but as the environmental responses of these mechanisms may be determined or affected differently within the same plant, it is necessary to consider their integration to provide an overview of transport systems in plants.

A prime objective of crop physiological research is to maximize the harvest index. Where environmental factors (i.e. water, mineral nutrients, light, temperature and atmospheric carbon dioxide) can be largely controlled, as in the protected crops industry, crop yields increase in proportion to total plant growth. However, for field crops, increases in yield have come mainly from changes in the partitioning of assimilates and nutrients by reducing the growth of non-harvestable sinks through plant breeding rather than through the optimization of growing conditions (Gifford, *et al.* 1984). The potential yield of field crops is ultimately regulated by light interception and canopy photosynthetic efficiency. However, there is little evidence that significant increases in maximum rates of photosynthesis (per unit leaf area) have occurred during plant selection (Gifford & Evans, 1981). As options for further increases in crop yield will necessitate improvements in dry matter partitioning and use, the role of the sources and sinks in controlling partitioning in whole plants should be better understood.

It is the aim of this chapter to re-assess the current concepts of source and sink and the possible mechanisms regulating source/sink interactions. Based on this assessment, the potentials or limitations for assimilate partitioning as a means of improving crop productivity can be defined. Also, we discuss how source and sink strengths are determined and how source, sink, the pathway and relevant environmental factors affect the performance of the sources and sinks, i.e. the *apparent* source or sink strength. Additionally, we examine how the synthesis and remobilization in the source organs and the use or storage of assimilates in the sink organs determines the intrinsic ability of the source or sink, i.e. the *potential* source or sink strength.

8.2 Whole plant source/sink relationships

8.2.1 *Source/sink concepts*

The application of the terms 'source' and 'sink' has gained considerable extension from their early use in translocation studies (Mason & Maskell, 1928). Generally, 'source' and 'sink' are functional descriptions of plant organs and tissues, recognizing their ability to supply or use a particular metabolic substance. Intrinsic to the concept is physical continuity between source and sink by vascular connection. Additional descriptive nomenclature is found in the literature (Fig. 8.1). As the terms are often used in several contexts (i.e. metabolism, morphology and transport), methods for their definition and quantification have been proposed (Warren Wilson,

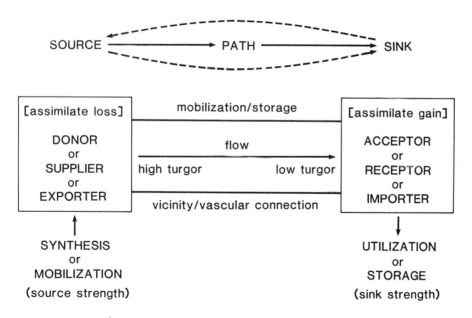

Fig. 8.1 Source/sink relationships for photoassimilates in a transport system. Both source and sink organs interact through the pathway to regulate the balance between supply and demand.

1967, 1972). Source and sink can be defined in terms of supply (losses) and demand (gains) of a particular substance in a particular plant part. Therefore, for transport, source/sink status can be described as the product of the differences between the gains (e.g. photosynthesis) and losses (e.g. respiration), by influx from and efflux to the plant environment and gains (import) and losses (export) by phloem translocation. However, the formal treatment and investigation of whole plant source/sink relations by extension of growth analysis techniques (Warren Wilson, 1972) has not been greatly developed. In this chapter we define source organs as net exporters and sink organs as net importers. Therefore, export from a leaf can be measured as:

$$T = (P\text{--}R) - \triangle C \tag{8.1}$$

where P is carbon gained by photosynthesis, R is carbon lost by respiration and $\triangle C$ is the net gain of carbon by the leaf (Ho, 1976a). Similarly, import into a fruit can be measured as

$$T = R + \triangle C \tag{8.2}$$

where $\triangle C$ is the net gain of carbon by the fruit (Walker & Ho, 1977a). The great 'plasticity' of plant systems in their ability to activate rapidly new sources and sinks to accommodate changes in their source/sink status, limits the categorical description of a plant organ unless a temporal parameter is introduced into source/sink definitions. However, there are intrinsic characteristics of source and sink organs which determine their capacities to export and import.

8.2.2 Source/sink status during plant development

The source/sink status of a plant is dynamic, since an organ may change from being a consumer (net importer) to a supplier (net exporter) or vice versa.

8.2.2.1 Seedlings and developing leaves

During plant development the supply of metabolites for growth and differentiation originates from various parts of the plant. Likewise, the demand for metabolites will change as plant parts grow, mature and senesce. A plant seed normally contains materials which are used during germination. These are frequently present in an endosperm, (e.g. tomato, maize and wheat) or in the cotyledons of the embryo (e.g. peas and lettuce). Changes in chemical composition in different parts of the seed during germination (Oota *et al.*, 1953) demonstrate a sequential mobilization of reserves. Seedlings are classified as epigeal, in which the cotyledons are above ground and usually photosynthetic (e.g. *Ricinus communis*), and hypogeal, in which the cotyledons remain below ground (e.g. *Phaseolus multiflorus*).

In the latter case the cotyledons are either themselves the source of reserves for the seedling, or receive materials from the endosperm. Eventually the first true leaves develop and these gradually attain a capacity for photosynthesis. As the young leaf grows, the physiological status of the organ changes from being a net consumer of assimilates (i.e. a sink) to being a net supplier (i.e. a source) and the general pattern of translocation is reversed (Wardlaw, 1968). The transition of a leaf from a sink to a source is associated with the development of the capacity to produce mobile assimilates and the development of a positive carbon balance between synthesis and use within the leaf (Turgeon & Webb, 1975; Giaquinta, 1978; Loescher *et al.*, 1982).

In the tomato, compound leaves become net exporters (sources) of carbon metabolites when they reach about 10% of their final area, individual leaflets being some 30% of their final leaf area (Ho & Shaw, 1977). A similar developmental period for leaf physiological transition has been reported for cucumber (Ho *et al.*, 1984), soybean (Thrower, 1967), and sugar-beet (Fellows & Geiger, 1974). In squash, sugar-beet and tobacco leaves (Turgeon & Webb, 1973) the capacity for importing ^{14}C-labelled photoassimilate is lost basipetally during expansion. In squash, import into the lamina tip occurred until the leaf was 10% expanded; import to the lamina base ceased when the blade was 45% expanded and net export occurred when the lamina was 35% expanded. There was a basipetal development of the capacity to export which immediately followed the loss of import capacity. In the tomato leaf (Ho & Shaw, 1977), photoassimilates exported from the mature terminal leaflets were initially redistributed to the immature basipetal leaflets. At the time of physiological transition both export and import of carbon may be occurring simultaneously (Isebrands & Larson, 1973; Ho & Shaw, 1977, 1979) with overlaps in physical and metabolic development within the leaf. In sugar-beet, for example, the onset of export is not correlated with any dramatic changes in the structural transformation within the phloem of the minor leaf veins prior to initiation of phloem loading (Fellows & Geiger, 1974). There is a gradual differentiation of phloem, commencing acropetally in the leaf, resulting in a large proportion of the minor vein sieve elements reaching maturity together. The decline in import and the onset of export may be regulated independently. In grafted 'albino' tobacco leaves the import of assimilates ceased at the same leaf age as in normal leaves despite the lack of photosynthesis and carbon export (Turgeon, 1984a). As the efflux of imported assimilates from the phloem tissue was very small, a physical interruption of the import pathway, or at the point of unloading, may cause the import to stop (Turgeon, 1984b; Blechschmidt-Schneider & Eschrich, 1985; see Ch. 6). Therefore the transition of a growing leaf from a sink to a source is mainly regulated by the physiological processes providing the translocate and by the anatomical development of the pathway for import or export.

8.2.2.2 Vegetative and reproductive growth

Prior to the transition from predominantly vegetative to reproductive growth, young leaves and roots compete proportionately for assimilates. However, once fruit set occurs, the additional demand for assimilates can severely affect dry matter distribution. Studies on the competition between fruiting and vegetative growth phases in the red raspberry (Waister *et al.*, 1977) demonstrated that, in a biennial cropping system, the absence of vegetative canes increased yields from fruiting canes. In the absence of fruiting canes, more vegetative canes were produced. Over five years, the annual removal of first-flush, vegetative canes increased the yield of fruit by 38% (Lawson & Wiseman, 1983). Increased yield of fruit by second-flush canes involved increases in the numbers of cropping nodes per cane and the production of more, and bigger, berries at each cropping node. In tomato, the rate of vegetative growth decreases to a minimum following the onset of fruiting (Salter, 1958). Similarly, root growth ceases four weeks after first anthesis and the growth of leaves decreases substantially at the time of maximum fruit growth rate (Hurd *et al.*, 1979). However, reproductive growth is not always at an advantage over vegetative growth; for example, under conditions of limited assimilate supply young leaves will develop while inflorescences abort (Kinet, 1977a). The competition for assimilates between vegetative and reproductive sinks is affected by the developmental stage of the reproductive sink.

Competition for assimilates also occurs between reproductive sinks. In wheat, the greater the relative size of grain and their proximity to a source leaf, the greater the amount of current assimilates received by those sinks (Cook & Evans, 1978). Similarly, in tomato, the position of fruits in a truss (Ho, 1980) and the induction sequence of fruit set (Bangerth & Ho, 1984) affect the import rate of assimilates into the fruits of the same truss. First-induced fruits are larger within the same truss and fruit thinning experiments suggest that fruits nearer the source had a higher potential to import assimilates.

8.2.3 Regulation of transport between sources and sinks

In the Münch osmotic pressure flow theory, solute secretion into the seive tubes at the source end of the translocation system provides the necessary driving force (i.e. osmotic pressure) for transport. Since water is transported along with solutes, its return via the xylem was proposed to complete a circulatory pathway. Elaborating the simple Münch osmotic pressure flow model to intact plants requires that, within the phloem transport system, a pressure gradient is maintained between regions of high osmotic pressure (sites of synthesis: sources) and regions of low osmotic pressure (sites of utilization: sinks). In soybean (Fisher, 1978) the maximum turgor pressure difference between leaves and roots was

400 kPa, which was sufficient to drive a pressure flow at an observed velocity of 8–9 mm min^{-1}. A similar pressure difference (500–700 kPa) between source and sink has been reported in sugar-beet (Fellows & Geiger, 1974) and in *Ricinus*, 400 kPa (Milburn, 1974). For larger herbs and trees, a range of gradients has been reported: 25 kPa m^{-1} for tall trees (Zimmermann, 1971), 35 kPa m^{-1} for *Quercus* (Hammel, 1968), and 550 kPa m^{-1} for *Fraxinus* (Dixon, 1933).

Without a continuing supply of assimilates through synthesis or remobilization in source regions, and conversely their use or compartmentation at sink regions, the Münch pressure flow model would ultimately reach equilibrium. When osmotic gradients between sources and sinks approached zero, transport would cease. This aspect of the model is considered fundamental to the source–sink concept as an explanation for whole plant assimilate distribution. Although it is widely accepted that the main transport mechanism driving pressure flow is osmosis through essentially passive sieve tubes (Weatherley & Johnson, 1968), active processes of phloem loading or unloading may also be involved (Ho & Baker, 1982). The pressure gradient between sources and sinks may also be regulated by phytochrome and plant growth regulators (PGRs) (Lüttge & Higinbotham, 1979). Potential control sites for PGR regulation of assimilate transport appear to be predominantly associated with sink tissues. PGRs have been shown to induce volume change in sink organs (Patrick & Wareing, 1980). In tomato plants grown in low light, assimilate supply to the aborting inflorescence was increased at the expense of the apical shoot by applying PGRs (a mixture of cytokinins and gibberellins) to the inflorescence (Leonard et al., 1983). As a result the development of these tomato plants was improved. Kinet (1977b) suggested that the promoting effect of BA was due to redirection of assimilate flow from the apex to the inflorescence while GA specifically stimulated inflorescence development.

However, in the absence of growth, PGRs may influence assimilate transport by altering cellular metabolism; inhibition of protein and nucleic acid synthesis prevented hormonal stimulation of assimilate transport in decapitated strawberry stolons (daCruz & Audus, 1978). Regulation of sugar-proton co-transport in sinks (i.e. pea stems, maize coleoptiles and roots) by the action of IAA and/or fusicoccin has also been reported (Colombo et al., 1978), and the possible role of ABA in sink tissues for assimilate unloading from the phloem has been proposed (Tanner, 1980). Studies on *in vitro* rates of sucrose uptake by PGR-treated bean stems (Patrick & Wareing, 1976) and estimations of *in vivo*, apoplastic sugar pool sizes (Patrick, 1982), have suggested that IAA and kinetin act principally on phloem transport processes, in contrast to GA$_3$ which exerts a dual action on phloem transport and sink metabolism.

Shifts in the hormonal balance within plants have been linked with changes in assimilate partitioning between leaves, stems and roots in soybean (Kasperbauer et al., 1984), tobacco (Kasperbauer, 1971) and

Phaseolus vulgaris (Downs, *et al.*, 1957). A comparison of dry matter partitioning in field-grown soybean (from spacing experiments) and cabinet-grown plants (from light quality experiments) suggested a correlation between phytochrome induction and endogenous PGR balance (Kasperbauer *et al.*, 1984); plants supplied with a red light treatment at the end of the photoperiod partitioned less dry matter to stems than to roots, when compared with plants receiving a far-red light treatment. The light effects were reversible if alternative light treatments were given sequentially. However, bulb formation in onion was apparently associated with increased far-red light treatment (Austin, 1972). Phytochrome is intimately associated with cellular membranes (Marmé, 1977) and several examples of the regulation of membrane permeability (i.e. polarization, depolarization) through red/far-red light treatments are reported (Lüttge & Higinbotham, 1978). The interaction between membrane transport and PGR regulation of dry matter distribution, through growth or metabolic adjustments in sources and sinks, remains controversial.

8.3 Regulation of carbon allocation in leaves

The ability of a leaf to export assimilates (i.e. source strength) can be seen as the balance between carbon assimilation and carbon use by the leaf. The potential source strength of a leaf will be determined mainly by the potential photosynthesis rate, and the apparent source strength will be affected by current environmental conditions and the developmental stage of the leaf.

This section discusses the fate of carbon assimilated by leaves, particularly the control of carbon allocation to the leaf (i.e. its growth and maintenance) and to the demand from the rest of the plant (i.e. export). By carbon allocation we mean the proportions of carbon fixed by the leaf which are allocated to storage in the leaf, to immediate export from the leaf and to respiration or growth of the leaf itself.

There have been a number of excellent recent reviews of the control of export and carbon allocation in leaves (Geiger, 1979; Geiger & Giaquinta, 1982; Geiger & Fondy, 1985; Geiger, 1986). These reviews will be updated and extended in an attempt to assess the role of carbon allocation in linking photosynthesis and export within the leaf.

8.3.1 Carbon allocation and photosynthesis

The pathway of carbon assimilated in photosynthesis has been described in detail and many of the biochemical controls are well understood (see Ch. 1; Geiger & Giaquinta, 1982; Horton, 1985). The control of triose phosphate export from the chloroplast and the control of sucrose synthesis have been partly elucidated and models presented (see Ch. 1). These models can account for the short-term control of carbon allocation (Stitt, 1986). The metabolic controls of starch synthesis and breakdown may also be involved in the control of export.

Once carbon has left the chloroplast it may be stored in the vacuole, often as soluble sugars, used for growth and maintenance within the leaf, or exported from the cell (and the leaf). In terms of carbon mass, the most important fraction in fully expanded leaves is that exported. Therefore, carbon export influences the proportion of carbon allocated to storage as starch or soluble sugars, The control of export may be achieved at a number of points in the pathway from chloroplast to sink:

1. Carbon may be retained in the chloroplast by the regulation of starch metabolism, preventing release of assimilates to the export pool.
2. The amount of sucrose available for export may be regulated by the rate of sucrose production.
3. The two processes of sucrose release from the mesophyll and its subsequent loading by the phloem may be regulated.
4. The allocation of carbon to other metabolic processes in the leaf may divert carbon from export.

8.3.1.1 Starch metabolism

Export may be affected by the rate of starch breakdown or the rate of starch synthesis.

8.3.1.1.1 Starch breakdown Some of the carbon exported from leaves during the night comes from the breakdown of starch. Thus the rate of export at night or in low light could be dependent on the processes of starch breakdown, particularly in plants which store large amounts of starch during the day. However, there is no suitable mechanism to explain the regulation of starch breakdown (Stitt, 1984).

Many leaves show a lag in starch breakdown at the start of the night which is apparently dependent on the level of soluble sugars in the leaf (Geiger & Batey, 1967; Gordon *et al.*, 1980; Fondy & Geiger, 1982; Grange, 1985a, b), such that the reserve sucrose (or hexose) content declines to a critical value before starch degradation can begin. There may be some control of breakdown related to the level of sugars. In sugar-beet (Fondy & Geiger, 1982), tomato, (Ho, 1976a, b) and pepper (Grange, 1985a, b) the rate of export at night, or in low light, is closely associated with the rate of starch mobilization. Thus, starch breakdown could be regulated to maintain approximately constant export rates throughout the night. It is difficult to determine whether the export rate determines starch breakdown rate or whether some control of starch breakdown rate determines the availability of sucrose for export. However, in pepper leaves grown in short days the rate of starch breakdown towards the end of the night declines as the amount of starch decreases, despite the presence of a significant amount of starch (Grange, 1985b). The relationship between starch breakdown and export suggests that starch breakdown rate can limit export rate.

Mechanisms for controlling starch synthesis are known (Priess, 1982; Stitt, 1984), so that starch content could be controlled by changing the synthesis rate against a continuous degradation rate. Chan & Bird (1960) and Stitt & Heldt (1981) reported starch breakdown in the light in tobacco leaves and in isolated spinach chloroplasts respectively. However, in the former experiment and in an earlier example (Porter *et al.*, 1959), there could have been endogenously regulated sequential synthesis and breakdown during the experimental period. Recently, Kruger *et al.*, (1983) were unable to demonstrate turnover during synthesis of starch in the light using a ^{14}C pulse labelling technique. Recent experiments with constant specific activity ^{14}C labelling of attached leaves (Fox & Geiger, 1984; Grange, 1984) showed no simultaneous synthesis and breakdown of starch unless the rate of leaf photosynthesis was reduced by a low CO_2 concentration. Thus, for attached leaves, the weight of evidence suggests an absence of degradation during synthesis of starch in the light and is evidence for some control of starch degradation. Nonetheless, starch can be degraded rapidly in the light if the demand for carbon export exceeds current carbon fixation (e.g. Ho, 1977; Grange, 1985a). Further work on the control of starch breakdown is needed to understand fully the implications for the control of export.

8.3.1.1.2   Starch synthesis The production of starch by the chloroplast prevents the release of a proportion of the assimilated carbon for export. Thus starch synthesis and export are in competition, and factors affecting the rate of starch synthesis will affect export. Environmental signals, notably daylength, appear to influence the rate of starch accumulation. Such signals might alter the rate of starch synthesis directly or indirectly by changing the activity of sinks or the processes of export.

Britz *et al.*, (1985) found changes in starch accumulation rate after only one inductive short day in the grass *Digitaria decumbens*. It is difficult to envisage developmental changes in sinks which could affect export rate in such a short time and thus indirectly alter the rate of starch accumulation. Moreover, Britz (1986) has also shown that the effects of daylength on diurnal patterns of net photosynthesis, starch and sugar accumulation in *Sorghum bicolor* are determined by three different circadian rhythms. Such photoperiodic responses are thought to be effected through changes in membranes (Marmé, 1977; Cleland, 1984; Friend, 1984). As many of the export processes in leaves are dependent on membrane transport (exchange of triose phosphate and inorganic phosphate at the chloroplast envelope, efflux from mesophyll cells, sugar loading at the sieve elements), it is possible that photoperiodic effects operate at these points. However, rapid and clear responses such as those reported above are unusual.

Nonetheless, many species have a higher starch accumulation rate when *grown* in short days (Hewitt *et al.*, 1985) and starch accumulation rates increase after *transfer* to short days (Chatterton & Silvius, 1979, 1980a, b; Sicher *et al.*, 1984). In these cases, it is more likely that the environmental changes affect processes further along the pathway from chloroplast to sink. Geiger *et al.*, (1985) suggested that plants grown in short days or gradually transferred to short days, behaved differently from plants abruptly transferred. They did not observe a sustained increase in starch accumulation and proposed that the changes in starch accumulation observed in transfer experiments resulted from stress. Disruption in the growth and development of sinks may change source leaf export and carbon allocation. Similarly, Grange (1985b) observed no significant effect of daylength on starch accumulation in pepper leaves grown in different daylengths at the same irradiance or with the same daily irradiance integral.

The average rate of starch accumulation over the light period is lower in long days, but the *maximum* rate during the day is similar (Challa, 1976; Chatterton & Silvius, 1979; Hewitt *et al.*, 1985). The effect of long days is to increase the 'lag' period in starch accumulation at the start of the day and to increase the period of decreased starch accumulation before dark. The early lag in starch accumulation is ascribed to the preferential allocation of carbon to soluble sugars, filling the 'export' pool before the starch reserves are augmented. In short-day plants both pools receive carbon simultaneously (Hewitt *et al.*, 1985). Long-day plants would export much of the carbon which is fixed early in the day and thus cause a lag in starch accumulation. Thus, daylength is probably affecting the rate of export rather than starch accumulation directly. If the export rate is decreased in short days without concomitant changes in assimilation then starch accumulation will increase. Moreover, an increase in soluble sugars in short days (Grange, 1985b; Hewitt *et al.*, 1985) suggests an oversupply of assimilates.

Thus starch metabolism is probably regulated indirectly through environmental and endogenous regulation of export capacity or through sink demand. The control of starch metabolism itself is unlikely to regulate export except in extreme circumstances, e.g. when starch is the main source of carbon for export during low light or in darkness.

8.3.1.2 The production of sucrose

The synthesis of sucrose depends on a number of processes. In light, where fixation exceeds the rate of export from the leaf, sucrose synthesis will be controlled by the pathway proposed by Stitt (1986) and Huber *et al.*, (1986); see Ch. 1.

Huber *et al.*, (1986) have reviewed the control features of sucrose phosphate synthase (SPS) and its interaction with other important enzymes in the metabolism of sucrose in the leaf. They proposed that SPS activity is regulated at the level of 'fine control' by metabolic effectors which bind to the enzyme and modify activity slightly, and at the level of 'coarse control' by endogenous rhythms which regulate the synthesis of SPS (Kerr & Huber, 1987). There seems no doubt that the sucrose synthesis is important in controlling export rates in many species. However, in the dark (Rufty & Huber, 1983; Hendrix & Huber, 1986), or in low light (Grange, 1985a), export may be governed by other events, such as the rate of starch degradation or the release of soluble reserves (e.g. vacuolar sugars).

Because sugars are compartmentalized in the leaf there are difficulties in interpreting changes of gross sugar content in the leaf with changing enzyme activities. The proportion of sugars contained in the vacuole has been estimated by labelling techniques (Geiger *et al.*, 1983), but determining compartmentation throughout the leaf is more difficult (Fisher, 1975; Fisher *et al.*, 1978). More importantly, enzyme studies should be accompanied by adequate estimates of total sucrose production, including that exported.

Moreover, there is evidence from tomato (Hammond *et al.*, 1984; Shaw *et al.*, 1986) that hexoses may have a role in controlling sucrose production. In tomato leaves, the sucrose pool is small and changes little when the plant is subjected to a range of treatments to alter the source/sink ratio. Little evidence was found of changes in sucrose in response to changes in the source/sink ratio after plants were surgically manipulated. Shaw (1985) proposed that hexose may have importance other than as a soluble reserve and quoted the observations of Stitt (1985) that glucose was more effective in stimulating the synthesis of fructose-2,6-bisphosphate and glucose-6-phosphate than either sucrose or sorbitol. Fructose-2,6-bisphosphate is a potent inhibitor of fructose bisphosphatase, the enzyme catalysing the first irreversible step in the synthesis of sucrose; see Ch. 1. In both tomato and pepper (Shaw *et al.*, 1986; Grange, 1985a, b, 1987) hexose levels change with a variety of treatments in ways which suggest a role other than simple 'overflow' storage. A diurnal rhythm has been observed in both these species and in cotton (Hendrix & Huber, 1986), whereby hexose levels rose during the morning and fell after the middle of the day. In addition, pepper leaves grown in short days, at high or low irradiance, show increased levels of hexoses and faster accumulation during the day. It is still not certain whether hexoses are derived entirely from recently synthesized sucrose or from other sources.

8.3.1.3 The loading of sucrose

As discussed in Ch. 5, there is ample evidence that sucrose loading is an important determinant of the rate of export. Export may be controlled by the loading process at two points: during the release of sucrose from the mesophyll cell and by uptake at the sieve elements. There is also evidence that sucrose transport may be symplastic (see Ch. 2), allowing further possibilities for controlling export.

8.3.1.3.1 *Release from the mesophyll cell*

If sucrose is exported to the free space before being loaded into the sieve elements, then the plasmalemma of mesophyll cells presents an important control point. Doman & Geiger (1979) have obtained evidence in sugar-beet that the release from the mesophyll can be influenced by potassium ions. Geiger (1979) has further suggested that the potassium status (and possibly also the phosphorus and nitrogen status) of the free space may be an important signal of the plant's nutrient status. This may control sucrose efflux and act as a 'feedforward' response to increase the rate of export and thus subsequent plant growth, made possible by abundant nutrient levels. However, further work (Conti & Geiger, 1982; Geiger & Conti, 1983) showed no effect of abundant K^+ on translocation. They suggested that increased K^+ accumulation by the sinks and mature leaves may cause a greater rate of synthetic metabolism, which in turn causes increased export from the source leaf. This was not consistent with the feedforward mechanism. Moreover, K^+ has been implicated in the loading process at the sieve element (Baker & Chaudry, 1986). It is still unclear whether there is any regulation of sucrose release into the mesophyll and whether it has any role in the control of export.

8.3.1.3.2 *Uptake at the sieve element – phloem loading*

Most of the work on the kinetics and transmembrane transport of phloem loading has been done on isolated tissues, protoplasts or unicellular organisms, and the mechanism of proton-sugar co-transport has been based on results from this material (see Ch. 5). Where heterogeneous plant tissues have been used there are difficulties in interpreting the results (Komor & Orlich, 1986). The weight of evidence favours proton-sucrose co-transport in many tissues but probably only in *Ricinus* cotyledons has such a system been unequivocally demonstrated, and even then only in certain cells. Some plant growth regulators (e.g. auxin, ABA, kinetin and gibberellin) have been shown to influence proton extrusion systems (Baker, 1985). There is also evidence that they have such a role in intact systems (Aloni *et al.*, 1986). Baker (1985) concluded that the loading process *per se* did not limit phloem transport and was capable of responding to a wide range of sucrose concentrations and changes in supply and demand.

The loading system may respond to sieve tube turgor (Geiger, 1979; Smith & Milburn, 1980). Osmotic potentials are much lower in minor vein cells than in mesophyll cells and transport of exogenously supplied sucrose was initially decreased, eventually recovered, in leaves where the free space was equilibrated with 800 mol m^{-3} (0.8 M) mannitol (Geiger *et al.*, 1974). The decreased turgor of the sieve tubes was purported to have initiated increased loading of the sieve tubes, thus restoring transport. The turgor-driven system proposed by Münch works equally well with any osmotically active solute (Lang, 1983); see Ch. 8. In fact, K$^+$ is osmotically more important than sucrose (Grange, 1978). Loading could be influenced both by sucrose and K$^+$ status independently of the effect of K$^+$ on sucrose efflux from mesophyll cells. If source leaves have abundant K$^+$, it may be loaded along with sucrose and unloaded at the sink or along the pathway, increasing turgor gradients and enhancing transport; K$^+$ gradients may be another mechanism by which the plant controls partitioning (Lang, 1983).

8.3.1.4 Allocation of carbon to other metabolic pathways

8.3.1.4.1 Dark respiration A considerable proportion of the carbon fixed during a day can be lost in dark respiration during the following night. In exporting leaves it can represent an important loss of total carbon or of the previous day's assimilation. The loss of carbon from a source leaf was found to be over 10% of that fixed during the previous light period, if leaves were subjected to high irradiance in short days (Grange, 1985b). Losses from leaves at moderate irradiance and in long days were less than 5%.

There is evidence that the rate of respiration may be dependent on the carbohydrate status of the leaf or the irradiance prevailing during the day (Ludwig *et al.*, 1975; Azcon-Bieto, 1984). This is generally termed maintenance respiration, and is that required to sustain vital metabolism even in mature leaves. Energy may also be required to operate the processes of loading and transport (Ho & Thornley, 1978). However, Grange (1985b) did not find evidence for such a relationship in pepper leaves. Some of the respiratory losses may be 'wasted' by poorly controlled metabolism operating when some soluble metabolites are in excess (Azcon-Bieto, 1984). The existence of such a pathway of respiration has important implications for the loss of potential export when the supply of assimilates is greater than the current demand.

8.3.1.4.2 Organic and amino acids The diurnal production of organic acids has been determined by Steer (1973) and Grange (1985a, b) in pepper and by Shaw (1985) in tomato. Malic acid is the largest fraction and is synthesized from newly fixed carbon. It appears to mix with an existing pool and is not exported to any significant extent during the dark period. The function of this compound is probably to balance the accumulated potassium ions and to maintain osmotic and pH balance in the leaf (Morgan, 1984). The production of malate is also involved in nitrate transport and reduction (Neyra & Hagemann, 1976; Blevins *et al.*, 1978).

The production of amino acids has been investigated in legumes (see Pate, 1980, 1986 and Ch. 4) where the exchange of newly fixed carbon (exported to the root) for microbially synthesized amino acids occurs rapidly. However, this is a special case and the factors governing the allocation of carbon in mature non-leguminous leaves has not been examined in detail. Some results from pulse-labelled leaves have been reported (Pearson & Steer, 1977; Shaw, 1985) which suggest that the amount of carbon used for this purpose is small. The diurnal course of amino acid production has been detailed by Steer (1973) in expanding pepper leaves. When growth is rapid in young leaves there is a considerable production of amino acids, with up to 40% of recently fixed carbon being diverted to amino acid synthesis. In slightly older leaves (Steer, 1974a, b) the proportion falls to less than 20% and in mature leaves it is below 10% of the currently fixed carbon. Thus, although amino acid production is important metabolically (for protein synthesis and maintenance) it does not greatly influence carbon export rates from mature source leaves.

8.3.2 The control of export

Many of the initial steps in carbon export are short-term controls to ensure efficient matching of carbon supply to export. The pathways of sucrose synthesis respond rapidly and sensitively to small fluctuations in the levels of intermediates and end-products. These systems ensure that the concentrations of the intermediates are held at optimum values and, more importantly, that sucrose supply is maintained under most of the environmental conditions to which the leaf is exposed. However, major persistent changes in the environment probably result in long-term adaptation; regulation is probably provided by the processes at the loading step. Changes here affect the previous chain of synthesis by feedback mechanisms and adjustments are made to optimize the production of sucrose and starch consistent with the new rate of export. There is good evidence for such a chain of regulation operating when changes in export rate are observed during changes in photosynthetic rate or sink demand.

8.3.2.1 The effect of photosynthetic rate on export

Export rate is proportional to photosynthetic rate in both short-term (Servaites & Geiger, 1974; Ho, 1976a; Gordon *et al.*, 1980; Grange, 1985a) and long-term experiments (Ho, 1978a, 1979). Most of the short-term relationships reported are for photosynthetic rates below or just above those measured in the growth environment. Although some experiments have included irradiances above that imposed during growth, this has not always resulted in significantly higher assimilation rates. This proportional relationship between export and fixation is probably controlled by the short-term mechanisms described above. The relationship breaks down only when fixation is low and reserves are mobilized, or when fixation suddenly exceeds the levels to which the leaf is acclimatized.

The data of Servaites & Geiger (1974), for sugar-beet, are often cited as supporting a direct relationship between export and assimilation, particularly as the source leaf assimilation rate was substantially increased. However, export from the source leaf was not measured directly but was inferred from the import rate of ^{14}C by a sink leaf during constant specific activity ^{14}C labelling of the source leaf. Recent work has shown that import rates may change independently of export rates (Geiger & Fondy, 1985). Moreover, such a response is not typical of other species. Silvius *et al.*, (1979) found no increase in export rate from soybean leaves over two days if plants previously grown in moderate irradiance were exposed to a high irradiance. The extra carbon fixed accumulated as starch. In contrast, the export rate from tomato leaves did increase when plants were transferred to a CO_2-enriched atmosphere (Ho, 1977, 1978a). However, the change was slow, taking about three days for the export rate to increase in proportion to the assimilation rate, even though the photosynthetic rate increased by 40% immediately on transfer. Huber *et al.*, (1984a) found no change in export rates from soybean leaves transferred to a CO_2-enriched atmosphere after even longer-term exposure. Again, the increased accumulation of carbon was stored as starch. Similarly, Bell & Incoll (1982) concluded that export from wheat leaves in the field could be described by saturation (Michaelis–Menten) kinetics. Recently, Grange (1987) showed that mature pepper leaves required about three days for a large increase in assimilation rate to cause a proportional increase in export rate.

Huber *et al.* (1984a) attributed the slow change in export rates to lack of adaptation in the activity of SPS. However, the treatments which increased the rate of carbon accumulation also caused the accumulation of sucrose in the leaf. Thus SPS activity was sufficient to produce sucrose for export but some process beyond sucrose synthesis was limiting. Grange (1987) also concluded that the limitation to export in pepper leaves was at the loading step because there were increases in sugar levels before the export rate became adapted to the increase in assimilation.

8.3.2.2 The effect of sink demand on export and carbon allocation

It is apparent from many experiments that the yield of most crop plants is source-limited. Slack & Calvert (1977) showed that removal of a truss from tomato plants resulted in yield increases in the remaining trusses. Nevertheless, there are many instances when dry matter production is in excess of that currently required by major sinks, and other plant organs may act as secondary storage sinks (Ho, 1979).

The demand of the sink organs affects export from leaves by feedback mechanisms acting through the processes discussed above. However, investigations of the interactions between individual sources and sinks are made difficult by the complex and extensive buffering capacity of most plant systems (Lang, 1978; Shaw, 1985). Moreover, the sink/source system is very plastic; when the demand from a particular sink is changed the redirection of assimilates to other sinks may be such that export is maintained constant. Indeed, it is only in longer-term experiments that real changes in export are seen in intact systems. Thorne & Koller (1974) increased the source/sink ratio by shading all but one of the source leaves on soybean plants. The export of labelled assimilates increased 8 d after treatment and it is uncertain how rapidly the source leaf responded. Further experiments by Fondy & Geiger (1980), using *Phaseolus*, showed that export rate did not always increase immediately and that, in any case, increases were small. In similar experiments on sugar-beet plants (Fondy & Geiger, 1980) the rate of export was not increased by darkening all but one source leaf. The distribution of assimilates amongst the sinks changed rapidly but both the amount of recently fixed assimilates and the absolute amount exported were maintained within narrow limits.

Treatments which remove sinks, particularly when they are anatomically close to the source, can reduce export from the source leaf within hours (Ho *et al.*, 1983; Hammond *et al.*, 1984). If the sink is further away, other sinks can accommodate the exported carbon and export takes longer to decrease. In fact, removal of the truss in a single source leaf/single truss system did not decrease export within the first 24 h, and the exported assimilates were found in the petiole and stem between the leaf and truss. Export declined after this time, but assimilation was unchanged, resulting in starch and sugar accumulation in the source leaf. Responses of photosynthesis to decreased export may also be a function of leaf age. In older leaves where starch levels are already high, reduced export may affect photosynthesis more rapidly (Shaw, 1985).

An important effect of removing sinks is the possible inhibition of photosynthesis and the resulting disturbances to source leaf physiology. The subject has been extensively reviewed (Neales & Incoll, 1968; Pinto, 1980; Geiger & Giaquinta, 1982) and will not be detailed here. Geiger & Giaquinta (1982) concluded that most of the experimental evidence did not favour a simple feedback response whereby decreased export caused the accumulation of photosynthetic products and an inhibition of photosynthesis. Changes in photosynthesis occurred slowly in response to decreases in export, with some species avoiding inhibition of photosynthesis by accumulating starch and sugars. Where changes in both export and photosynthesis did occur there were other explanations for the correlation.

Presumably the slow response of export rate to increased assimilation or changes in sink demand is an adaptation to prevent inefficient excess of loading capacity. If an increase in export is required then the other systems within the leaf have time to adjust (Thorne & Koller, 1974).

8.3.3 *Diurnal allocation*

Much of what we know about the allocation of carbon in leaves has been gained using relatively few species, notably Geiger's work on sugar-beet, with other workers concentrating on soybean (Huber & Israel, 1982; Huber *et al.*, 1984a, b), barley (Gordon *et al.*, 1980, 1982; Farrar & Farrar, 1985) and tomato (Ho, 1976a, b, 1977, 1978a, b, 1979). Occasionally, when other plants are examined, it appears that different species adopt widely different strategies of export and storage, particularly diurnally, e.g. sweet pepper (Grange 1985a,b, 1987), strawberry (Shaw & Grange, 1984) and a number of other species (Gordon, 1986).

Species can be divided into those that immediately export a large proportion of recently fixed carbon (e.g. C_4 grasses) and those which store a large proportion for export during the night (Gordon, 1986). Indeed, a given species may change its strategy during development (e.g. pepper, Hammond & Burton, 1983; soybean, Huber *et al.*, 1986). In addition, there is variation in the stored compound. Most dicotyledons which retain assimilates do so as starch with additional reserves as soluble sugars. Many grasses store little starch but large amounts of sugars, usually sucrose. Many species are intermediate, storing both sucrose and starch, e.g. strawberry (Shaw & Grange, 1984) and spinach (Huber *et al.*, 1986). Some species, e.g. tomato, store mainly hexoses and starch. A number of important crop species, notably cucurbits (Richardson *et al.*, 1982) and fruit trees (Bieleski & Redgwell, 1977; Yamaki, 1982), store sugar alcohols such as stachyose or sorbitol. There has been little research into why these different strategies are adopted, nor what the consequences for the control of export and storage are likely to be. Such research may be important, for there is evidence that some species may have different mechanisms controlling carbon allocation and export (Richardson *et al.*, 1984); see Ch. 2.

These different strategies are probably a result of the ecological demands on the plants. Many of the grasses are perennial and grow in intermittently cold or arid environments. Storing soluble sugar rather than starch may help to maintain turgor, through decreased osmotic potential (Morgan, 1984), or avoid a possible reduction of starch breakdown rate during cold nights. This may also explain why some herbaceous perennials (e.g. strawberry, spinach) also store large amounts of sucrose, for they are similarly exposed to chilling during their development. Plants in temperate, moist environments can grow rapidly at night. Indeed, there is a good reason for maintaining transport throughout the night, as water stress is minimal and temperatures can remain moderate compared with those of the day, so that plant growth can then be maximized. This requires a large store of fixed carbon, provided by starch. The use of starch may be energetically and osmotically more efficient and less available for exploitation by pests or pathogens than high concentrations of soluble sugars.

The control of carbon allocation in source organs is complex and difficult to describe by a unified overall scheme of control. It is likely that temporary changes in the environment elicit short-term controls in the triose-P shuttle or sucrose synthesis pathway. Changes in the loading system or in the gross levels of important enzymes are probably responses to longer-term changes in the environment. Furthermore, genetic and developmental differences between species have led to different control mechanisms to respond to these environmental changes. The synthesis of a common mechanism for carbon allocation in leaves may be unattainable but a thorough understanding of the major steps is important in research directed at improving crop yield.

8.4 Regulation of carbon compartmentation in sink organs

8.4.1 Sink strength and carbon partitioning

The partitioning of dry matter, mainly photoassimilates, into various sink organs of the same plant differs substantially during the course of plant development (see Sect. 8.2). However, it is not clear whether assimilates are partitioned to individual sink organs by being: (a) 'pushed' by the source organs according to the availability of assimilates (i.e. source strength) and/or the vicinity of the source to the sinks; or (b) 'pulled' by individual sink organs according to their assimilate demand (i.e. sink strength). Whether the partitioning of dry matter into individual sink organs is due to a sink-passive mode or a sink-active mode, the capacity of the sink organ to receive assimilates (i.e. sink size) and the location of the sink organ in relation to the source organ (i.e. path), are the constraints on sink strength (Peel & Ho, 1970; Cook & Evans, 1983). If partitioning of assimilates is mainly a sink-active process, the metabolic activity of the sink organs (i.e. sink activity) will be an important component of sink strength.

In an attempt to quantify the sink strength, Warren Wilson (1967) proposed that the strength of each sink is given by the product of its size (dry weight) and its relative rate of change in dry weight. Subsequently, Warren Wilson (1972) interpreted this relationship as:

sink strength = sink size × sink activity

which can be experimentally measured on a sink organ as:

absolute growth rate = dry weight × relative growth rate

Whilst this description of sink strength is conceptually correct it does not consider the respiration of the sink organs and competition between sink organs for a limited supply of assimilates when the sink strength is esti-

CONDITIONS   SINK STRENGTHS   PARAMETERS

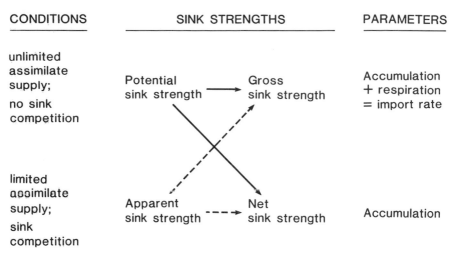

Fig. 8.2 Definitions of sink strength according to the conditions and parameters used for measurement.

mated (Wareing & Patrick, 1975). As the respiratory losses of imported assimilates in the sink organ can be substantial (Farrar, 1985), the absolute growth rate or the net accumulation of dry matter in a sink organ underestimates the actual strength of a sink and is a measure of the 'net sink strength'. Thus, the import rate by a sink organ, measured as the sum of the net gain of carbon and respiratory loss of carbon (Walker & Ho, 1977a), should give a more appropriate estimate of the sink strength and can be termed 'gross sink strength'.

In practice, most measurements of sink strength are made when more than one sink organ competes for a limited supply of assimilates within a plant. Thus, the import rate or the absolute growth rate may be restricted by the availability of assimilate supply to that specific sink organ and will be lower than the real sink strength. For this reason, estimates obtained with a limited supply of assimilates should be termed 'apparent sink strength' and estimates obtained with an unlimited supply of assimilates should be termed 'potential sink strength'. The difference between apparent sink strength and potential sink strength, in tomato fruit, has been demonstrated under non-limiting assimilate supply following fruit-thinning treatments (Ho, 1978b). The relationships between different types of sink strengths and their estimation are illustrated in Fig. 8.2.

To define the roles of sink size and sink activity in determining sink strength, measurements on the import rate or absolute growth rate alone do not reveal the underlying mechanisms controlling assimilate partitioning. Sink size and sink activity should be defined as the physical and physiological constraints of sink strength, respectively. For instance, among cultivars, the final grain size of wheat is closely related to both the number of endosperm cells and the number of starch granules per endosperm cell (Gleadow *et al.*, 1982). Close relationships between grain size and endosperm cell number have been found among grains within the same ear (Singh & Jenner, 1984). These observations demonstrate that the accumulation of starch, the principal form of dry matter in wheat grains, is potentially constrained by the plastid number in the grains. Therefore, the sink size of a wheat grain, measured as the number of plastids or endosperm cells early in grain development, appears to be a more appropriate estimation than the dry weight of the grain, for the latter is the resultant rather than the determinant of the sink strength of the grain.

If sink size is a measure of the capacity of the sink organ to receive the imported assimilates, then sink activity is a measure of the ability of the sink organ to attract assimilates. Therefore sink activity should be estimated as the rate of the rate-limiting processes which include the unloading of assimilates from conducting tissue, transport of assimilates across the plasmalemma or tonoplast, and chemical conversion and use of assimilates (Ho, 1986); see Ch. 6. In comparison to sink size, sink activity may be affected more by environmental factors during sink organ development. For instance, during the rapid growth period after the final fruit cell number has been fixed, the rate of import into tomato fruit can be readily altered by changing the temperature of the fruit itself (Walker & Ho, 1977b). Based on the inverse relationship between the import rate and the sucrose concentration in the fruit at different temperatures, a reduction in the import rate at low temperatures might be due to less sucrose hydrolysis (Walker & Thornley, 1977). In this case, the change of sink strength is mainly due to the change of sink activity, which can be measured by the acid invertase activity in the fruit (Walker *et al.*, 1978; Eschrich, 1980). However, the sink strength of a tomato fruit must be determined by both the sink size and sink activity, as the growth rates of parthenocarpic tomato fruits induced by different growth regulators are related to both the cell number and cell size (Bunger-Kibler & Bangerth, 1983), although their relative importance changes during fruit development. The importance of cell size in determining sink strength has also been reported in corn kernels (Reddy & Daynard, 1983) and in soybean seeds (Egli *et al.*, 1981).

The discussion above suggests a new approach to the estimation of sink strength, with the emphasis on the underlying mechanisms. Instead of measuring the overall growth rate or the resultant weight of a sink organ, sink strength can be estimated by its morphological determinants, e.g. the number of cells, or the biochemical determinants, e.g. the activity of key enzymes in the rate-limiting processes. However, neither the rate of starch synthesis in wheat grains (Jenner, 1986) nor the rate of sucrose hydrolysis in tomato fruit (Ho, 1986) has been unequivocally proven to be the rate-limiting process for the import of assimilates by these organs. Estimations of the sink strength of a wheat grain or a tomato fruit by their cell number in the endosperm or the pericarp, and their rates of starch synthesis or sucrose hydrolysis, respectively, remain an attractive proposition.

8.4.2 Characteristics of sink organs

In general, the import of assimilates by a sink organ consists of: (1) the movement of sucrose, the principal translocated assimilate, along the phloem conducting tissue; (2) the unloading or transfer of sucrose to the intercellular free space or sink cells, respectively; (3) the retrieval of sucrose or its metabolites (e.g. hexoses) by the sink cell from the cell wall; and (4) the metabolism or compartmentation of sugars inside the sink cell. Although the mechanisms for controlling import are not certain, the import rate of individual sink organs can be regulated by the osmotic potential gradient between the source and sink assuming that there is a pressure flow (Milburn, 1974; Ho, 1979); see Ch 7. The low sucrose concentration at the unloading site within the sink, which is required to maintain a steep osmotic potential gradient, can be obtained by: (a) chemical conversion of sucrose to hexoses (to prevent reloading) or to starch (to reduce osmotic concentration); (b) physical compartmentation of sucrose into the vacuole; and (c) utilization of sucrose for growth in biosynthesis or respiration (Thorne, 1985). Studies of these processes in a number of sink organs have demonstrated that sink organs can be classified into groups.

Primarily, sink organs can be classified into those concerned with utilization or storage. Most of the meristem tissues of growing organs are utilization sinks, whilst the storage tissue of growing organs are the storage sinks.

8.4.2.1 Utilization sink

In utilization sinks, sucrose would most likely be transferred from the sieve elements to the utilization sink cells via intercellular cytoplasmic connections. Sufficient plasmodesmata are found in young leaves (Turgeon & Webb, 1975; Schmalstig & Geiger, 1985) and root tips (Dick & ap Rees, 1975; Giaquinta *et al.*, 1983) to facilitate such a transfer without involving the hydrolysis of sucrose, which would normally take place in the cell wall free space (Giaquinta, 1977). Most of the imported sucrose in the cytosol of the utilization sink cell would be used immediately for respiration or for structural growth and only a small proportion of the sucrose stored temporarily. For instance, the fibrous roots of barley use 40% and 55% of the imported sucrose for respiration and structural growth, respectively (Farrar, 1985) and the stored sugar only accounts for 1% of the root weight, with a turnover time of 0.5 h for the sugar pool (Farrar, 1981). Therefore, it is envisaged that a sucrose concentration gradient between the sieve elements and the sink cell can be sustained efficiently by the use of the sucrose in the sink cell, thus facilitating further transport (Geiger & Fondy, 1980). If the use of sucrose is the regulatory step for the control of import rate in the utilization sink organs, factors affecting the rates of biosynthesis of structural materials, such as proteins and cellulose, or of respiration, should also affect the rate of import. The relative growth rate of the root of chickpea was found to be linearly related to the respiration of the root (Kallarackal & Milburn, 1985), and the import of assimilates into the nodules (as well as the respiration of the nodules) of soybean was simultaneously retarded by low oxygen (Gordon *et al.*, 1985), but there is as yet no direct evidence that the import rate by the utilization sinks is primarily regulated by the rate of respiration or growth.

8.4.2.2 Storage sink

In the storage sink, a substantial proportion of the imported assimilates is stored rather than being used for concurrent respiration or growth. Although the transfer of sucrose from the sieve elements to the adjacent sink cells may be via symplastic connections, an apoplastic route of transfer has frequently been identified between maternal and embryonic tissue of the reproductive sinks, e.g. wheat grains (Cook & Oparka, 1983) and bean seeds (Thorne, 1981; Offler & Patrick, 1984), or in vegetative sinks, e.g. sugar-cane stalk (Glasziou & Gayler, 1972). As sucrose or its metabolites have to be retrieved by the sink cells from the apoplast, membrane transport of sugars is involved. The imported sucrose is stored as such, or as its metabolites, through either physical compartmentation or chemical conversion. Based on the pathway of accumulation, the storage sinks can be classified into three types: sugar sink, sugar/starch sink, and starch sink (Ho & Baker, 1982).

8.4.2.2.1 Sugar sink Imported sucrose is stored mainly as sucrose, hexoses or fructans. For sugar sinks storing mainly sucrose, the imported sucrose may be stored in the vacuole, without hydrolysis along the path, as in sugar-beet tap-root (Giaquinta 1977; Wyse, 1979), or with hydrolysis at the cell wall or free space prior to the resynthesis of sucrose for storage, as in sugar-cane stalks (Glasziou & Gayler, 1972).

In the tap-roots of sugar-beet, the transport of sucrose from the free space to the cytosol may be down a concentration gradient (Saftner *et al.*, 1983) and is then confined entirely to the vacuole (Leigh *et al.*, 1979). However, the uptake of sucrose by the vacuole from the cytosol is against a steep gradient with a more than 10-fold increase in sucrose concentration. The transport of sucrose into the vacuole consists of both active uptake and passive permeation of sucrose (Willenbrink & Doll, 1979). The active uptake may be a proton-sucrose antiport with the proton motive force energized by the tonoplast-bound ATPase (Doll *et al.*, 1979; Briskin *et al.*, 1985). However, the imported sucrose in the cell wall space of the sugar-cane stalk is hydrolysed by the acid invertase (Bowen & Hunter, 1972) and the resultant hexoses are taken up selectively by the sink cell (Komor *et al.*, 1981). Hexoses may be transported either by a sugar-proton antiport into the vacuole and then resynthesized as sucrose for storage

(Thom & Komor, 1984), or they may be transported by UDP-glucose-dependent, tonoplast-bound group translocators, with sucrose being subsequently made from the UDP-glucose inside the vacuole (Thom & Maretzki, 1985). Neither of these mechanisms are unequivocally proven.

In many fruits (e.g. strawberry, orange, grape and cucumber), the imported sucrose or stachyose is stored mainly as hexoses. In grapes, the accumulation of hexoses in the vacuole is preceded in sequence by: the hydrolysis of sucrose, phosphorylation of hexoses, synthesis and then hydrolysis of sucrose phosphate, synthesis and then hydrolysis of sucrose (Hawker, 1969). The sucrose phosphate may be synthesized at the tonoplast by group translocators and then used for synthesis of hexoses inside the vacuole (Brown & Coombe, 1982).

Fructans with varying degrees of polymerization are stored in substantial amounts in the leaf bases of onions (Darbyshire & Henry, 1978), the stems of temperate grasses (Pollock & Jones, 1979) or in the tubers of Jerusalem artichokes (Edelman & Jefford, 1968). As the synthesis and remobilization of fructans is closely linked with the metabolism of sucrose and starch (Pollock, 1986), the role of fructan accumulation in regulating import in these organs is not discussed here.

The form of sugars stored may change during sink organ development. For instance, during the period of rapid growth, sweet melon accumulates only a small amount of hexoses with hardly any sucrose. Later accumulation is entirely as sucrose (Lester & Dunlop, 1985). This change in the form of sugars for storage is most probably due to the changes of related enzyme activity in the sink organ during development.

8.4.2.2.2 Sugar/starch sink In some storage sinks, starch and sugars are equally important for storage at certain stages of sink development. When a tomato fruit is growing most rapidly, imported sucrose is stored in similar amounts as starch and hexoses (Ho *et al.*, 1983). Later, starch is depleted and hexoses account for more than half of the fruit dry weight. A correlation between the starch content in green fruit and the total soluble solids in the ripe fruit has been found in a number of cultivars (Dinar & Stevens, 1981). However, the absolute amount of starch accumulated by a tomato fruit is small in comparison to the amount of hexoses. Enhancing the accumulation of starch, by altering the water relations in the plant, did not increase the import rate (Ehret & Ho, 1986). Therefore, hydrolysis of sucrose, rather than starch accumulation, is likely to be a rate-limiting process in controlling the import of sucrose into the tomato fruit (Ho, 1986).

8.4.2.2.3 Starch sink Starch is the principal form of stored assimilate in many fruits (e.g. cereal kernels), stem tubers (e.g. potato) and root tubers (e.g. sweet potato). The imported sucrose is either converted directly into starch (e.g. wheat grain) or hydrolysed into hexoses prior to starch synthesis (e.g. corn kernels).

Table 8.1 Classification of sink organs in plants according to the route of transport within the sink organ and the critical processes related to the fate of imported assimilates

Sink organs	Routes	Critical processes
I. Utilization sinks (meristem tissue)	Symplastic	Biosynthesis of cellular structural material; respiration
II. Storage sinks (storage tissue)	Symplastic and apoplastic	Regulation of transport and compartmentation of assimilates
1. Sugar sinks		
(a) Sucrose sinks (sugar-beet tap-root)		H^+/sucrose antiport across the tonoplast
(sugar-cane stalk)		Hydrolysis of sucrose in the cell wall and the transport of hexoses across the tonoplast
(b) Hexose sinks (grape berry)		Group translocators for sucrose phosphate prior to hexose storage in the vacuole
(c) Fructan sinks		Unidentified
2. Sugar starch sinks (tomato fruit)		Hydrolysis of sucrose and accumulation of hexoses
3. Starch sinks (wheat grain)		Retrieval of sucrose from the cavity for starch synthesis in the endosperm
(corn kernel)		Hydrolysis of sucrose in the pedicel apoplast prior to starch synthesis

In wheat grains, sucrose is transported from the conducting tissue to the endosperm down a concentration gradient (Jenner, 1974). Although the retrieval of sucrose from the endosperm cavity by the endosperm cells may partly involve a membrane-carrier-mediated process (Ho & Gifford, 1984), the most likely determinants of sink strength are the number of endosperm cells (Gleadow *et al.*, 1982) and the capacity of the endosperm cells to convert sucrose to starch (Jenner, 1986); see Sect. 8.4.1. The enzyme activity of starch synthesis declines as the grain approaches maturity (Turner, 1969) and the synthesis of starch cannot be enhanced by increasing the sucrose supply (Jenner, 1970; Barlow, *et al.*, 1983). The conversion of sucrose to starch, rather than the supply of sucrose, is likely to be the process limiting grain filling in wheat.

In corn kernels, sucrose is hydrolysed in the pedicel free space (Shannon, 1972) prior to the uptake of hexoses by the endosperm transfer cells (Felker & Shannon, 1980) before starch synthesis in the endosperm. The unloading of sucrose into the pedicel free space may be regulated osmotically by the sugar concentration in the free space (Porter *et al.*,

1985). However, the amount of starch in the endosperm is genetically determined. The low level of starch in the shrunken mutants of corn is due to mutations in either *sh* or *sh-2* loci, causing a severe reduction in starch synthesis enzymes such as sucrose synthetase (Chourey & Nelson, 1976) or ADPG pyrophosphorylase, respectively (Tsai & Nelson, 1966). In this case the lack of starch synthesis is the rate-limiting factor causing low import by the kernels.

The general classification of sink organs summarized in Table 8.1 is a guide to the possible controls of import of assimilates in different types of sinks, according to the routes of transport and the physical and biochemical compartmentation of assimilates within sink organs. As there is such a diversity of these characteristics, the control of import may differ greatly amongst sink organs.

8.4.3 Regulation of sink competition

The partitioning of assimilates to various sink organs within a plant appears to follow an order of priority which may change with plant development. In fact, the partitioning of assimilates into various sinks changes more readily than the rate of assimilate export when the demand of one of the sinks is altered (see Sect. 8.3.2.2). In the tomato, the developing inflorescence may abort if the available supply of assimilates is sufficient to meet only the demand from the meristems at the apex and the roots (Kinet, 1977a). When fruit growth rate reached a maximum, the growth of leaves was reduced to a minimum and root growth stopped. Furthermore, the initiation of new trusses was delayed when the assimilate supply was limited, because the plant already had a heavy load of fruit in early trusses (Salter, 1958; Hurd *et al.*, 1979). Therefore, the priority for assimilate partitioning in flowering tomato plants decreases in the order of roots > young leaves > inflorescence. In the fruiting plant, the priority order is reversed, such that fruit > young leaves > flower > roots.

A direct competition for assimilates between the initiating first inflorescence and the adjacent young leaves or roots in tomato has been well recognized in horticultural practice. By restricting either root growth (Cooper, 1964) or removing the adjacent young leaves in the winter (Leopold & Lam, 1960), the development of the first truss can be secured. There is an anatomical basis for such competition, as leaves 1,3,6 and 8 (numbered from the base of the plant), are the main suppliers of assimilates for the developing inflorescence, as well as for the apex and roots in the young flowering plant (Russell & Morris, 1983). However, it is not certain why the developing inflorescence should compete so poorly against the apex and roots. The assimilate supply to the inflorescence decreased immediately when the irradiance was reduced. However, the inflorescence received extra assimilate supply only after a time lag when the irradiance was increased. Thus the lower priority of the inflorescence to receive assimilates must be determined by its sink strength relative to other sinks, rather than by the availability of the assimilate supply. Furthermore, an aborting inflorescence, induced. by insufficient irradiance, contains much less cytokinin (Leonard & Kinet, 1982) and the floral development can be restored by applying cytokinin and gibberellin to the inflorescence (Kinet, 1977b). Apparently, the added growth regulators restored cell division of the inflorescence (Kinet *et al.*, 1986) and diverted the assimilates from the apex to the inflorescence without affecting the rate of export from the leaves (Leonard *et al.*, 1983).

These observations suggest that competition for assimilates between two utilization sinks may be mainly determined by their cell division activity which is regulated by endogenous PGR levels. A close relationship between cell division activity and PGR levels has been demonstrated in the comparison between the seeded and parthenocarpic isogenic lines of tomato, where a higher cell division is associated with a higher cytokinin level (Mapelli *et al.*, 1978). In contrast, the competition between rapidly growing fruits and young leaves at the apex of tomato may represent competition between storage sinks and utilization sinks, respectively. Whether the greater sink strength of the fruit is due to its larger sink size or higher sink activity remains to be determined. Nevertheless, there is some indication that the sink strength of the stem and the ear in wheat is genetically determined. The dwarfing gene *Gai/Rht2* is responsible for reducing stem growth in the dwarf genotype of winter wheat and, consequently, the grain yield as well as the harvest index increases (Brooking & Kirby, 1981). In this sense, sink competition is a genetic expression.

Apart from sink size, competition between sinks can also be regulated by sink activity. Although the competition between fruits in the same truss may be due to their respective cell numbers, which is related to their positions or fruit-setting sequence in the truss (Bangerth & Ho, 1984), fruit growth rate can also be altered by fruit temperature alone. It has also been reported that at a constant air temperature of 18°C, warming the roots to 25°C or 30°C increased root growth but decreased fruit growth. When the root was cooled to 15°C, root growth was reduced while dry matter accumulation in fruit increased (Holder, 1984). By altering the sink activity and thus the sink strength of individual sinks, the order of priority of competing sinks may change.

In essence, the regulation of sink competition determines assimilate partitioning in plants. The potential priority of competing sinks may be genetically determined and both the potential sink size and sink activity may be regulated by PGRs. When the factors determining the sink strength and the regulating mechanism between sources and sinks are thoroughly investigated, the regulation of partitioning will then be a powerful means for improving crop productivity.

8.5 Concluding remarks

Partitioning of assimilates within a plant is an integrated transport phenomenon. Recent advances in research on the loading and unloading processes have allowed the study of the links between assimilation and use of dry matter in the whole plant, but the controlling points in the transport system are far from clear. While more work is needed to investigate the biophysical and biochemical aspects of the transport processes, the complexity of this integrated transport phenomenon can only be understood fully when the regulation of the interaction between sources and sinks is known.

Clearly, there are source-limiting and sink-limiting situations for plant growth. Continuous improvement of crop yield can only be achieved by overcoming the rate-limiting processes in the source and sink organs alternately. Further improvement of crop productivity can result from an improvement in carbon assimilation and carbon partitioning. For the latter, research on the determination of sink strength and the regulation of sink competition is most important. Advances in agriculture will be brought about by plant breeding and the optimization of growing environment to ensure desirable gene expression. While genetic engineering has been regarded as a most promising tool for future agricultural improvement, the need to identify controlling factors in the various processes associated with assimilate partitioning, and the mechanism integrating these processes, remains an important priority.

Acknowledgements

We would like to thank Dr Alun Rees and Dr Bob Thompson for their helpful comments on the manuscript.

References

Aloni, B., Daie, J. & Wyse, R. E. (1986). Enhancement of [^{14}C]sucrose export from source leaves of *Vicia faba* by gibberellic acid. *Plant Physiology* **82**, 962–6.

Austin, R. B. (1972). Bulb formation in onions as affected by photoperiod and spectral quality of light. *Journal of Horticultural Science* **47**, 493–504.

Azcon-Bieto, J. (1984). The effect of carbohydrate status on the photosynthetic, stomatal and respiratory physiology of wheat leaves. In Sybesma, C. (ed.). *Advances in Photosynthesis Research*, Vol. III, pp. 905–8. The Hague, M. Nijhoff/Dr W. Junk.

Baker, D. A. (1985). Regulation of phloem loading. In Jeffcoat, B., Hawkins,

A. F. & Stead, A. D. (eds.). *Regulation of Sources and Sinks in Crop Plants*, Monograph 12, pp. 163–76. Bristol, British Plant Growth Regulator Group.

Baker, D. A. & Chaudry, S. B. (1986). Active sucrose transport in the castor bean: Effect of monovalent cations. In Cronshaw, J., Lucas, W. J. & Giaquinta, R. T. (eds.). *Plant Biology, Vol. 1, Phloem Transport*, pp. 77–88. New York, Alan R. Liss.

Bangerth, F. & Ho, L. C. (1984). Fruit position and fruit set sequence in a truss as factors determining final size of tomato fruits. *Annals of Botany* **53**, 315–9.

Barlow, E. W. R., Donovan, G. R. & Lee, J. W. (1983). Water relations and composition of wheat ears grown in liquid culture: Effect of carbon and nitrogen. *Australian Journal of Plant Physiology* **10**, 99–108.

Bell, C. J. & Incoll, L. D. (1982). Translocation from the flag leaf of winter wheat in the field. *Journal of Experimental Botany* **33**, 896–909.

Bieleski, R. L. & Redgwell, R. J. (1977). Synthesis of sorbitol in apricot leaves. *Australian Journal of Plant Physiology* **4**, 1–10.

Blechschmidt-Schneider, S. & Eschrich, W. (1985). Microautoradiographic localisation of imported ^{14}C-photosynthate in induced sink leaves of two dicotyledonous C_4 plants in relation to phloem unloading. *Planta* **163**, 439–49.

Blevins, D. G., Barnett, N. M. & Frost, W. B. (1978). Role of potassium and malate in nitrate upake and translocation by wheat seedlings. *Plant Physiology* **62**, 784–8.

Bowen, J. C. & Hunter, J. E. (1972). Sugar transport in immature internodal tissue of sugarcane. II. Mechanism of sucrose transport. *Plant Physiology* **49**, 789–93.

Briskin, D. P., Thornley, W. R. & Wyse, R. E. (1985). Membrane transport in isolated vesicles from sugar beet taproot. II. Evidence for a Sucrose/H antiport. *Plant Physiology* 78, 871–5.

Britz, S. J. (1986). The role of circadian rhythms in the photoperiodic response of photosynthate partitioning in *Sorghum* leaves: A progress report. In Cronshaw, J., Lucas W. J. & Giaquinta, R. T. (eds.). *Plant Biology, Vol. 1, Phloem Transport*, pp. 527–34. New York, Alan R. Liss.

Britz, S. J., Hungerford, W. E. & Lee, D. R. (1985). Photoperiodic regulation of photosynthate partitioning in leaves of *Digitaria decumbens* Stent. *Plant Physiology* **78**, 710–14.

Brooking, I. R. & Kirby, E. J. M. (1981). Interrelationships between stem and ear development in winter wheat: the effects of a Norin 10 dwarfing gene, *Gai/Rht2*. *Journal of Agricultural Science, Cambridge* **97**, 373–81.

Brown, S. C. & Coombe, B. G. (1982). Sugar transport by an enzyme complex at the tonoplast of grape pericarp cells. *Naturwissenschaften* **69**, 43–5.

Bunger-Kibler, S. & Bangerth, F. (1983). Relationship between cell number, cell size and fruit size of seeded fruits of tomato (*Lycopersicon esculentum* Mill.) and those induced parthenocarpically by the application of plant growth regulators. *Plant Growth Regulation* **1**, 143–54.

Challa, H. (1976). *An analysis of the diurnal course of growth, carbon dioxide exchange and carbohydrate reserve content of cucumber.* Agricultural Research Report 861. Wageningen, the Netherlands, Centre for Agrobiological Research.

Chan, T. T. & Bird, I. F. (1960). Starch dissolution in tobacco leaves in the light. *Journal of Experimental Botany* **11**, 335–40.

Chatterton, N. J. & Silvius, J. E. (1979). Photosynthate partitioning into starch in soybean leaves. I. Effects of photoperiod versus photosynthetic period duration. *Plant Physiology* **64**, 749–53.

Chatterton, N. J. & Silvius, J. E., (1980a). Photosynthate partitioning into leaf starch as affected by daily photosynthetic period duration in six species. *Physiologia Plantarum* **49**, 141–4.

Chatterton, N. J. & Silvius, J. E. (1980b). Acclimation of photosynthate partitioning and photosynthetic rates to changes in length of the daily photosynthetic period. *Annals of Botany* **46**, 739–45.

Chourey, P. S. & Nelson, O. E. (1976). The enzymatic deficiency conditioned by Shrunken-1 mutation in maize. *Biochemical Genetics* **14**, 1041–55.

Cleland, C. F. (1984). Biochemistry of induction – the immediate action of light. In Vince-Prue, D., Thomas, B. & Cockshull, K. E. (eds.). *Light and the Flowering Process*, pp. 123–36. London, Academic Press.

Colombo R., de Michelis M. I. & Lado P. (1978). 3-0-Methyl glucose uptake stimulation by auxin and by fusicoccin in plant materials and its relationship with proton extrusion, *Planta* **138**, 249–56.

Conti, T. R. & Geiger, D. R. (1982). Potassium nutrition and translocation in sugar beet. *Plant Physiology* **70**, 168–72.

Cook, M. G. & Evans, L. T. (1978). Effect of relative size and distance of competing sinks on the distribution of photosynthetic assimilate in wheat. *Australian Journal of Plant Physiology* **5**, 495–509.

Cook, M. G. & Evans, L. T. (1983). The roles of sink size and location in the partitioning of assimilates in wheat ears. *Australian Journal of Plant Physiology* **10**, 313–27.

Cook, H. & Oparka, K. J. (1983). Movement of fluorescein into isolated caryopses of wheat and barley. *Plant, Cell and Environment* **6**, 239–42.

Cooper, A. J. (1964). A study of the development of the first inflorescence of glasshouse tomatoes. *Journal of Horticultural Science* **39**, 92–7.

daCruz, G. S. & Audus, L. J. (1978). Studies of hormone directed transport in decapitated stolons of *Saxifraga sarmentosa*. *Annals of Botany* **42**, 1009–27.

Darbyshire, B. & Henry, R. J. (1978). The distribution of fructan in onions. *New Phytologist* **81**, 29–34.

Dick, P. S. & ap Rees, T. (1975). The pathway of sugar transport in roots of *Pisum sativum*. *Journal of Experimental Botany* **26**, 305–14.

Dinar, M. & Stevens, M. A. (1981). The relationship between starch accumulation and soluble solids content of tomato fruits. *Journal of the American Society of Horticultural Science* **106**, 415–8.

Dixon H. H. (1933). Bast sap. *Proceedings of the Royal Dublin Society* **20**, 487–94.

Doll, S., Rodier, F. & Willenbrink, J. (1979). Accumulation of sucrose in vacuoles isolated from red beet tissue. *Planta* **144**, 407–11.

Doman D. C. & Geiger D. R. (1979). Effects of exogenously supplied foliar potassium on phloem loading in *Beta vulgaris* L. *Plant Physiology* **64**, 528–33.

Downs R. J., Hendricks, S. B. & Borthwick, H. A. (1957). Photoreversible control of elongation of pinto beans and other plants. *Botanical Gazette* **118**, 199–208.

Edelman, J. & Jefford, T. G. (1968). The mechanism of fructosan metabolism in higher plants as exemplified in *Helianthus tuberosus*. *New Phytologist* **67**, 517–31.

Egli, D. B., Fraser, J., Legget, J. E. & Poneleit, C. G. (1981). Control of seed growth in soybean (*Glycine max* L. Merrill). *Annals of Botany* **48**, 171–6.

Ehret, D. L. & Ho, L. C. (1986). Effects of salinity on dry matter partitioning and fruit growth in tomatoes grown in nutrient film culture. *Journal of Horticultural Science* **61**, 361–7.

Eschrich, W. (1980). Free space invertase, its possible role in phloem unloading. *Berichte der Deutschen Botanischen Gesellschaft* **93**, 363–78.

Farrar, J. F. (1981). Respiration rate of barley roots: its relation to growth, substrate supply and the illumination of the shoot. *Annals of Botany* **48**, 53–63.

Farrar, J. F. (1985). Fluxes of carbon in roots of barley plants. *New Phytologist* **99**, 57–69.

Farrar, S. C. & Farrar, J. F. (1985). Fluxes of carbon compounds in leaves and roots of barley plants. In Jeffcoat, B., Hawkins, A. F. & Stead, A. D. (eds.). *Regulation of Sources and Sinks in Crop Plants*, Monograph 12, pp. 67–84. Bristol, British Plant Growth Regulator Group.

Felker, F. C. & Shannon, J. C. (1980). Movement of ¹⁴C-labelled assimilates into kernels of *Zea mays* L. III. An anatomical examination and microautoradiographic study of assimilate transfer. *Plant Physiology* **65**, 864–70.

Fellows, R. J. & Geiger, D. R. (1974). Structural and physiological changes in sugar beet leaves during sink to source conversion. *Plant Physiology* **54**, 877–85.

Fisher, D. B. (1975). Translocation kinetics of photosynthates. In Aronoff, S., Dainty, J., Gorham, P. R., Srivastava, L. M. & Swanson, C. A. (eds.). *Phloem Transport*, pp. 495–520. NATO Advanced Study Institute Series. London, Plenum Press.

Fisher, D. B. (1978). An evaluation of the Münch hypothesis for phloem transport in soybean. *Planta* **139**, 25–8.

Fisher, D. B., Housley T. C. & Christy A. L. (1978). Source pool kinetics for ¹⁴C-photosynthate translocation in Morning Glory and Soybean. *Plant Physiology* **61**, 291–5.

Fondy, B. R. & Geiger, D. R. (1980). Effect of rapid changes in sink-source ratio on export and distribution of products of photosynthesis in leaves of *Beta vulgaris* L. and *Phaseolus vulgaris* L. *Plant Physiology* **66**, 945–9.

Fondy, B. R. & Geiger, D. R. (1982). Diurnal patterns of translocation and carbohydrate metabolism in source leaves of *Beta vulgaris* L. *Plant Physiology* **70**, 671–6.

Fondy, B. R. & Geiger D. R. (1985). Diurnal changes in allocation of newly fixed carbon in exporting sugar beet leaves. *Plant Physiology* **78**, 753–7.

Fox, T. C. & Geiger, D. R. (1984). Effects of decreased net carbon exchange on carbohydrate metabolism in sugar beet source leaves. *Plant Physiology* **76**, 763–8.

Friend, D. J. C. (1984). The interaction of photosynthesis and photoperiodism in induction. In: Vince-Prue, D., Thomas, B. & Cockshull, K. E. (eds.). *Light and the Flowering Process*, pp. 257–76. London, Academic Press.

Geiger, D. R. (1979). Control of partitioning and export of carbon in leaves of higher plants. *Botanical Gazette* **140**, 241–8.

Geiger, D. R. (1986). Processes affecting carbon allocation and partitioning among sinks. In Cronshaw, J., Lucas, W. J. & Giaquinta, R. T. (eds.). *Plant Biology, Vol. 1, Phloem Transport*, pp. 375–88. New York, Alan R. Liss.

Geiger, D. R. & Batey, J. W. (1967). Translocation of ¹⁴C sucrose in sugar beet during darkness. *Plant Physiology* **42**, 1743–9.

Geiger, D. R. & Conti, T. R. (1983). Relation of increased potassium nutrition to photosynthesis and translocation of carbon. *Plant Physiology* **71**, 141–4

Geiger, D. R. & Fondy, B. R. (1980). Phloem loading and unloading: Pathways and mechanisms. *What's New in Plant Physiology* **11**, 25–8.

Geiger, D. R. & Fondy, B. R. (1985). Responses of export and partitioning to internal and environmental factors. In Jeffcoat, B., Hawkins, A. F. & Stead, A. D. (eds.). *Regulation of Sources and Sinks in Crop Plants*, Monograph 12, pp. 177–94. Bristol, British Plant Growth Regulator Group.

Geiger, D. R. & Giaquinta, R. T. (1982). Translocation of photosynthate. In *Photosynthesis: Development, Carbon Metabolism and Plant Productivity*, Vol. II, pp. 345–86. Academic Press.

Geiger, D. R., Jablonski, L. M. & Ploeger, B. J. (1985). Significance of carbon allocation to starch in growth of *Beta vulgaris* L. In: Heath, R. L. & Priess, J. (eds.). *Regulation of Carbon Partitioning in Photosynthetic Tissues*, Riverside Symposium 8, pp. 289–308. Baltimore, American Society of Plant Physiologists.

Geiger, D. R., Ploeger, B. J., Fox, T. C. & Fondy, B. R. (1983). Sources of sucrose translocated from illuminated sugar beet source leaves. *Plant Physiology* **72**, 964–70.

Geiger, D. R., Sovonick, S. A., Shock, T. L. & Fellows, R. J. (1974). Role of free space in translocation in sugar beet. *Plant Physiology* **54**, 892–8.

Giaquinta, R. T. (1977). Sucrose hydrolysis in relation to phloem translocation in *Beta vulgaris*. *Plant Physiology* **60**, 339–43.

Giaquinta R. T. (1978). Source and sink leaf metabolism in relation to phloem translocation – carbon partitioning and enzymology. *Plant Physiology* **61**, 380–5.

Giaquinta, R. T., Lin, W., Sadler, N. L. & Franseschi, V. R. (1983). Pathway of phloem unloading of sucrose in corn roots. *Plant Physiology* **72**, 362–7.

Gifford, R. M. & Evans, L. T. (1981). Photosynthesis, carbon partitioning and yield. *Annual Review of Plant Physiology* **32**, 485–509.

Gifford, R. M., Thorne, J. H., Hitz, W. D. & Giaquinta, R. T. (1984). Crop productivity and photoassimilate partitioning. *Science* **225**, 801–8.

Glasziou, K. T. & Gayler, K. R. (1972). Sugar accumulation in sugar cane. Role of cell walls in sucrose transport. *Plant Physiology* **49**, 912–13.

Gleadow, R. M., Dalling, M. J. & Halloran, G. M. (1982). Variation in endosperm characteristics and nitrogen content in six wheat lines. *Australian Journal of Plant Physiology* **9**, 539–51.

Gordon, A. J. (1986). Diurnal patterns of photosynthate allocation and partitioning among sinks. In Cronshaw, J., Lucas, W. J. & Giaquinta, R. T. (eds.). *Plant Biology, Vol. 1, Phloem Transport*, pp. 499–517. New York, Alan R. Liss.

Gordon, A. J., Ryle, G. J. A., Mitchell, D. F. & Powell, C. E. (1982). The dynamics of carbon supply from leaves of barley plants grown in long or short days. *Journal of Experimental Botany* **33**, 241–50.

Gordon, A. J., Ryle, G. J. A., Mitchell, D. F. & Powell, C. E. (1985). The flux of ^{14}C-labelled photosynthate through soybean root nodules during N_2 fixation. *Journal of Experimental Botany* **46**, 756–9.

Gordon, A. J., Ryle, G. J. A. & Webb, G. (1980). The relationship between sucrose and starch during 'dark' export from leaves of uniculm barley. *Journal of Experimental Botany* **31**, 845–50.

Grange, R. I. (1978). Evidence for solution flow in the phloem of willow. *Planta* **138**, 15–23.

Grange, R. I. (1984). The extent of starch turnover in mature pepper leaves in the light. *Annals of Botany* **54**, 289–91.

Grange, R. I. (1985a). Carbon partitioning and export in mature leaves of pepper (*Capsicum annuum*). *Journal of Experimental Botany* **36**, 734–44

Grange, R. I. (1985b). Carbon partitioning in mature leaves of pepper. Effects of daylength. *Journal of Experimental Botany* **36**, 1749–59.

Grange, R. I. (1987). Carbon partitioning in mature pepper leaves. Effects of transfer to low or high irradiance. *Journal of Experimental Botany* **38**, 77–83.

Hammel, H. T. (1968). Measurement of turgor pressure and its gradient in the phloem of oak. *Plant Physiology* **43**, 1042–8.

Hammond, J. B. W. & Burton, K. S. (1983). Leaf starch metabolism during the growth of pepper (*Capsicum annuum*) plants. *Plant Physiology* **73**, 61–5.

Hammond, J. B. W., Burton, K. S., Shaw, A. F. & Ho, L. C. (1984). Source-sink relationships and carbon metabolism in tomato leaves. 2. Carbohydrate pools and catabolic enzymes. *Annals of Botany* **53**, 307–14.

Hawker, J. S. (1969). Changes in the activities of enzymes concerned with sugar metabolism during the development of grape berries. *Phytochemistry* **8**, 9–17.

Hendrix, D. L. & Huber, S. C. (1986). Diurnal fluctuations in cotton leaf carbon exchange rate, sucrose synthesising enzymes, leaf carbohydrate content and carbon export. In Cronshaw, J., Lucas, W. J. & Giaquinta, R. T. (eds.). *Plant Biology, Vol. 1, Phloem Transport*, pp. 369–73. New York, Alan R. Liss.

Hewitt, J. D., Casey, L. L. & Zobel, R. W. (1985). Effect of daylength and night temperature on starch accumulation and degradation in soybean. *Annals of Botany* **56**, 513–22.

Ho, L. C. (1976a). The relationship between the rates of carbon transport and of photosynthesis in tomato leaves. *Journal of Experimental Botany* **27**, 87–97.

Ho, L. C. (1976b). The effects of current photosynthesis on the origin of translocates in old tomato leaves. *Annals of Botany* **40**, 1153–62.

Ho, L. C. (1977). Effects of CO_2 enrichment on the rates of photosynthesis and translocation in tomato leaves. *Annals of Applied Biology* **87**, 191–200.

Ho, L. C. (1978a). The regulation of carbon transport and the carbon balance of mature tomato leaves. *Annals of Botany* **42**, 155–64.

Ho, L. C. (1978b). Translocation of assimilates in the tomato plant. In *Opportunities for Chemical Plant Growth Regulation*, BCPC Monograph 21, pp. 159–66. UK, Boots.

Ho, L. C. (1979). Regulation of assimilate translocation between leaves and fruits in the tomato. *Annals of Botany* **43**, 437–48.

Ho, L. C. (1980). Control of import into tomato fruits. *Berichte der Deutschen Botanischen Gesellschaft* **93**, 315–25.

Ho, L. C. (1984). Partitioning of assimilates in fruiting tomato plants. *Plant Growth Regulator* **2**, 277–85.

Ho, L. C. (1986). Metabolism and compartmentation of translocates in sink organs. In Cronshaw, J., Lucas, W. J. & Giaquinta, R. T. (eds.). *Plant Biology, Vol. 1, Phloem Transport*, pp. 317–24. New York, Alan R. Liss.

Ho L. C. & Baker D. A. (1982). Regulation of loading and unloading in long distance transport systems. *Physiologia Plantarum* **56**, 225–30.

Ho, L. C. & Gifford, R. M. (1984). Accumulation and conversion of sugars by developing wheat grains. V. The endosperm apoplast and apoplastic transport. *Journal of Experimental Botany* **35**, 58–73.

Ho, L. C., Hurd, R. G., Ludwig, L. J., Shaw, A. F., Thornley, J. H. M. & Withers, A. C. (1984). Changes in photosynthesis, carbon budget and mineral content during the growth of the first leaf of cucumber. *Annals of Botany* **54**, 87–101.

Ho, L. C. & Shaw, A. F. (1977). Carbon economy and translocation of ^{14}C in leaflets of the seventh leaf of tomato during leaf expansion. *Annals of Botany* **41**, 833–48.

Ho L. C. & Shaw A. F. (1979). Net accumulation of water and minerals and the carbon budget in an expanding leaf of tomato. *Annals of Botany* **43**, 45–54.

Ho, L. C., Shaw, A. F., Hammond, J. B. W. & Burton, K. S. (1983). Source-sink relationships and carbon metabolism in tomato leaves. 1. ^{14}C assimilate compartmentation. *Annals of Botany* **52**, 365–72.

Ho, L. C., Sjut, V. & Hoad, G. V. (1983). The effect of assimilate supply on fruit growth and hormone levels in tomato plants. *Plant Growth Regulator* **1**, 155–71.

Ho, L. C. & Thornley, J. H. M. (1978). Energy requirements for assimilate translocation from mature tomato leaves. *Annals of Botany* **42**, 481–3.

Holder, R. (1984). Root temperature and the growth of tomato plants and the feasibility of energy saving in tomato production. PhD Thesis, Liverpool University, UK.

Horton, P. (1985). Regulation of photochemistry and its interaction with carbon metabolism. In Jeffcoat, B., Hawkins, A. F. & Stead, A. D. (eds.). *Regulation of Sources and Sinks in Crop Plants*, Monograph 12, pp. 19–33. Bristol, British Plant Growth Regulator Group.

Huber, S. C. & Israel, D. W. (1982). Biochemical basis for partitioning of photosynthetically fixed carbon between starch and sucrose in soybean (*Glycine max* Merr.) leaves. *Plant Physiology* **69**, 691–6.

Huber, S. C., Kerr, P. S. & Kalt-Torres, W. (1986). Biochemical control of allocation of carbon for export and storage in source leaves. In Cronshaw, J., Lucas, W. J. and Giaquinta, R. T. (eds.). *Plant Biology, Vol. 1, Phloem Transport*, pp. 355–67. New York, Alan R. Liss.

Huber, S. C., Rogers, H. H. & Israel, D. W. (1984a). Effects of CO_2 enrichment on photosynthate partitioning in soybean (*Glycine max*) leaves. *Physiologia Plantarum* **62**, 95–101.

Huber, S. C, Rogers, H. H. & Mowry, F. L. (1984b). Effects of water stress on photosynthesis and carbon partitioning in soybean (*Glycine max* L. Merr.) plants grown in the field at different CO_2 contents. *Plant Physiology* **76**, 244–9.

Hurd, R. G., Gay, A. P. & Mountifield, A. C. (1979). The effect of partial flower removal on the relation between root, shoot and fruit growth in the indeterminate tomato. *Annals of Applied Biology* **93**, 77–89.

Isebrands J. G. & Larson P. R. (1973). Anatomical changes during leaf ontogeny in *Populus deltoides*. *American Journal of Botany* **60**, 199–208.

Jenner, C. F. (1970). Relationship between levels of soluble carbohydrate and starch synthesis in detached ears of wheat. *Australian Journal of Biological Sciences* **23**, 991–1003.

Jenner, C. F. (1974). Factors in the grain regulating the accumulation of starch. In Bieleski, R. L., Ferguson, A. R. & Cresswell, M. M. (eds.). *Mechanisms of Regulation of Plant Growth*. The Royal Society of New Zealand, Bull. 12, pp. 901–8.

Jenner, C. F. (1986). End product storage in cereals. In Cronshaw, J., Lucas, W. J. & Giaquinta, R. T. (eds.). *Plant Biology, Vol. 1, Phloem Transport*, pp. 561–72. New York, Alan R. Liss.

Kallarackal, J. & Milburn, J. A. (1985). Respiration and phloem translocation in the roots of chickpea (*Cicer arietinum*). *Annals of Botany* **56**, 211–18.

Kasperbauer, M. J. (1971). Spectral distribution of light in a tobacco canopy and effects of end of day light quality on growth and development. *Plant Physiology* **47**, 775–8.

Kasperbauer, M. J., Hunt, P. G. & Sojka, R. E. (1984). Photosynthate partitioning and nodule formation in soybean plants that received red or far-red light at the end of the photosynthetic period. *Physiologia Plantarum* **61**, 549–54.

Kerr, P. S. & Huber, S. C. (1987). Coordinate control of sucrose formation in soybean leaves by sucrose-phosphate synthase and fructose-2,6-bisphosphate. *Planta* **170**, 197–204.

Kinet, J. M. (1977a). Effect of light condition on the development of the inflorescence in tomato. *Scientia Horticulturae* **6**, 15–26.

Kinet, J. M. (1977b). Effect of defoliation and growth substances on the development of the inflorescence in tomato. *Scientia Horticulturae* **6**, 27–35.

Kinet, J. M., Zime, V, Linotte, A., Jacqmond, A. & Bernier, G. (1986). Resumption of cellular activity induced by cytokinin and gibberellin treatments in tomato flowers targeted for abortion in unfavourable light conditions. *Physiologia Plantarum* **64**, 67–73.

Komor, E. & Orlich, G. (1986). Sugar-proton symport: from single cells to phloem loading. In Cronshaw, J., Lucas, W. J. & Giaquinta, R. T. (eds.). *Plant Biology, Vol. 1, Phloem Transport*, pp. 53–65. New York, Alan R. Liss.

Komor, E., Thom, M. & Maretzki, A. (1981). The mechanism of sugar uptake by sugarcane suspension cells. *Planta* **153**, 181–92.

Kruger, N. J., Bulpin, P. V. & ap Rees, T. (1983). The extent of starch degradation in the light in pea leaves. *Planta* **157**, 271–3.

Lang, A. (1978). Interactions between source, path and sink in determining phloem translocation rate. *Australian Journal of Plant Physiology* **5**, 665–74.

Lang, A. (1983). Turgor-regulated translocation. *Plant, Cell and Environment* **6**, 683–9.

Lawson, H. M. & Wiseman, J. S. (1983). Techniques for the control of cane vigour in red raspberry in Scotland; effects of timing and frequency of cane removal treatments on growth and yield in cv. Glen Clova. *Journal Horticultural Science* **58**, 247–60.

Leigh, R. A., ap Rees, T., Fuller, W. A. & Banfield, J. (1979). The location of acid invertase activity and sucrose in vacuoles isolated from storage roots of red beet (*Beta vulgaris* L.). *Biochemical Journal* **178**, 539–47.

Leonard, M. & Kinet, J. M. (1982). Endogenous cytokinin and gibberellin levels in relation to inflorescence development in tomato. *Annals of Botany* **50**, 127–30.

Leonard, M., Kinet, J. M., Bodson, M & Bernier, G. (1983). Enhanced inflorescence development in tomato by growth substance treatments in relation to ^{14}C-assimilate distribution. *Physiologia Plantarum* **57**, 85–9.

Leopold, A. C. & Lam, S. L. (1960). A leaf factor influencing tomato earliness. *Proceedings of the American Society for Horticultural Science* **76**, 543–7.

Lester, G. E. & Dunlop, J. R. (1985). Physiological changes during development and ripening of 'Perlita' muskmelon fruits. *Scientia Horticulturae* **26**, 323–31.

Loescher, W. H., Marlow, G. C. & Kennedy, R. A. (1982). Sorbitol metabolism and sink-source interconversion in developing apple leaves. *Plant Physiology* **70**, 335–9.

Ludwig, J., Charles-Edwards, D. A. & Withers, A. C. (1975). Tomato leaf photosynthesis and respiration in various light and carbon dioxide environments. In Marcelle, R. (ed.). *Environmental and Biological Control of Photosynthesis*. The Hague, Dr W. Junk.

Lüttge, U. & Higinbotham, N. (1979). *Transport in Plants*. New York, Springer-Verlag.

Mapelli, S., Frova, C., Torti, G. & Soressi, G. P. (1978). Relationship between set, development and activities of growth regulators in tomato berries. *Plant and Cell Physiology* **19**, 1281–8.

Marmé, D. (1977). Phytochrome: membranes as possible sites of primary action. *Annual Review of Plant Physiology* **28** 173–98.

Mason, T. G. & Maskell, J. (1928). Studies on the transport of carbohydrates in the cotton plant. II. The factors determining the rate and the direction of movement of sugars. *Annals of Botany*, O.S. **42**, 571–636.

Milburn, J. A. (1974). Phloem transport in *Ricinus*: concentration gradients between source and sink. *Planta* **117**, 303–19.

Morgan, J. (1984). Osmoregulation and water stress in higher plants. *Annual Review of Plant Physiology* **35**, 299–319.

Neales, T. F. & Incoll, L. D. (1968). The control of leaf photosynthesis rate by assimilate concentration in the leaf: A review of the hypotheses. *Botanical Review* **34**, 107–25.

Neyra, C. A. & Hageman, R. H. (1976). Relationships between carbon dioxide, malate and nitrate accumulation and reduction in corn (*Zea mays* L.) seedlings. *Plant Physiology* **58**, 726–30.

Offler, C. E. & Patrick, J. W. (1984). Cellular structures, plasma membrane surface areas and plasmodesmatal frequencies of seed coats of *Phaseolus vulgaris* L. in relation to photosynthate transfer. *Australian Journal of Plant Physiology* **11**, 79–99.

Oota Y., Fujii, R. & Osawa, S. (1953). Changes in chemical constituents during the germination stage of a bean, *Vigna sesquipedalis*. *Journal of Biochemistry, Tokyo* **40**, 649–61.

Pate, J. S. (1980). Transport and partitioning of nitrogenous solutes. *Annual Review of Plant Physiology* **31**, 313–40.

Pate, J. S. (1986). Xylem-to-phloem transfer – vital component of the nitrogen-partitioning system of a nodulated legume. In Cronshaw, J., Lucas, W. J. & Giaquinta, R. T. (eds.). *Plant Biology, Vol. 1, Phloem Transport*, pp. 445–62. New York, Alan R. Liss.

Patrick J. W. (1982). Hormonal control of assimilate transport. In Wareing, P. F. (ed.). *Plant Growth Substances*, pp. 569–678. Academic Press.

Patrick, J. W. & Wareing, P. F. (1976). Auxin-promoted transport of metabolites in stems of *Phaseolus vulgaris* L. *Journal of Experimental Botany* **27**, 969–82.

Patrick, J. W. & Wareing, P. F. (1980). Hormonal control of assimilate movement and distribution. In Jeffcoat, B. (ed.). *Aspects and Prospects of Plant Growth Regulators*, Monograph 6, pp. 65–84. Joint DPGRG and BPGRG Symposium.

Pearson, C. J. & Steer, B. T. (1977). Daily changes in nitrate uptake and metabolism in *Capsicum annuum*. *Planta* **137**, 107–12.

Peel, A. J. & Ho, L. C. (1970). Colony size of *Tuberolachnus salignus* Gmelin in relation to mass transport of ^{14}C-labelled assimilates from the leaves in willow. *Physiologia Plantarum* **23**, 1033–8.

Pinto, C. M. (1980). Régulation de la photosynthèse par le demande d'assimilates: mécanismes possibles. *Photosynthetica* **14**, 611–37.

Pollock, C. J. (1986). Fructans and the metabolism of sucrose in vascular plants. *New Phytologist* **104**, 1–24.

Pollock, C. J. & Jones, T. (1979). Seasonal patterns of fructan metabolism in forage grasses. *New Phytologist* **83**, 9–15.

Porter, G. A., Knievel, D. P. & Shannon, J. C. (1985). Sugar efflux from maize (*Zea mays* L.) pedicel tissue. *Plant Physiology* **77**, 524–31.

Porter, M. H., Martin, R. V. & Bird, I. F. (1959). Synthesis and dissolution of starch labelled with 14-carbon in tobacco leaf tissue. *Journal of Experimental Botany* **10**, 264–76.

Preiss, J. (1982). Regulation of the biosynthesis and degradation of starch. *Annual Review of Plant Physiology* **33**, 431–54.

Reddy, V. M. & Daynard, T. B. (1983). Endosperm characteristics associated with properties of grain filling and kernel size in corn. *Maydica* **28**, 339–55.

Richardson, P. T., Baker, D. A. & Ho, L. C. (1982). The chemical composition of cucurbit vascular exudates. *Journal of Experimental Botany* **33**, 1239–47.

Richardson, P. T., Baker, D. A. & Ho, L. C. (1984). Assimilate transport in Cucurbits. *Journal of Experimental Botany* **35**, 1575–81.

Rufty, T. W. & Huber, S. C. (1983). Changes in starch formation and activities of sucrose phosphate synthase and cytoplasmic fructose-1,6-bisphosphate in response to source-sink alterations. *Plant Physiology* **72**, 474–80.

Russell, C. R. & Morris, D. A. (1983). Patterns of assimilate distribution and source-sink relationships in young reproductive tomato plants (*Lycopersicon esculentum* Mill.). *Annals of Botany* **52**, 357–63.

Saftner, R. A., Daie, J. & Wyse, R. E. (1983). Sucrose uptake and compartmentation in sugar beet root tissue discs. *Plant Physiology* **72**, 1–6.

Salter, P. J. (1958). The effect of different water-regimes on the growth of plants under glass. IV. Vegetative growth and fruit development in the tomato. *Journal of Horticultural Science* **33**, 1–12.

Schmalstig, J. & Geiger, D. R. (1985). Phloem unloading in developing leaves of sugar beet. I. Evidence for pathway through the symplast. *Plant Physiology* **79**, 237 41.

Servaites, J. C. & Geiger, D. R. (1974). Effects of light intensity and oxygen on photosynthesis and translocation in sugar beet. *Plant Physiology* **54**, 575–8.

Shannon, J. C. (1972). Movement of ^{14}C-labelled assimilates into kernels of *Zea mays* L. I. Pattern and rate of sugar movement. *Plant Physiology* **49**, 198–202.

Shaw, A. F. (1985). Source leaf carbon metabolism and export in tomato. PhD Thesis University of Sussex, UK.

Shaw, A. F. & Grange. R. I. (1984). Regulation of the partitioning of leaf carbon during the light period. *Annual Report of the Glasshouse Crops Research Institute for 1982*, pp. 41–5.

Shaw, A. F., Grange, R. I. & Ho, L. C. (1986). The regulation of source leaf assimilate compartmentation. In Cronshaw, J., Lucas, W. J. & Giaquinta, R. T. (eds.). *Plant Biology, Vol. 1, Phloem Transport*, pp. 391–8. New York, Alan R. Liss.

Sicher, R. C., Kremer, D. F. & Harris, W. G. (1984). Diurnal carbohydrate metabolism of barley primary leaves. *Plant Physiology*, 76, 165–9.

Silvius, J. E., Chatterton, N. J. & Kremer, D. F. (1979). Photosynthate partitioning in soybean leaves at two irradiance levels. Comparative responses of acclimated and unacclimated leaves. *Plant Physiology* **64**, 872–5.

Singh, B. K. & Jenner, C. F. (1984). Factors controlling endosperm cell number and grain dry weight in wheat: Effects of shading on intact plants and of variation in nutritional supply to detached, cultured ears. *Australian Journal of Plant Physiology* **11**, 151–63.

Slack, G. & Calvert, A. (1977). The effect of truss removal on the yield of early sown tomatoes. *Journal of Horticultural Science* **52**, 309–15.

Smith, J. A. C. & Milburn, J. A. (1980). Phloem turgor and the regulation of sucrose loading in *Ricinus communis* L. *Planta* **148**, 42–8.

Steer, B. T. (1973). Diurnal variations in photosynthetic products and nitrogen metabolism in exanding leaves. *Plant Physiology* **51**, 744–8.

Steer, B. T. (1974a). Control of diurnal variations in photosynthetic products. I. Carbon metabolism. *Plant Physiology* **54**, 758–61.

Steer, B. T. (1974b). Control of diurnal variations in photosynthetic products. II. Nitrate reductase activity. *Plant Physiology* **54**, 762–5.

Stitt, M. (1984). Degradation of starch in chloroplasts: a buffer to sucrose metabolism. In Lewis, D. H. (ed.). *Storage Carbohydrates in Vascular Plants: Distribution, Physiology and Metabolism*, Society for Experimental Biology Seminar Series, Vol. 19, pp. 205–29. Cambridge University Press.

Stitt, M. (1985). Control of photosynthetic sucrose synthesis by fructose-2,6-bisphosphate: comparative studies in C_3 and C_4 species. In: Jeffcoat, B., Hawkins, A. F. & Stead, A. D. (eds.). *Regulation of Sources and Sinks in Crop Plants*, Monograph 12, Bristol, British Plant Growth Regulator Group.

Stitt, M. (1986). Regulation of photosynthetic sucrose synthesis: Integration, adaptation and limits. In Cronshaw, J., Lucas W. J. & Giaquinta, R. T. (eds.). *Plant Biology, Vol. 1, Phloem Transport*, pp. 331–47. New York, Alan R. Liss.

Stitt, M. & Heldt, H. W. (1981). Simultaneous synthesis and degradation of starch in spinach chloroplasts. *Biochimica et Biophysica Acta* **6638**, 1–11.

Tanner, W. (1980). On the possible role of ABA in phloem unloading. *Berichte der Deutschen Botanischen Gesellschaft* **93**, 349–51.

Thom, M. & Komor, E. (1984). H^+-sugar antiport as the mechanism of sugar uptake by sugarcane vacuoles. *FEBS Letters* **173**, 1–4.

Thom, M. & Maretzki, A. (1985). Group translocation as a mechanism for sucrose transfer into vacuoles from sugarcane cells. *Proceedings of the National Academy of Sciences, USA* **82**, 4697–701.

Thorne, J. H. (1981). Morphology and ultrastructure of maternal seed tissues of soybean in relation to the import of photosynthate. *Plant Physiology* **67**, 1016–25.

Thorne, J. H. (1985). Phloem unloading of C and N assimilates in developing seeds. *Annual Review of Plant Physiology* **36**, 317–43.

Thorne, J. H. & Koller, H. R. (1974). Influence of assimilate demand on photosynthesis, diffusive resistances, translocation and carbohydrate levels. *Plant Physiology* **54**, 201–7.

Thrower, S. L. (1967). The pattern of translocation during leaf ageing. In Woolhouse, H. W. (ed.). *Aspects of the Biology of Ageing*, SEB Symposium 21, pp. 483–506. Cambridge University Press.

Tsai, C. Y. & Nelson, O. E. (1966). Starch deficient maize mutants lacking adenosine diphosphate glucose pyrophoshorylase activity. *Science* **151**, 341–3.

Turgeon, R. (1984a). Termination of nutrient import and development of vein loading capacity in albino tobacco leaves. *Plant Physiology* **76**, 45–8.

Turgeon, R. (1984b). Efflux of sucrose from minor veins of tobacco leaves. *Planta* **161**, 120–8.

Turgeon R. & Webb J. A. (1973). Leaf development and phloem transport in *Cucurbita pepo*: transition from import to export. *Planta* **113**, 179–91.

Turgeon, R. & Webb, J. A. (1975). Leaf development and phloem transport in *Cucurbita pepo*. Maturation of minor veins. *Planta* **129**, 265–9.

Turner, J. F. (1969). Starch synthesis and changes in uridine diphosphate glucose pyrophosphorylase and adenosine diphosphate glucose pyrophosphorylase in developing wheat grain. *Australian Journal of Biological Science* **22**, 1321–7.

Waister, P. D., Cormack, M. R. & Sheets, W. A. (1977). Competition between fruiting and vegetative phases in the red raspberry. *Journal of Horticultural Science* **52**, 75–85.

Walker, A. J. & Ho, L. C. (1977a). Carbon translocation in the tomato: Carbon import and fruit growth. *Annals of Botany* **41**, 813–23.

Walker, A. J. & Ho, L. C. (1977b). Carbon translocation in the tomato: Effect of fruit temperature on carbon metabolism and the rate of translocation. *Annals of Botany* **41**, 825–32.

Walker, A. J., Ho, L. C. & Baker, D. A. (1978). Carbon translocation in the tomato: Pathways of carbon metabolism in the fruit. *Annals of Botany* **43**, 437–48.

Walker, A. J. & Thornley, J. H. M. (1977). The tomato fruit: Import, growth, respiration and carbon metabolism at different fruit sizes and temperature. *Annals of Botany* **41**, 977–85.

Wardlaw, I. F. (1968). Carbohydrates in plants. *Botanical Review* **34**, 79–105.

Wareing, P. F. & Patrick, J. (1975). Source-sink relations and the partition of assimilates in the plant. In Cooper, J. P. (ed.). *Photosynthesis and Productivity in Different Environments*, pp. 481–99.

Warren Wilson, J. (1967). Ecological data on dry-matter production by plants and plant communities. In Bradley, E. F. & Denmead, O. T. (eds.). *The Collection and Processing of Field Data*, pp. 77–123. New York, Wiley.

Warren Wilson, J. (1972). Control of crop processes. In Rees, A. R., Cockshull, K. E., Hand D. W. & Hurd, R. G. (eds.). *Crop Processes in Controlled Environments*, pp. 7–30. London and New York, Academic Press.

Weatherley, P. E. & Johnson, R. P. (1968). The form and function of sieve tubes: A problem of reconciliation. *International Review of Cytology* **24**, 149–92.

Willenbrink, J. & Doll, S. (1979). Characteristics of sucrose uptake system of vacuoles isolated from red beet tissue: Kinetics and specificity of the sucrose uptake system. *Planta* **147**, 159–62.

Wyse, R. E. (1979). Sucrose uptake by sugar beet taproot tissue. *Plant Physiology* **64**, 837–41.

Yamaki, S. (1982). Distribution of sorbitol, neutral sugars, free amino acids, malic acid and some hydrolytic enzymes in vacuoles of apple cotyledons. *Plant and Cell Physiology* **23**, 881–9.

Zimmermann, M. H. (1971). Transport in the phloem. In Zimmermann, M. H. & Brown, C. C. *Trees, Structure and Function*, pp. 221–79. Springer-Verlag.

Appendix   Physico-chemical aspects of phloem sap

J. A. Milburn and D. A. Baker

In the recent past our knowledge of phloem sap composition has advanced considerably. The tables in this section are intended to provide guidance in phloem sap characteristics and to indicate the reliability of measurements e.g. at different degrees of maturation or seasonal changes.

Techniques for the extraction of phloem sap were originally through repeated incisions but the number of species capable of *sustained* exudation was only relatively small e.g. palms and Agaves. Many species have now been added (*Yucca, Cucurbita, Ricinus, Lupinus* etc). The technique of extraction via aphid stylets has been extended since the 1950s. The great attraction of this technique is that it samples single sieve tubes: its disadvantage is that the sap may be augmented by saliva or phytoalexin-like chemicals. Extraction using EDTA wound-sustained exudation, has been used to increasing effect since 1974. It appears to be exceptionally useful in extracting phloem sap from cereals e.g. rice. This technique remains somewhat problematic in that it is difficult to assess the extent of contamination from parenchymatous cells adjacent to the wound and also affected by the EDTA.

Tables I–III give general sap analyses, IV gives an indication of sugars and sugar alcohols in different families. Tables V and VI indicate how variable the amino acid composition may be at different times. Many additional papers could be advanced to further support this view. Tables VII, VIII and IX indicate polyamine, protein and hormone levels. The high protein value for *Cucurbita* protein may well be anomalous.

Speed and Specific Mass Transfer measurements in both intact and exuding systems are given in Tables X and XI. The differences between them are very marked showing that phloem sieve tubes seldom achieve their full conducting potential in the intact unwounded state. Finally Table XII gives values for the percentage of sieve tubes in phloem. This has been

345

a vexed question for many years but great strides have been possible using fluorescent stains and ultraviolet microscopy.

The intention is to provide a useful comparative check. Further bibliographic details should be consulted in the original tabulation references.

Table Ia The composition of the exudate obtained from incisions made in the bark of *Ricinus* plants. Some values for xylem exudate are included for comparison. Concentrations are expressed in mg cm^{-3} and also in eq m^{-3} or mol m^{-3} where relevant (from Hall & Baker (1972) and Hall *et al.* (1971)).

	Phloem Sap (mg cm^{-3})		**Xylem Sap**
Dry matter	100–125		2 mg cm^{-3}
Sucrose	80–106		0
Reducing sugars	Absent		0
Protein	1.45–2.20		
Amino acids	5.2 (as glutamic acid)	37.4 mol m^{-3}	
Keto acids	2.0–3.2 (as malic acid)	30–47 eq m^{-3}	
Phosphate	0.35–0.55	7.4–11.4 eq m^{-3}	
Sulphate	0.024–0.048	0.5–1.0 eq m^{-3}	
Chloride	0.355–0.675	10–19 eq m^{-3}	
Nitrate	Absent		
Bicarbonate	0.010	1.7 eq m^{-3}	
Potassium	2.3–4.4	60–112 eq m^{-3}	
Sodium	0.046–0.276	2–12 eq m^{-3}	7–8 eq m^{-3}
Calcium	0.020–0.092	1.0–4.6 eq m^{-3}	3–4 eq m^{-3}
Magnesium	0.109–0.122	9–10 eq m^{-3}	
Ammonium	0.029	1.6 eq m^{-3}	
Auxin	10.5 × 10^{-6}	0.60 × 10^{-4} mol m^{-3}	
Gibberellin	2.3 × 10^{-6}	0.67 × 10^{-5} mol m^{-3}	
Cytokinin	10.8 × 10^{-6}	0.52 × 10^{-4} mol m^{-3}	
ATP	0.24–0.36	0.40–0.60 mol m^{-3}	
pH	8.0–8.2		6.0
Conductance	1.32 mS m^{-1} at 18°C		
Solute potential	−1.42 to −1.52 MPa		
Viscosity	1.34 × 10^{-3} N s m^{-2} at 20°C		

Table Ib The amino acid composition of phloem exudate obtained from *Ricinus* plants. From Hall & Baker (1972).

Amino Acid	Concentration	
	(mol m^{-3})	(%)
Glutamic acid	13.0	34.76
Aspartic acid	8.8	23.53
Threonine	5.4	14.44
Glycine	2.4	6.42
Alanine	2.0	5.35
Serine	1.6	4.28

Amino Acid	Concentration	
	(mol m^{-3})	(%)
Valine	1.6	4.28
Isoleucine	1.0	2.67
Phenylalanine	0.6	1.60
Histidine	0.4	1.07
Leucine	0.4	1.07
Lysine	0.2	0.53
Arginine	Trace	Trace
Methionine	Trace	Trace
Total amino acids	37.4	100.00

Table Ic Phloem sap analysed from 9-week-old *Ricinus* plants grown on NO_3^-–N Long Ashton solution. Figures are mean ± S.E. for 6 plants or values for pooled samples. For calculation of the approximate theoretical contribution of the solutes to sap ψ_s it was assumed that the monovalent and divalent cations were balanced by monovalent and divalent anions, respectively. The organic acids concentration value refers to mol m^{-3} charge equivalents. Adapted from Smith & Milburn (1980a).

	Concentration (mol m^{-3})	Theoretical contribution to sap ψ_s to (MPa)
Neutral compounds		
sucrose	259 ± 21	−0.71
reducing sugars	trace	0
amino acids + amides bearing no net charge	86.5	−0.22
Cations		
K$^+$	68.1 ± 4.1	
Na$^+$	6.7 ± 1.9	−0.36
amino acids (+ve) + NH$_4^+$	2.7	
Mg^{2+}	3.9 ± 0.3	−0.04
Ca^{2+}	1.3 ± 0.4	
Anions		
Cl$^-$	8.9 ± 3.4	
H$_2$PO$_4^{-4}$/HPO$_4^{2-}$	7.6 ± 0.3	
NO$_3^-$	3.0	
organic acids	20.7	
amino acids (−ve)	23.8	
pH of sap = 7.3–7.6		

Total		−1.33
Observed sap ψ_S =		−1.36 ± 0.10
Sap ψ_s accounted for ± 98%		

Table IIa Properties and concentration of solutes in phloem exudate at *Yucca flaccida*. Data adapted from Tammes & Van Die (1975).

pH	8.0–8.2
Total dry matter	17.1–19.2%
Sucrose	150–180 mg cm^{-3}
Fructose	2–4 mg cm^{-3}
Glucose	2–4 mg cm^{-3}
Hexose phosphates	Trace
Total amino acids and amides	0.05–0.08 M
Total protein	0.5–0.8 mg cm^{-3}
Allantoin, allantoic acid and urea	Slight trace
Invertase activity	Absent
Malic acid	ca. 2.7 eq m^{-3}
Oxalic acid	ca. 1.4 eq m^{-3}
Other known di- and tricarboxylic acids	ca. 2.9 eq m^{-3}

Table IIb Free amino acids and amides (mole percentages) of three exudate samples from *Yucca flaccida*.

Glutamine and glutamic acid	54.9%	56.1%	58.5%
Asparagine and aspartic acid	9.4	9.2	7.4
Valine	8.0	8.0	5.5
Proline + threonine	9.2	4.4	6.5
Serine	5.7	5.4	6.1
Lysine	4.0	3.6	2.7
Isoleucine	2.8	2.8	1.8
Leucine	2.4	2.2	1.9
Glycine	1.5	1.6	1.0
Alanine	1.3	1.0	1.5
Phenylalanine	2.0	1.9	1.6
Ornithine	0.3	0.3	Trace
Tyrosine	Trace	Trace	Trace

Table III Properties and concentrations of solutes in phloem exudates from three samples of exudate from sugarbeet roots (*Beta vulgaris*). Adapted from Fife *et al.* (1962).

Measurement (units)	I	II	III
pH	8.02	6.12	6.53
Electrical conductivity (mmho cm^{-1})	10.79	2.30	4.30
Solute potential (MPa)	1.804	0.163	2.156
Viscosity (relative to H_2O)	1.58	0.98	2.18
Surface tension* (10^{-3}N m^{-1})	50.3	57.6	43.4
Density (g cm^{-3})	1.059	1.004	1.082
Electrolytes (as % of total solutes)	21.5	50.3	7.7
Total solids (%)	16.0	0.8	20.4

Table III (continued)

Measurement (units)	I	II	III
Sucrose (%)	8.8	0.10	16.5
Reducing sugars (%)	0.33	0.03	0.76
Soluble nitrogen, excluding nitrates (g dm^{-3})	5.65	0.10	1.16
Total amino acids, calculated as glutamic acid (%)	1.10		0.44
Protein nitrogen, heat-coagulable (g dm^{-3})	0.48		0.55

Table IV Phloem sap composition and sugar alcohols by family. Code: Absent = not detected; 0 = trace amounts; 1 = 0.5–2%; 2 = 2–10%; 3 = 10–20% and 4 = 20–30% (w/v) sucrose equivalent. Condensed from Zimmermann & Ziegler (1975)

Family	Sugars					Sugar Alcohols			
	SUCROSE	RAFFINOSE	STACHYOSE	VERBASCOSE	AJUGOSE	MANNITOL	SORBITOL	DULCITOL	INOSITOL
Aceraceae	4	1	0						0
Anacardiaceae	3	1	1	1					1
Annonaceae	3	0	0						
Apiaceae	3		1						1
Apocynaceae	2	1	0	1					1
Aquifoliaceae	3	1	1						1
Araliaceae	2								
Asteraceae	2	1							1
Berberidaceae	2	1							1
Betulaceae (see Corylaceae)									
Bignoniaceae	2	1	3	1	1				
Bombacaceae	2								
Boraginaceae	3	1							1
Buddleiaceae	3	3	2	1	1				1
Burseraceae	3	1							
Buxaceae	4	2	2	1		0			1
Caesalpiniaceae	3		0						1
Calycanthaceae	3	1	1						2
Caprifoliaceae	3	2	1	0					1
Casuarinaceae	3								
Celastraceae	2	1	3	0				3	0
Cercidiphyllaceae	4	2	1						0
Cistaceae	1	1							0
Clethraceae	4	4	4	4	1				1

Family	Sugars					Sugar Alcohols			
	SUCROSE	RAFFINOSE	STACHYOSE	VERBASCOSE	AJUGOSE	MANNITOL	SORBITOL	DULCITOL	INOSITOL
Cneoraceae	3	0	0	0					1
Combretaceae	3	2	1	0		2			
Compositae (see Asteraceae)									
Coriariaceae	3	1	1						0
Cornaceae	4	1	1						0
Corylaceae	3	2	1	1					0
Dichapetalaceae	3	0	1						
Dilleniaceae	2	0	1						
Ebenaceae	4	2	1	1					1
Elaeagnaceae	4	1	0	0					0
Elaeocarpaceae	3								
Ericaceae	3	2	1						1
Euphorbiaceae	2	2	0	1					1
Fabaceae	3	0	0						0
Fagaceae	4	0	0	0	0				1
Flacourtiaceae	3	0	1						
Garryaceae	2	1							1
Hammamelidaceae	3	2	1	1					0
Hernandiaceae	3								
Hippocastanaceae	4	0	0						0
Hyericaceae	3								
Juglandaceae	4	2	2	0	1				1
Lamiaceae	2	1	2	0					2
Lardizabalaceae	3	2	0	0					1
Laurcaceae	3	1	1	1					1
Lecythidaceae	3								
Leguminosae (see Caesalpiniaceae, Fabaceae, Mimosaceae)									
Lythraceae	3	3	1	0					0
Magnoliaceae	3	0	1	0					1
Malpighiaceae	3	0	1						
Malvaceae	3	0		0					0
Meliaceae	4	0	0	0					1
Melianthaceae	4		0	0					1
Mimosaceae	3	0	0						1
Monimiaceae	1								1
Moraceae	4	2	2	0					1
Moringaceae	3								
Myrsinaceae	3	0							
Myrtaceae	2	2	0	0					1
Nyctaginaceae	3	0							1

Family	Sugars					Sugar Alcohols			
	SUCROSE	RAFFINOSE	STACHYOSE	VERBASCOSE	AJUGOSE	MANNITOL	SORBITOL	DULCITOL	INOSITOL
Nyssaceae	3		0						
Olaceae	2	2	3	1	1	3			1
Onagraceae	2	3	2						
Oxalidaceae	3								
Papilionaceae (see Fabaceae)									
Pittosporaceae	2	0	1						1
Polygalaceae	1	0							0
Polygonaceae	3								
Proteaceae	4	1							0
Prunoideae	2	1	0				3		1
Punicaceae	4	3	3						1
Rhamnaceae	3	1	0						1
Rosaceae (Chrysobalanoideae)	3	0	0						
Rosaceae (Pomoideae)	3	0	0	1			3		1
Rosaceae (Rosoideae)	3	1	1	0					
Rubiaceae	2	0	0						
Rutaceae	3	0	1	0					1
Salicaceae	4	1	1	1					1
Santalaceae	3								
Sapindaceae	3	0	0						1
Sapotaceae	3	1							0
Saxifragaceae	2	0	1	0					1
Scrophulariaceae	4	3	4	1	0				1
Simarubaceae	3								
Solanaceae	1		0						1
Sonneratiaceae	2								
Spiroideae	3	1	0	1			3		0
Staphylcaccac	2	1	0	0					1
Sterculiaceae	3	0							0
Styracaceae	4	1	1						1
Tamaricaceae	3	1							1
Thymeleaceae	2								1
Tiliaceae	3	1	1	0					1
Ulmaceae	4	1	1	1					1
Umbelliferae (see Apiaceae)									
Verbenaceae	2	1	3	0					1
Vitaceae	3	2	2	1					0
Zygophyllaceae	3								

Table V Proportion of amino acids collected from leaves of wheat (*Triticum aestivum* L). Time after anthesis is given in days. 5 mol m^{-3} disodium EDTA at pH 7.0 served as a 10 min pretreatment. It was followed by 7 hours collection in darkness at 80% relative humidity into a new solution – six leaves per 4 ml with five replicates. Adapted from Simpson & Dalling (1981).

Amino acid	Concentration of amino acids (mol leaf^{-1})		
	5 (days)	15(d)	25(d)
Glutamic acid	26.2	34.5	9.5
Asparagine	20.2	9.5	7.1
Glutamine	1.1	4.8	26.2
Serine	9.5	9.5	7.1
Alanine	9.5	8.3	8.3
Glycine	7.1	8.3	4.8
Threonine	4.8	4.8	3.6
Valine	5.0	4.8	8.3
Leucine	5.9	4.8	7.2
Isoleucine	3.6	4.8	4.8
Other	7.1	5.9	13.1
Extraction Rate nmol leaf hr^{-1}	7.0	12.0	20.6

Table VI The volume of phloem sap collected from *Salix alba L* using stylets of the willow aplid *Tuberolachnus salignus* (Gmelin) and the concentration of individual amino acids and amids. + = presence of small amount; − = absence. (Adapted from Leckstein & Llewellyn, 1975).

	May	June	July	August	September	October
Aspartic acid Threonine	2.66	0.07	+	0.31	0.2	0.43
Asparagine Glutamine	0.35	+	2.92	0.15	0.25	0.14
Serine	−	0.04	0.89	1.34	0.56	0.89
Glutamic acid	1.92	1.56	2.06	0.19	0.68	0.09
Glycine	0.12	0.07	0.17	0.04	0.28	0.22
Alanine	0.10	0.04	0.11	+	0.11	0.09
Valine	+	+	+	0.11	−	+
Cysteine	−	+	0.15	−	0.16	+
Methionine	−	+	−	+	+	+
Iso-leucine	−	−	0.12	0.10	0.14	+
Leucine	+	+	0.16	0.14	0.10	+
Tyrosine	−	+	+	+	+	−
Phenylalanine	+	0.05	0.21	0.05	0.11	+
Lysine	+	0.003	0.16	0.17	0.008	0.10
Histidine	+	+	0.17	0.1	0.15	+
Arginine	+	+	+	−	−	+
Tryptophan	−	+	0.14	+	+	−
Vol. of sap collected in mm^3	17.3	17.3	38.2	40.9	24.9	21.8

Table VII Polyamine concentrations collected from basal stems in 200 litres of 20 mM sodium – EDTA for 3 hours. The exudate was believed to be phloem exudate. Controls in water gave only trace amounts (except sunflower 13 days which gave 430 ± 70 p mol^{-1} soln). Each experiment was the average of two experiments analysed in duplicate. Adapted from Friedman *et al.* (1986).

Plant	Age	Polyamine concentration	
Genus Species	(days)	Putrescine (pmol ml^{-1} soln)	Spermidine (pmol ml^{-1} soln)
Sunflower			
Helianthus annuus	8	4490 ± 530	960 ± 115
	13	430 ± 78	Trace
Mungbean			
Vigna radiata	8	1255 ± 231	1550 ± 215

Table VIII Protein concentrations in phloem exudates of some species (adapted from Kursanov, 1984).

Species	Time of exudation	Protein-content	Worker (Year)
Tilia platyphyllos Scop.	Oct.	0.31	Kennecke *et al.*, (1971)
Carpinus betulus L.	Oct.	0.60	Kennecke *et al.*, (1971)
Fraxinus americana L.	Oct.	1.12	Kennecke *et al.*, (1971)
Quercus borealis Michx.	Oct.	1.06	Kennecke *et al.*, (1971)
Salix viminalis L.	Oct.	1.30	Kennecke *et al.*, (1971)
Robinia pseudoacacia	Sept.	1.95–2.10	Ziegler (1956)
Ricinus communis L.		1.45–2.20	Hall & Baker (1972)
Yucca flaccida Haw.		0.5–0.8	Tammes & Van Die (1964)
Cucurbita maxima DuCH.		9.8	Kollmann *et al.*, (1970)

Table IX An example of hormone transport in sieve tubes of Soybean leaves. Sap was extracted by the EDTA technique. It is obvious that IAA and ABA are both mobile in the phloem in free or combined form. There are effects from pod removal and also the direction of exudate flow. Adapted from Hein *et al.*, (1984).

	(ng/gfwt leaf)	
	Control	Depodded
ABA Basipetal	10–20	12–30
Acropetal	0–5	0–5
IAA Ester	4–19%	1–7%
IAA (Free)	0.3%	0.4%

Table X Speed of assimilate transport in the phloem of plant species analysed by differing techniques. Based on a tabulation by Kursanov (1984).

Method	Species	Organ	Mean velocity (cm h^{-1})	Worker & Year
Intact Systems				
From dry weight increment of the sink	*Solanum tuberosum*	Stolons	50	Dixon & Ball (1922)
	Dioscorea alata	Stalk	88	Mason & Lewin (1926)
	Tilia parvifolis	Trunk	31	Crafts (1939)
	Carpinus betulus	Trunk	73	Crafts (1939)
	Quercus rubra	Trunk	52	Münch (1930)
	Acer platanoides	Trunk	42	Münch (1930)
	Pyrus communis	Fruitstalk	10	Münch (1930)
	Prunus domestica	Fruitstalk	10	Münch (1930)
	Cucumis sativus	Fruitstalk	24	Crafts (1932)
From transport of ^{14}C compounds with $^{14}CO_2$ assimilation	*Gossypium hirsutum*	Stalk	40	Kursanov & Vyskrebentseva (1954)
	Vitis vinifera	Stalk	60	Swanson & El-Shishiny, (1958)
	Solanum tuberosum	Stalk	50	Mokromosov & Bubenshchikova (1961)
	Soja hispida	Stalk	63	Moorby et al., (1963)
	Macrocystis	Thallus	70	Parker (1963)
	Beta vulgaris	Petiole	85	Mortimer (1965)
	Dioscorea sp.	Stalk	70	Lawton (1967)
	Helianthus annuus	Stalk	70	Whittle (1971)
	Marsilea quadrifolia	Petiole	60–150	Gahnke et al., (1981)
From transport of ^{14}C compounds	*Beta vulgaris*	Petiole	80	Kursanov et al., (1953)
	Ricinus communis	Petiole-stem	80–84	Hall et al., (1971)
	Ricinus communis	Fruit stalk	118	Kallarackal & Milburn (1983)
From transport of ^{11}C	*Zea mays*	Leaf	162–219	Troughton et al. (1974)
From concentration wave speed	*Quercus borealis*	Trunk	up to 460	Huber et al., (1937)
	Quercus rubra	Trunk	250	Huber et al., (1937)
	Fraxinus americana	Trunk	up to 150	Zimmermann (1958)

Exuding Systems

Stems & Fruitstalks

Plant	Organ		Author
Arenga	Fruit stalk	1140	Tammes (1952)
Ricinus communis	Stem	100+	Milburn (1972)
Ricinus communis	Stem	1584	Hall *et al*., (1971)
Ricinus communis	Fruit stalk		Kallarackal & Milburn (1983)
Cucumis sativus	Stem	300	Crafts (1932)
Cocos nucifera	Fruit stalk	1140	Milburn & Zimmermann (1972)
Ricinus communis	Stem	63–200	Smith & Milburn (1980)

Aphid mouthparts

Plant	Organ		Author
Salix acutifolia	Stem	500	Mittler (1952)

Table XI Specific mass transfer of dry matter via phloem from published data. Table adapted from Canny (1973).

INTACT SYSTEMS

Organ utilized	Author (Date)	Plant	Specific mass transfer g h^{-1} cm^{-2} phloem
Stems, tubers fruit stalks & roots			
	Dixon & Ball (1922)	*Solanum tuberosum* tuber	4.93
	Mason & Lewin (1926)	*Dioscorea alata* tuber	4.45
	Crafts (1933)	*Solanum tuberosum* tuber	2.1
	Clements (1940)	*Kigelia africana* fruit	2.6
	Crafts & Lorenz (1944)	Connecticut field pumpkin	3.3
		Early prolific straightneck summer squash	0.85 to 3.77 mean 2.68
	Colwell (1942)	*Cucurbita*	4.8
	Evans *et al*., (1970)	Wheat peduncle	4.4
	Münch (1930)	*Pyrus communis* fruit	0.59
		Prunus domestica fruit	0.57
		Fagus sylvatica fruit	0.61
		Quercus pedunculata fruit	0.59
		Quercus pedunculata	0.47
		δ × *sessiliflora* fruit	

Table XI

Organ utilized	Author (Date)	Plant	Specific mass transfer g hr^{-1} cm^{-2} phloem
	Kallarackal & Milburn (1984)	*Ricinus communis*	6.18
	Kallarackal & Milburn (1985)	*Cicer arietinum*	6.5
Petioles	Birch-Hischfeld (1920)	*Phaseolus multiflorus*	0.56
	Dixon (1932)	*Tropaeolum majus*	1.4
	Crafts (1931a)	*Tropaeolum majus*	0.51
		Phaseolus	0.5
	Sachs (1884)	*Helianthus annuus*	4.6
	Geiger et al., (1969)	*Beta vulgaris*	0.13
Tree trunks	Münch (1930)	*Pinus sylvestris I*	1.07
		Pinus sylvestris II	1.18
		Acer pseudoplatanus	6.33
		Acer platanoides	5.15
		Tilia parvifolia	2.92
		Carpinus betulus	6.25
		Quercus rubra	4.50
	Crafts (1931a)	Pear tree (*Pyrus*)	0.9
Exuding Systems Stems, tubers & fruit stalks	Tammes (1933, 1952)	*Arenga* peduncle	95.5
	Milburn & Zimmermann (1977)	*Cocos* peduncle	71.2
	Kallarackal & Milburn (1984)	*Ricinus* peduncle	112.7

Table XII Percentage of sieve tubes in phloem transverse sections. It is assumed that phloem fibres are normally excluded though the status of immature fibres can be quite ambiguous. Some workers have excluded ray tissue and others have specified 'sieve tubes' only – usually identified by callose deposits from reconstructed sections of macerated material. All were determined by transverse sectioning except those marked (*) which were longitudinal partial-macerations.

Worker (Year)	Subject material	Percentage sieve tubes	Details of phloem if given
Münch (1930)	Trees (many species) (Trunks)	66	Excludes rays and fibres
Crafts (1931)	Cucurbita and Solanum (Stem)	20	
Crafts (1933)	Solanum (Potato stolon)	37	
Geiger et al. (1969)	Beta (Sugar beet petioles)	29 ± 3	
Evans et al. (1970)	Triticum (Wheat peduncle)	33	Sieve tube luminae
Canny	Cucurbita ficifolia (Petiole) Ext. phl.	48	Sieve tubes only
	Cucurbita ficifolia (Petiole) Int. phl.	52	Sieve tubes only
	Solanum (Potato stolon)	52	Excluding rays
	Solanum (Potato stolon)	40	Including rays
	Beta (Sugar beet petiole)	78	Including rays
Milburn & Zimmermann (1971)	Cocos nucifera (Fruit stalk)	45	Excl. fibres
Smith & Milburn (1980)	Ricinus communis (Stem)	ca 10	Excl. fibres
Milburn & Kallarackal (1984)	Musa sapientum (Stem pseudo)	18.4	Including fibres
	Musa sapientum (Petiole)	18.8	Including fibres
	Musa sapientum (Fruitstalk)	4.8	Including fibres
	Cucumis myriocarpus (Stem)	24.0	Including fibres
	Cucumis myriocarpus (Petiole)	20.7	Including fibres
	Cucumis myriocarpus (Fruitstalk)	20.0	Including fibres
	Ricinus communis (Stem)	5.9	Including fibres
	(Petiole)	8.9	Including fibres
	(Fruitstalk)	6.8	Including fibres
	(Stem*)	6.1	Including fibres
	(Petiole*)	10.5	Including fibres
	(Fruitstalk*)	6.1	Including fibres

References

Canny, M. J. (1973). *Phloem Translocation*. Cambridge University Press.

Fife, J. M., Price, C. & Fife, D. C. (1962). Some properties of phloem exudate, collected from roots of sugar beet. *Plant Physiology*. **37**, 791–2.

Friedman, R., Levin, N. & Altman, A. (1986). Presence and identification of polyamines in xylem and phloem exudates of plants. *Plant Physiology*, **82**, 1154–7.

Hall, S. M. & Baker, D. A. (1972). The chemical compositon of *Ricinus* phloem exudate. *Planta* **106**, 131–40.

Hall, S. M., Baker, D. A. & Milburn, J. A. (1971). Phloem transport of ^{14}C-labelled assimilates in *Ricinus*. *Planta* **100**, 200–7.

Heine, M. B., Brenner, M. L. & Brun, W. A. (1984). Effects of pod removal on the transport and accumulation of abscisic acid and indole-3-acetic acid in soybean leaves. *Plant Physiology* **76**, 955–8.

Kallarackal, J. & Milburn, J. A. (1983). Studies on phloem sealing mechanism in *Ricinus* fruit stalks. *Australian Journal of Plant Physiology* **10**, 561–8.

Kallarackal, J. & Milburn, J. A. (1984). Specific mass transfer and sink-controlled phloem translocation in castor bean. *Australian Journal of Plant Physiology*, **56**, 211–18.

Kallarackal, J. & Milburn, J. A. (1985). Respiration and phloem translocation in the roots of chickpea (*Cicer aretinum*). *Annals of Botany* **56**, 211–18.

Kursanov, A. L. (1984). *Assimilate Transport in Plants*. Amsterdam, New York, Oxford, Elsevier.

Leckstein, P. M. & Llewellyn, M. (1975). Quantitative analysis of seasonal variation in the amino acids in phloem sap of *Salix alba* L. *Planta*. **124**, 89–91.

Milburn, J. A. (1972). Phloem transport in *Ricinus*. *Pesticide Science* **3**, 653–65.

Milburn, J. A. & Zimmermann, M. H. (1977). Preliminary studies on sap flow in *Cocos nucifera*. II. Phloem transport. *New Phytologist* **79**, 543–58.

Milburn, J. A. & Kallarackal, J. (1984). Quantitative determination of sieve-tube dimensions in *Ricinus, Cucumis* and *Musa*. *New Phytologist* **96**, 383–95.

Simpson, R. J. & Dalling, M. J. (1981). Nitrogen redistribution during grain growth in wheat (*Triticum aestivum* L.) III. Enzymology and transport of amino acids from senescing flag leaves. *Planta* **151**, 447–56.

Smith, J. A. C. & Milburn, J. A. (1980a). Osmoregulation and the control of phloem sap composition in *Ricinus communis* L. *Planta* **148**, 28–34.

Smith, J. A. C. & Milburn, J. A. (1980b). Phloem transport, solute flux and the kinetics of sap exudation in *Ricinus communis* L. *Planta* **148**, 35–41.

Troughton, J. H., Moorby, J. & Currie, B. G. (1974). Investigations of carbon transport in plants. I. The use of carbon-11 to estimate various parameters of the translocation process. *Journal of Experimental Botany* **25**, 684–94.

Van Die, J. & Tammes, P. M. L. (1975). Phloem exudation from monocotyledons axes. In Zimmermann, M. H. and Milburn, J. A. (eds.) *Phloem Transport. Encyclopedia of Plant Physiology*. New Series. Vol. I. pp. 196–222. Heidelberg, Springer-Verlag.

Zimmermann, M. H. and Ziegler, H. (1975). List of sugars of sugar alcohols in sieve-tube exudates. In Zimmermann, M. H. and Milburn, J. A. (eds) *Phloem Transport. Encyclopedia of Plant Physiology*. New Series. Vol. I. pp. 480–503. Heidelberg, Springer-Verlag.

Author index

Numbers in italics refer to authors listed in the References at ends of chapters.

Subject index

abscisic acid (ABA), 193, 230–9, 243, 272, 311, 353
  content, 238
  in sieve tube sap, 243
  stimulated, 255
abscission layers, 225, 230
Abutilon, 55
  megapotamicum, 51
  nectary, 55
Acer, 230, 281
acetyl coenzyme, 292
Achlys triphylla, 102
acid
  fuchsin, 160
  invertase, 219, 327
  phosphatases, 102, 127
  trap mechanism, 193, 196
actin-like, 112
action potentials, 235, 255
adenine nucleotide transport, 17
adenosine-5'-diphosphate (ADP), 17
  glucose pyrophosphorylase, 22, 32, 33, 330
adenosine-5'-triphosphate (ATP), 7, 10, 17, 18, 118, 274
  generated, 2, 3
  supply, 195
  -ase, 253, 255
  -ase activity, 127, 173
  -ase inhibitors, 176
  -energized proton efflux, 27, 235
  /ADP ratio, 235
adenylates, 17
Aegilops comosa, 127
aerial,
  root, 213

stems, 88
Agave,
  americana, 220, 280
alanine, 16, 17, 26, 163, 227
  amino transferase, 163
  percentage distribution of, 161
  ^{14}C labelling of, 159
  ^{15}N labelling of, 158
Albizia julibrissin, 235
albuminous cells, 128
aldolase condensation, 21
aleurone layer, 243, 244
alfalfa mosaic virus, 62
allantoic acid, 156
  ^{14}C labelling of, 159
allantoin, 156
  ^{14}C labelling of, 159
allantoinase, 163
Allium, 55
  ascalonicum, 231
  leaf, 64
Alnus, 281
Aloe arborescens, 35
Amaranthus caudatus, 248
  edulis, 15
  retroflexis, 53
amide-N protein, percentage distribution of, 161
amino acids, 151, 171, 172, 177–9, 184, 187, 239, 273, 285, 318, 319
  acidic, 184
  amphoteric, 207
  balance, 149
  basic, 184
  components of phloem, 156, 352
  concentrations of, 147